Contemporary Abstract Algebra

EIGHTH EDITION

Joseph A. Gallian
University of Minnesota Duluth

Australia • Brazil • Japan • Korea • Mexico • Singapore • Spain • United Kingdom • United States

To my wonderful grandson, Joey Yapel.

Notations

(The number after the item indicates the page where the notation is defined.)

SET THEORY

$\cap_{i \in I} S_i$	intersection of sets S_i, $i \in I$
$\cup_{i \in I} S_i$	union of sets S_i, $i \in I$
$[a]$	$\{x \in S \mid x \sim a\}$, equivalence class of S containing a, 18
$\lvert s \rvert$	number of elements in the set of S

SPECIAL SETS

Z	integers, additive groups of integers, ring of integers
Q	rational numbers, field of rational numbers
Q^+	multiplicative group of positive rational numbers
F^*	set of nonzero elements of F
$\mathbf{R}$	real numbers, field of real numbers
$\mathbf{R}^+$	multiplicative group of positive real numbers
$\mathbf{C}$	complex numbers

FUNCTIONS AND ARITHMETIC

f^{-1}	inverse of the function f
$t \mid s$	t divides s, 3
$t \nmid s$	t does not divide s, 3
$\gcd(a, b)$	greatest common divisor of the integers a and b, 4
$\operatorname{lcm}(a, b)$	least common multiple of the integers a and b, 6
$\lvert a + b \rvert$	$\sqrt{a^2 + b^2}$, 13
$\phi(a)$	image of a under ϕ, 20
$\phi: A \to B$	mapping of A to B, 20
gf, $\alpha\beta$	composite function, 21

ALGEBRAIC SYSTEMS

D_4	group of symmetries of a square, dihedral group of order 8, 33
D_n	dihedral group of order $2n$, 34
e	identity element, 43
Z_n	group $\{0, 1, \ldots, n - 1\}$ under addition modulo n, 44
$\det A$	the determinant of A, 45
$U(n)$	group of units modulo n (that is, the set of integers less than n and relatively prime to n under multiplication modulo n), 46
$\mathbf{R}^n$	$\{(a_1, a_2, \ldots, a_n) \mid a_1, a_2, \ldots, a_n \in \mathbf{R}\}$, 47
$SL(2, F)$	group of 2×2 matrices over F with determinant 1, 48
$GL(2, F)$	2×2 matrices of nonzero determinants with coefficients from the field F (the general linear group), 48
g^{-1}	multiplicative inverse of g, 51
$-g$	additive inverse of g, 52
$\lvert G \rvert$	order of the group G, 60
$\lvert g \rvert$	order of the element g, 60
$H \leq G$	subgroup inclusion, 61
$H < G$	subgroup $H \neq G$, 61
$\langle a \rangle$	$\{a^n \mid n \in Z\}$, cyclic group generated by a, 65
$Z(G)$	$\{a \in G \mid ax = xa$ for all x in $G\}$, the center of G, 66

$C(a)$ — $\{g \in G \mid ga = ag\}$, the centralizer of a in G, 68

$\langle S \rangle$ — subgroup generated by the set S, 71

$C(H)$ — $\{x \in G \mid xh = hx \text{ for all } h \in H\}$, the centralizer of H, 72

$\phi(n)$ — Euler phi function of n, 84

$N(H)$ — $\{x \in G \mid xHx^{-1} = H\} = \{x \in G \mid Hx = xH\}$, the normalizer of H in G, 95

$\mathrm{cl}(a)$ — conjugacy class of a, 95

G^n — $\{g^n \mid g \in G\}$, 96

S_n — group of one-to-one functions from $\{1, 2, \ldots, n\}$ to itself, 101

A_n — alternating group of degree n, 110

$G \approx \overline{G}$ — G and $\overline{G}$ are isomorphic, 128

ϕ_a — mapping given by $\phi_a(x) = axa^{-1}$ for all x, 135

$\mathrm{Aut}(G)$ — group of automorphisms of the group G, 136

$\mathrm{Inn}(G)$ — group of inner automorphisms of G, 136

aH — $\{ah \mid h \in H\}$, 144

aHa^{-1} — $\{aha^{-1} \mid h \in H\}$, 144

$|G:H|$ — the index of H in G, 148

HK — $\{hk \mid h \in H, k \in K\}$, 150

$\mathrm{stab}_G(i)$ — $\{\phi \in G \mid \phi(i) = i\}$, the stabilizer of i under the permutation group G, 151

$\mathrm{orb}_G(i)$ — $\{\phi(i) \mid \phi \in G\}$, the orbit of i under the permutation group G, 151

$G_1 \oplus G_2 \oplus \cdots \oplus G_n$ — external direct product of groups $G_1, G_2, \ldots, G_n$, 162

$U_k(n)$ — $\{x \in U(n) \mid x \bmod k = 1\}$, 166

G' — commutator subgroup, 181

$H \triangleleft G$ — H is a normal subgroup of G, 185

G/H — factor group, 187

$H \times K$ — internal direct product of H and K, 196

$H_1 \times H_2 \times \cdots \times H_n$ — internal direct product of $H_1, \ldots, H_n$, 197

$\mathrm{Ker}\,\phi$ — kernel of the homomorphism ϕ, 208

$\phi^{-1}(g')$ — inverse image of g' under ϕ, 210

$\phi^{-1}(\overline{K})$ — inverse image of $\overline{K}$ under ϕ, 211

$Z[x]$ — ring of polynomials with integer coefficients, 246

$M_2(Z)$ — ring of all 2×2 matrices with integer entries, 246

$R_1 \oplus R_2 \oplus \cdots \oplus R_n$ — direct sum of rings, 247

nZ — ring of multiples of n, 249

$Z[i]$ — ring of Gaussian integers, 249

$U(R)$ — group of units of the ring R, 251

$\mathrm{char}\, R$ — characteristic of R, 258

$\langle a \rangle$ — principal ideal generated by a, 268

$\langle a_1, a_2, \ldots, a_n \rangle$ — ideal generated by $a_1, a_2, \ldots, a_n$, 268

R/A — factor ring, 268

$A + B$ — sum of ideals A and B, 275

AB — product of ideals A and B, 275

$\mathrm{Ann}(A)$ — annihilator of A, 277

$N(A)$ — nil radical of A, 277

$F(x)$ — field of quotients of $F[x]$, 291

$R[x]$ — ring of polynomials over R, 298

Contents

Preface

Dear Sir or Madam, will you read my book, it took me years to write, will you take a look?

John Lennon and Paul McCartney, *"Paperback Writer," single, 1966**

Although I wrote the first edition of this book more than 25 years ago, my goals for it remain the same. I want students to receive a solid introduction to the traditional topics. I want readers to come away with the view that abstract algebra is a contemporary subject—that its concepts and methodologies are being used by working mathematicians, computer scientists, physicists, and chemists. I want students to see the connections between abstract algebra and number theory and geometry. I want students to be able to do computations and to write proofs. I want students to enjoy reading the book. And I want to convey to the reader my enthusiasm for this beautiful subject.

Educational research has shown that an effective way of learning mathematics is to interweave worked-out examples and practice problems. Thus, I have made examples and exercises the heart of the book. The examples elucidate the definitions, theorems, and proof techniques. The exercises facilitate understanding, provide insight, and develop the ability of the students to do proofs. The exercises often foreshadow definitions, concepts, and theorems to come. Many exercises focus on special cases and ask the reader to generalize. Generalizing is a skill that students should develop but rarely do. Even if an instructor chooses not to spend class time on the applications in the book, I feel that having them there demonstrates to students the utility of the theory.

Changes for the eighth edition include 200 new exercises, new examples, and a freshening of the quotations, historical notes, and biographies. These changes accentuate and enhance the hallmark features that have made previous editions of the book a comprehensive, lively, and engaging introduction to the subject:

- Extensive coverage of groups, rings, and fields, plus a variety of nontraditional special topics

- A good mixture of nearly 2000 computational and theoretical exercises appearing in each chapter and in Supplementary Exercise sets that synthesize concepts from multiple chapters
- Back-of-the-book skeleton solutions and hints to the odd-numbered exercises
- Worked-out examples—now totaling nearly 300—ranging from routine computations to quite challenging problems
- Computer exercises, which utilize interactive software available on my website, that stress guessing and making conjectures
- A large number of applications from scientific and computing fields, as well as from everyday life
- Numerous historical notes and biographies that spotlight the people and events behind the mathematics
- Lines from popular songs, poems, and quotations
- Scores of photographs, hundreds of figures, numerous tables and charts, and reproductions of stamps and currency that honor mathematicians
- Annotated suggested readings and media for interesting further exploration of topics

To make room for the new material, the computer exercises from previous editions are available at **www.d.umn.edu/~jgallian** or through Cengage's book companion site at **www.cengage.com/math/gallian**. The first website also offers a wealth of additional online resources supporting the book, including:

- True/false questions with comments
- Flash cards
- Essays on learning abstract algebra, doing proofs, and reasons why abstract algebra is a valuable subject to learn
- Links to abstract algebra–related websites and software packages and much, much more

Additionally, Cengage offers the following student and instructor ancillaries to accompany the book:

- A *Student Solutions Manual*, available for purchase separately, with detailed solutions to the odd-numbered exercises in the book (ISBN:978-1-133-60853-0)
- Solution Builder, an online instructor database that offers complete, worked-out solutions to all exercises in the text, which allows you to create customized, secure solutions printouts (in PDF format) matched exactly to the problems you assign in class. Sign up for access at www.cengage.com/solutionbuilder.

- An *Instructor's Solutions Manual* with solutions to all the exercises in the book and additional test questions and solutions
- An online laboratory manual, written by Julianne Rainbolt, with exercises designed to be done with the free computer algebra system software GAP
- Online instructor answer keys to the book's computer exercises and the exercises in the GAP lab manual

Special thanks go to my copy editor for this edition, Jeff Anderson, and the accuracy checker, Roger Lipsett. I am grateful to each for their careful attention to the manuscript. My appreciation also goes to Molly Taylor, Shaylin Hogan, and Alex Gontar from Cengage Learning, as well as Katie Costello and the Cengage production staff.

The thoughtful input of the following people, who served as reviewers for the eighth edition, is also sincerely appreciated: Homer Austin, Salisbury University; David Barth-Hart, Rochester Institute of Technology; Bret Benesh, College of St. Benedict and St. John's University; Daniel Daly, Southeast Missouri State University; Paul Felt, University of Texas of the Permian Basin; Donald Hartig, California Polytechnic State University, San Luis Obispo; Nancy Ann Neudauer, Pacific University; Bingwu Wang, Eastern Michigan University; Dana Williams, Dartmouth College; and Norbert Youmbi, Saint Francis University.

Over the years, many faculty and students have kindly sent me valuable comments and suggestions. They have helped to make each edition better. I owe many thanks to my UMD colleague Robert McFarland for giving me numerous exercises and comments that have been included in this edition. Douglas Dunham, another UMD colleague, has generously provided the spectacular cover image for this edition. For an explanation of the mathematics underlying this image see **www.d.umn.edu/~jgallian/Dunhamimage**. Please send any comments and suggestions you have to me at **jgallian@d.umn.edu**

Joseph A. Gallian

PART 1

Integers and Equivalence Relations

For online student resources, visit this textbook's website at www.CengageBrain.com

0 Preliminaries

The whole of science is nothing more than a refinement
of everyday thinking.

ALBERT EINSTEIN, *Physics and Reality*

Properties of Integers

Much of abstract algebra involves properties of integers and sets. In this chapter we collect the properties we need for future reference.

An important property of the integers, which we will often use, is the so-called Well Ordering Principle. Since this property cannot be proved from the usual properties of arithmetic, we will take it as an axiom.

Well Ordering Principle

Every nonempty set of positive integers contains a smallest member.

The concept of divisibility plays a fundamental role in the theory of numbers. We say a nonzero integer t is a *divisor* of an integer s if there is an integer u such that $s = tu$. In this case, we write $t \mid s$ (read "t divides s"). When t is not a divisor of s, we write $t \nmid s$. A *prime* is a positive integer greater than 1 whose only positive divisors are 1 and itself. We say an integer s is a *multiple* of an integer t if there is an integer u such that $s = tu$ or, equivalently, if t is a divisor of s.

As our first application of the Well Ordering Principle, we establish a fundamental property of integers that we will use often.

Theorem 0.1 Division Algorithm

Let a and b be integers with $b > 0$. Then there exist unique integers q and r with the property that $a = bq + r$, where $0 \leq r < b$.

PROOF We begin with the existence portion of the theorem. Consider the set $S = \{a - bk \mid k \text{ is an integer and } a - bk \geq 0\}$. If $0 \in S$, then b

divides a and we may obtain the desired result with $q = a/b$ and $r = 0$. Now assume $0 \notin S$. Since S is nonempty [if $a > 0$, $a - b \cdot 0 \in S$; if $a < 0$, $a - b(2a) = a(1 - 2b) \in S$; $a \neq 0$ since $0 \notin S$], we may apply the Well Ordering Principle to conclude that S has a smallest member, say $r = a - bq$. Then $a = bq + r$ and $r \geq 0$, so all that remains to be proved is that $r < b$.

If $r \geq b$, then $a - b(q + 1) = a - bq - b = r - b \geq 0$, so that $a - b(q + 1) \in S$. But $a - b(q + 1) < a - bq$, and $a - bq$ is the *smallest* member of S. So, $r < b$.

To establish the uniqueness of q and r, let us suppose that there are integers q, q', r, and r' such that

$$a = bq + r, \quad 0 \leq r < b, \quad \text{and} \quad a = bq' + r', \quad 0 \leq r' < b.$$

For convenience, we may also suppose that $r' \geq r$. Then $bq + r = bq' + r'$ and $b(q - q') = r' - r$. So, b divides $r' - r$ and $0 \leq r' - r \leq r' < b$. It follows that $r' - r = 0$, and therefore $r' = r$ and $q = q'$. ■

The integer q in the division algorithm is called the *quotient* upon dividing a by b; the integer r is called the *remainder* upon dividing a by b.

■ **EXAMPLE 1** For $a = 17$ and $b = 5$, the division algorithm gives $17 = 5 \cdot 3 + 2$; for $a = -23$ and $b = 6$, the division algorithm gives $-23 = 6(-4) + 1$. ■

Definitions Greatest Common Divisor, Relatively Prime Integers

The *greatest common divisor* of two nonzero integers a and b is the largest of all common divisors of a and b. We denote this integer by gcd(a, b). When gcd(a, b) = 1, we say a and b are *relatively prime*.

The following property of the greatest common divisor of two integers plays a critical role in abstract algebra. The proof provides an application of the division algorithm and our second application of the Well Ordering Principle.

■ Theorem 0.2 GCD Is a Linear Combination

For any nonzero integers a and b, there exist integers s and t such that gcd(a, b) = as + bt. Moreover, gcd(a, b) is the smallest positive integer of the form as + bt.

PROOF Consider the set $S = \{am + bn \mid m, n \text{ are integers and } am + bn > 0\}$. Since S is obviously nonempty (if some choice of m

and n makes $am + bn < 0$, then replace m and n by $-m$ and $-n$), the Well Ordering Principle asserts that S has a smallest member, say, $d = as + bt$. We claim that $d = \gcd(a, b)$. To verify this claim, use the division algorithm to write $a = dq + r$, where $0 \le r < d$. If $r > 0$, then $r = a - dq = a - (as + bt)q = a - asq - btq = a(1 - sq) + b(-tq) \in S$, contradicting the fact that d is the smallest member of S. So, $r = 0$ and d divides a. Analogously (or, better yet, by symmetry), d divides b as well. This proves that d is a common divisor of a and b. Now suppose d' is another common divisor of a and b and write $a = d'h$ and $b = d'k$. Then $d = as + bt = (d'h)s + (d'k)t = d'(hs + kt)$, so that d' is a divisor of d. Thus, among all common divisors of a and b, d is the greatest. ∎

The special case of Theorem 0.2 when a and b are relatively prime is so important in abstract algebra that we single it out as a corollary.

Corollary

If a and b are relatively prime, then there exist integers s and t such that $as + bt = 1$.

EXAMPLE 2 $\gcd(4, 15) = 1$; $\gcd(4, 10) = 2$; $\gcd(2^2 \cdot 3^2 \cdot 5, 2 \cdot 3^3 \cdot 7^2) = 2 \cdot 3^2$. Note that 4 and 15 are relatively prime, whereas 4 and 10 are not. Also, $4 \cdot 4 + 15(-1) = 1$ and $4(-2) + 10 \cdot 1 = 2$. ∎

The next lemma is frequently used. It appeared in Euclid's *Elements*.

Euclid's Lemma $p \mid ab$ **Implies** $p \mid a$ **or** $p \mid b$

If p is a prime that divides ab, then p divides a or p divides b.

PROOF Suppose p is a prime that divides ab but does not divide a. We must show that p divides b. Since p does not divide a, there are integers s and t such that $1 = as + pt$. Then $b = abs + ptb$, and since p divides the right-hand side of this equation, p also divides b. ∎

Note that Euclid's Lemma may fail when p is not a prime, since $6 \mid (4 \cdot 3)$ but $6 \nmid 4$ and $6 \nmid 3$.

Our next property shows that the primes are the building blocks for all integers. We will often use this property without explicitly saying so.

Theorem 0.3 Fundamental Theorem of Arithmetic

Every integer greater than 1 *is a prime or a product of primes. This product is unique, except for the order in which the factors appear. That is, if* $n = p_1p_2 \cdots p_r$ *and* $n = q_1q_2 \cdots q_s$, *where the p's and q's are primes, then* $r = s$ *and, after renumbering the q's, we have* $p_i = q_i$ *for all i.*

We will prove the existence portion of Theorem 0.3 later in this chapter (Example 11). The uniqueness portion is a consequence of Euclid's Lemma (Exercise 31).

Another concept that frequently arises is that of the least common multiple of two integers.

Definition Least Common Multiple

The *least common multiple* of two nonzero integers a and b is the smallest positive integer that is a multiple of both a and b. We will denote this integer by $\text{lcm}(a, b)$.

We leave it as an exercise (Exercise 10) to prove that every common multiple of a and b is a multiple of $\text{lcm}(a, b)$.

EXAMPLE 3 $\text{lcm}(4, 6) = 12$; $\text{lcm}(4, 8) = 8$; $\text{lcm}(10, 12) = 60$; $\text{lcm}(6, 5) = 30$; $\text{lcm}(2^2 \cdot 3^2 \cdot 5, 2 \cdot 3^3 \cdot 7^2) = 2^2 \cdot 3^3 \cdot 5 \cdot 7^2$. ∎

Modular Arithmetic

Another application of the division algorithm that will be important to us is modular arithmetic. Modular arithmetic is an abstraction of a method of counting that you often use. For example, if it is now September, what month will it be 25 months from now? Of course, the answer is October, but the interesting fact is that you didn't arrive at the answer by starting with September and counting off 25 months. Instead, without even thinking about it, you simply observed that $25 = 2 \cdot 12 + 1$, and you added 1 month to September. Similarly, if it is now Wednesday, you know that in 23 days it will be Friday. This time, you arrived at your answer by noting that $23 = 7 \cdot 3 + 2$, so you added 2 days to Wednesday instead of counting off 23 days. If your electricity is off for 26 hours, you must advance your clock 2 hours, since $26 = 2 \cdot 12 + 2$. Surprisingly, this simple idea has numerous important applications in mathematics and computer science. You will see a few of them in this section. The following notation is convenient.

When $a = qn + r$, where q is the quotient and r is the remainder upon dividing a by n, we write $a \bmod n = r$. Thus,

$$
\begin{aligned}
3 \bmod 2 &= 1 \text{ since } 3 = 1 \cdot 2 + 1, \\
6 \bmod 2 &= 0 \text{ since } 6 = 3 \cdot 2 + 0, \\
11 \bmod 3 &= 2 \text{ since } 11 = 3 \cdot 3 + 2, \\
62 \bmod 85 &= 62 \text{ since } 62 = 0 \cdot 85 + 62, \\
-2 \bmod 15 &= 13 \text{ since } -2 = (-1)15 + 13.
\end{aligned}
$$

In general, if a and b are integers and n is a positive integer, then $a \bmod n = b \bmod n$ if and only if n divides $a - b$ (Exercise 7).

In our applications, we will use addition and multiplication mod n. When you wish to compute $ab \bmod n$ or $(a + b) \bmod n$, and a or b is greater than n, it is easier to "mod first." For example, to compute $(27 \cdot 36) \bmod 11$, we note that $27 \bmod 11 = 5$ and $36 \bmod 11 = 3$, so $(27 \cdot 36) \bmod 11 = (5 \cdot 3) \bmod 11 = 4$. (See Exercise 9.)

Modular arithmetic is often used in assigning an extra digit to identification numbers for the purpose of detecting forgery or errors. We present two such applications.

EXAMPLE 4 The United States Postal Service money order shown in Figure 0.1 has an identification number consisting of 10 digits together with an extra digit called a *check*. The check digit is the 10-digit number modulo 9. Thus, the number 3953988164 has the check digit 2, since

Figure 0.1

3953988164 mod 9 = 2.[†] If the number 39539881642 were incorrectly entered into a computer (programmed to calculate the check digit) as, say, 39559881642 (an error in the fourth position), the machine would calculate the check digit as 4, whereas the entered check digit would be 2. Thus, the error would be detected. ∎

∎ EXAMPLE 5 Airline companies, the United Parcel Service, and the rental-car companies Avis and National use the mod 7 values of identification numbers to assign check digits. Thus, the identification number 00121373147367 (see Figure 0.2) has the check digit 3 appended

Figure 0.2

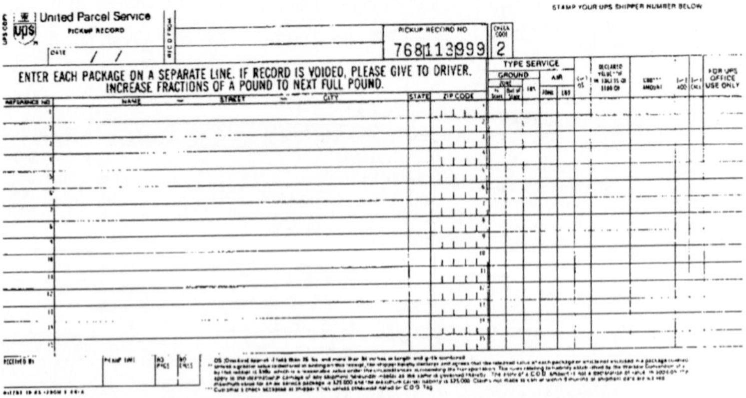

United Parcel Service
PICKUP RECORD

PICKUP RECORD NO 768113999 | CHECK CODE 2

ENTER EACH PACKAGE ON A SEPARATE LINE. IF RECORD IS VOIDED, PLEASE GIVE TO DRIVER.
INCREASE FRACTIONS OF A POUND TO NEXT FULL POUND.

REFERENCE NO | NAME | STREET | CITY | STATE | ZIP CODE | TYPE SERVICE: GROUND, AIR | DECLARED VALUE | FOR UPS OFFICE USE ONLY

STAMP YOUR UPS SHIPPER NUMBER BELOW

Figure 0.3

[†]The value of N mod 9 is easy to compute with a calculator. If $N = 9q + r$, where r is the remainder upon dividing N by 9, then on a calculator screen $N \div 9$ appears as $q.rrrrr\ldots$, so the first decimal digit is the check digit. For example, $3953988164 \div 9 = 439332018.222$, so 2 is the check digit. If N has too many digits for your calculator, replace N by the sum of its digits and divide that number by 9. Thus, 3953988164 mod 9 = 56 mod 9 = 2. The value of 3953988164 mod 9 can also be computed by searching Google for "3953988164 mod 9."

to it because 121373147367 mod 7 = 3. Similarly, the UPS pickup record number 768113999, shown in Figure 0.3, has the check digit 2 appended to it. ■

The methods used by the Postal Service and the airline companies do not detect all single-digit errors (see Exercises 41 and 45). However, detection of all single-digit errors, as well as nearly all errors involving the transposition of two adjacent digits, is easily achieved. One method that does this is the one used to assign the so-called Universal Product Code (UPC) to most retail items (see Figure 0.4). A UPC identification number has 12 digits. The first six digits identify the manufacturer, the next five identify the product, and the last is a check. (For many items, the 12th digit is not printed, but it is always bar-coded.) In Figure 0.4, the check digit is 8.

Figure 0.4

To explain how the check digit is calculated, it is convenient to introduce the dot product notation for two k-tuples:

$$(a_1, a_2, \ldots, a_k) \cdot (w_1, w_2, \ldots, w_k) = a_1w_1 + a_2w_2 + \cdots + a_kw_k.$$

An item with the UPC identification number $a_1a_2 \cdots a_{12}$ satisfies the condition

$$(a_1, a_2, \ldots, a_{12}) \cdot (3, 1, 3, 1, \ldots, 3, 1) \bmod 10 = 0.$$

To verify that the number in Figure 0.4 satisfies this condition, we calculate

$$(0 \cdot 3 + 2 \cdot 1 + 1 \cdot 3 + 0 \cdot 1 + 0 \cdot 3 + 0 \cdot 1 + 6 \cdot 3 + 5 \cdot 1 \\ + 8 \cdot 3 + 9 \cdot 1 + 7 \cdot 3 + 8 \cdot 1) \bmod 10 = 90 \bmod 10 = 0.$$

The fixed k-tuple used in the calculation of check digits is called the *weighting vector*.

Now suppose a single error is made in entering the number in Figure 0.4 into a computer. Say, for instance, that 021000958978 is entered (notice that the seventh digit is incorrect). Then the computer calculates

$$0 \cdot 3 + 2 \cdot 1 + 1 \cdot 3 + 0 \cdot 1 + 0 \cdot 3 + 0 \cdot 1 + 9 \cdot 3 \\ + 5 \cdot 1 + 8 \cdot 3 + 9 \cdot 1 + 7 \cdot 3 + 8 \cdot 1 = 99.$$

Since 99 mod 10 $\neq 0$, the entered number cannot be correct.

In general, any single error will result in a sum that is not 0 modulo 10.

The advantage of the UPC scheme is that it will detect nearly all errors involving the transposition of two adjacent digits as well as all errors involving one digit. For doubters, let us say that the identification number given in Figure 0.4 is entered as 021000658798. Notice that the last two digits preceding the check digit have been transposed. But by calculating the dot product, we obtain 94 mod 10 $\neq 0$, so we have detected an error. In fact, the only undetected transposition errors of adjacent digits a and b are those where $|a - b| = 5$. To verify this, we observe that a transposition error of the form

$$a_1a_2 \cdots a_ia_{i+1} \cdots a_{12} \rightarrow a_1a_2 \cdots a_{i+1}a_i \cdots a_{12}$$

is undetected if and only if

$$(a_1, a_2, \ldots, a_{i+1}, a_i, \ldots, a_{12}) \cdot (3, 1, 3, 1, \ldots, 3, 1) \bmod 10 = 0.$$

That is, the error is undetected if and only if

$$(a_1, a_2, \ldots, a_{i+1}, a_i, \ldots, a_{12}) \cdot (3, 1, 3, 1, \ldots, 3, 1) \bmod 10 \\ = (a_1, a_2, \ldots, a_i, a_{i+1}, \ldots, a_{12}) \cdot (3, 1, 3, 1, \ldots, 3, 1) \bmod 10.$$

This equality simplifies to either

$$(3a_{i+1} + a_i) \bmod 10 = (3a_i + a_{i+1}) \bmod 10$$

or

$$(a_{i+1} + 3a_i) \bmod 10 = (a_i + 3a_{i+1}) \bmod 10,$$

depending on whether i is even or odd. Both cases reduce to $2(a_{i+1} - a_i) \bmod 10 = 0$. It follows that $|a_{i+1} - a_i| = 5$, if $a_{i+1} \neq a_i$.

In 2005, United States companies began to phase in the use of a 13th digit to be in conformance with the 13-digit product identification numbers used in Europe. The weighting vector for 13-digit numbers is (1, 3, 1, 3, . . . , 3, 1).

Identification numbers printed on bank checks (on the bottom left between the two colons) consist of an eight-digit number $a_1a_2 \cdots a_8$ and a check digit a_9, so that

$$(a_1, a_2, \ldots, a_9) \cdot (7, 3, 9, 7, 3, 9, 7, 3, 9) \bmod 10 = 0.$$

As is the case for the UPC scheme, this method detects all single-digit errors and all errors involving the transposition of adjacent digits a and b except when $|a - b| = 5$. But it also detects most errors of the form $\cdots abc \cdots \to \cdots cba \cdots$, whereas the UPC method detects no errors of this form.

In Chapter 5, we will examine more sophisticated means of assigning check digits to numbers.

What about error correction? Suppose you have a number such as 73245018 and you would like to be sure that if even a single mistake were made in entering this number into a computer, the computer would nevertheless be able to determine the correct number. (Think of it. You could make a mistake in dialing a telephone number but still get the correct phone to ring!) This is possible using two check digits. One of the check digits determines the magnitude of any single-digit error, while the other check digit locates the position of the error. With these two pieces of information, you can fix the error. To illustrate the idea, let us say that we have the eight-digit identification number $a_1a_2 \cdots a_8$. We assign two check digits a_9 and a_{10} so that

$$(a_1 + a_2 + \cdots + a_9 + a_{10}) \bmod 11 = 0$$

and

$$(a_1, a_2, \ldots, a_9, a_{10}) \cdot (1, 2, 3, \ldots, 10) \bmod 11 = 0$$

are satisfied.

Let's do an example. Say our number before appending the two check digits is 73245018. Then a_9 and a_{10} are chosen to satisfy

$$(7 + 3 + 2 + 4 + 5 + 0 + 1 + 8 + a_9 + a_{10}) \bmod 11 = 0 \qquad (1)$$

and

$$\begin{aligned}(7 \cdot 1 + 3 \cdot 2 + 2 \cdot 3 + 4 \cdot 4 + 5 \cdot 5 + 0 \cdot 6 \\ + 1 \cdot 7 + 8 \cdot 8 + a_9 \cdot 9 + a_{10} \cdot 10) \bmod 11 = 0.\end{aligned} \qquad (2)$$

Since $7 + 3 + 2 + 4 + 5 + 0 + 1 + 8 = 30$ and $30 \bmod 11 = 8$, Equation (1) reduces to

$$(8 + a_9 + a_{10}) \bmod 11 = 0. \qquad (1')$$

Likewise, since $(7 \cdot 1 + 3 \cdot 2 + 2 \cdot 3 + 4 \cdot 4 + 5 \cdot 5 + 0 \cdot 6 + 1 \cdot 7 + 8 \cdot 8) \bmod 11 = 10$, Equation (2) reduces to

$$(10 + 9a_9 + 10a_{10}) \bmod 11 = 0. \qquad (2')$$

Since we are using mod 11, we may rewrite Equation (2′) as

$$(-1 - 2a_9 - a_{10}) \bmod 11 = 0$$

and add this to Equation (1′) to obtain $7 - a_9 = 0$. Thus $a_9 = 7$. Now substituting $a_9 = 7$ into Equation (1′) or Equation (2′), we obtain $a_{10} = 7$ as well. So, the number is encoded as 7324501877.

Now let us suppose that this number is erroneously entered into a computer programmed with our encoding scheme as 7824501877 (an error in position 2). Since the sum of the digits of the received number mod 11 is 5, we know that some digit is 5 too large (assuming only one error has been made). But which one? Say the error is in position i. Then the second dot product has the form $a_1 \cdot 1 + a_2 \cdot 2 + \ldots + (a_i + 5)i + a_{i+1} \cdot (i + 1) + \ldots + a_{10} \cdot 10 = (a_1, a_2, \ldots, a_{10}) \cdot (1, 2, \ldots, 10) + 5i$. So, $(7, 8, 2, 4, 5, 0, 1, 8, 7, 7) \cdot (1, 2, 3, 4, 5, 6, 7, 8, 9, 10) \bmod 11 = 5i \bmod 11$. Since the left-hand side mod 11 is 10, we see that $i = 2$. Our conclusion: The digit in position 2 is 5 too large. We have successfully corrected the error.

Modular arithmetic is often used to verify the validity of statements about divisibility regarding all positive integers by checking only finitely many cases.

EXAMPLE 6 Consider the statement, "The sum of the cubes of any three consecutive integers is divisible by 9." This statement is equivalent to checking that the equation $(n^3 + (n + 1)^3 + (n + 2)^3) \bmod 9 = 0$ is true for all integers n. Because of properties of modular arithmetic, to prove this, all we need do is check the validity of the equation for $n = 0, 1, \ldots, 8$. ■

Modular arithmetic is occasionally used to show that certain equations have no rational number solutions.

EXAMPLE 7 We use mod 3 arithmetic to show that there are no integers a and b such that $a^2 - 6b = 2$. To see this, suppose that there are such integers. Then, taking both sides modulo 3, there is an integer solution to $a^2 \bmod 3 = 2$. But trying $a = 0, 1$, and 2 we obtain a contradiction. ■

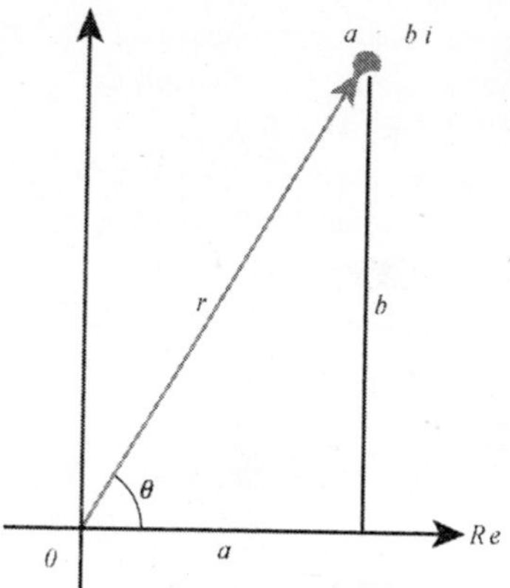

Figure 0.5

Complex Numbers

Recall that complex numbers are expressions of the form $a + b\sqrt{-1}$, where a and b are real numbers. The number $\sqrt{-1}$ is defined to have the property $\sqrt{-1}^2 = -1$. It is customary to use i to denote $\sqrt{-1}$. Then, $i^2 = -1$. Complex numbers written in the form $a + bi$ are said to be in *standard* form. In some instances it is convenient to write a complex number $a + bi$ in another form. To do this we represent $a + bi$ as the point (a,b) in a plane coordinatized by a horizontal axis called the *real* axis and a vertical i axis called the *imaginary* axis. The distance from the point $a + bi$ to the origin is $r = \sqrt{a^2 + b^2}$ and is often denoted by $|a + bi|$. If we draw the line segment from the origin to $a + bi$ and denote the angle formed by the line segment and the positive real axis by θ, we can write $a + bi$ as $r(\cos\theta + i\sin\theta)$ (see Figure 0.5). This form of $a + bi$ is called the *polar* form. An advantage of the polar form is demonstrated in parts 5 and 6 of Theorem 0.4.

Theorem 0.4 Properties of Complex Numbers

1. Closure under addition: $(a + bi) + (c + di) = (a + c) + (b + d)i$
2. Closure under multiplication: $(a + bi)(c + di) = (ac) + (ad)i + (bc)i + (bd)i^2 = (ac - bd) + (ad + bc)i$
3. Closure under division $(c + di \neq 0)$: $\dfrac{(a + bi)}{(c + di)} = \dfrac{(a + bi)}{(c + di)}\dfrac{(c - di)}{(c - di)} = \dfrac{(ac + bd) + (bc - ad)i}{c^2 + d^2} = \dfrac{(ac + bd)}{c^2 + d^2} + \dfrac{(bc - ad)}{c^2 + d^2}i$

4. Complex conjugation: $(a + bi)(a - bi) = a^2 + b^2$
5. Inverses: For every nonzero complex number $a + bi$ there is a complex number $c + di$ such that $(a + bi)(c + di) = 1$. (That is, $(a + bi)^{-1}$ exists in C.)
6. Powers: For every complex number $a + bi = r(\cos\theta + i\sin\theta)$ and every positive integer n, we have $(a + bi)^n = [r(\cos\theta + i\sin\theta)]^n = r^n(\cos n\theta + i\sin n\theta)$.
7. Radicals: For every complex number $a + bi = r(\cos\theta + i\sin\theta)$ and every positive integer n, we have $(a + bi)^{\frac{1}{n}} = [r(\cos\theta + i\sin\theta)]^{\frac{1}{n}} = r^{\frac{1}{n}}(\cos\frac{\theta}{n} + i\sin\frac{\theta}{n})$.

PROOF Parts 1 and 2 are definitions. Part 4 follows from part 2. Part 6 is proved in Example 10 in the next section of this chapter. Part 7 follows from Exercise 25 in this chapter. ∎

The next example illustrates properties of complex numbers.

EXAMPLE 8 $(3 + 5i) + (-5 + 2i) = -2 + 7i$;
$(3 + 5i)(-5 + 2i) = -25 + (-19)i = -25 - 19i$;

$$\frac{3 + 5i}{-2 + 7i} = \frac{3 + 5i}{-2 + 7i}\,\frac{-2 - 7i}{-2 - 7i} = \frac{29 - 31i}{53} = \frac{29}{53} + \frac{-31}{53}i;$$

$(3 + 5i)(3 - 5i) = 9 + 25 = 34$;

$$(3 + 5i)^{-1} = \frac{3}{34} - \frac{5}{34}i.$$

To find $(3 + 5i)^4$ and $(3 + 5i)^{\frac{1}{4}}$ we first note that if $\theta = \arctan\frac{5}{3}$, then $\cos\theta = \frac{3}{\sqrt{34}}$ and $\sin\theta = \frac{5}{\sqrt{34}}$. Thus, $(3 + 5i)^4 = ((\sqrt{34}(\cos\theta + i\sin\theta))^4 = \sqrt{34}^4(\cos 4\theta + i\sin 4\theta)$ and $(3 + 5i)^{\frac{1}{4}} = (\sqrt{34}(\cos\theta + i\sin\theta))^{\frac{1}{4}} = \sqrt{34}^{\frac{1}{4}}(\cos\frac{\theta}{4} + i\sin\frac{\theta}{4})$.

Mathematical Induction

There are two forms of proof by mathematical induction that we will use. Both are equivalent to the Well Ordering Principle. The explicit formulation of the method of mathematical induction came in the 16th century. Francisco Maurolico (1494–1575), a teacher of Galileo, used it in 1575 to prove that $1 + 3 + 5 + \cdots + (2n - 1) = n^2$, and Blaise Pascal (1623–1662) used it when he presented what we now call Pascal's triangle for the coefficients of the binomial expansion. The term *mathematical induction* was coined by Augustus De Morgan.

Theorem 0.5 First Principle of Mathematical Induction

> *Let S be a set of integers containing a. Suppose S has the property that whenever some integer $n \geq a$ belongs to S, then the integer $n + 1$ also belongs to S. Then, S contains every integer greater than or equal to a.*

PROOF The proof is left as an exercise (Exercise 33). ∎

So, to use induction to prove that a statement involving positive integers is true for every positive integer, we must first verify that the statement is true for the integer 1. We then *assume* the statement is true for the integer n and use this assumption to prove that the statement is true for the integer $n + 1$.

Our next example uses some facts about plane geometry. Recall that given a straightedge and compass, we can construct a right angle.

EXAMPLE 9 We use induction to prove that given a straightedge, a compass, and a unit length, we can construct a line segment of length $\sqrt{n}$ for every positive integer n. The case when $n = 1$ is given. Now we assume that we can construct a line segment of length $\sqrt{n}$. Then use the straightedge and compass to construct a right triangle with height 1 and base $\sqrt{n}$. The hypotenuse of the triangle has length $\sqrt{n + 1}$. So, by induction, we can construct a line segment of length $\sqrt{n}$ for every positive integer n. ∎

EXAMPLE 10 DeMOIVRE'S THEOREM We use induction to prove that for every positive integer n and every real number θ, $(\cos\theta + i\sin\theta)^n = \cos n\theta + i\sin n\theta$, where i is the complex number $\sqrt{-1}$. Obviously, the statement is true for $n = 1$. Now assume it is true for n. We must prove that $(\cos\theta + i\sin\theta)^{n+1} = \cos(n + 1)\theta + i\sin(n + 1)\theta$. Observe that

$$\begin{aligned}(\cos\theta + i\sin\theta)^{n+1} &= (\cos\theta + i\sin\theta)^n(\cos\theta + i\sin\theta)\\ &= (\cos n\theta + i\sin n\theta)(\cos\theta + i\sin\theta)\\ &= \cos n\theta\cos\theta + i(\sin n\theta\cos\theta\\ &\quad + \sin\theta\cos n\theta) - \sin n\theta\sin\theta.\end{aligned}$$

Now, using trigonometric identities for $\cos(\alpha + \beta)$ and $\sin(\alpha + \beta)$, we see that this last term is $\cos(n + 1)\theta + i\sin(n + 1)\theta$. So, by induction, the statement is true for all positive integers. ∎

In many instances, the assumption that a statement is true for an integer n does not readily lend itself to a proof that the statement is true for the integer $n + 1$. In such cases, the following equivalent form of induction may be more convenient. Some authors call this formulation the *strong form* of induction.

Theorem 0.6 Second Principle of Mathematical Induction

Let S be a set of integers containing a. Suppose S has the property that n belongs to S whenever every integer less than n and greater than or equal to a belongs to S. Then, S contains every integer greater than or equal to a.

PROOF The proof is left to the reader. ■

To use this form of induction, we first show that the statement is true for the integer a. We then *assume* that the statement is true for *all* integers that are greater than or equal to a and less than n, and use this assumption to prove that the statement is true for n.

EXAMPLE 11 We will use the Second Principle of Mathematical Induction with $a = 2$ to prove the existence portion of the Fundamental Theorem of Arithmetic. Let S be the set of integers greater than 1 that are primes or products of primes. Clearly, $2 \in S$. Now we assume that for some integer n, S contains all integers k with $2 \leq k < n$. We must show that $n \in S$. If n is a prime, then $n \in S$ by definition. If n is not a prime, then n can be written in the form ab, where $1 < a < n$ and $1 < b < n$. Since we are assuming that both a and b belong to S, we know that each of them is a prime or a product of primes. Thus, n is also a product of primes. This completes the proof. ■

Notice that it is more natural to prove the Fundamental Theorem of Arithmetic with the Second Principle of Mathematical Induction than with the First Principle. Knowing that a particular integer factors as a product of primes does not tell you anything about factoring the next larger integer. (Does knowing that 5280 is a product of primes help you to factor 5281 as a product of primes?)

The following problem appeared in the "Brain Boggler" section of the January 1988 issue of the science magazine *Discover.**

EXAMPLE 12 The Quakertown Poker Club plays with blue chips worth \$5.00 and red chips worth \$8.00. What is the largest bet that cannot be made?

---

*"Brain Boggler" by Maxwell Carver. Copyright © 1988 by *Discover Magazine*. Used by permission.

To gain insight into this problem, we try various combinations of blue and red chips and obtain 5, 8, 10, 13, 15, 16, 18, 20, 21, 23, 24, 25, 26, 28, 29, 30, 31, 32, 33, 34, 35, 36, 37, 38, 39, 40. It appears that the answer is 27. But how can we be sure? Well, we need only prove that every integer greater than 27 can be written in the form $a \cdot 5 + b \cdot 8$, where a and b are nonnegative integers. This will solve the problem, since a represents the number of blue chips and b the number of red chips needed to make a bet of $a \cdot 5 + b \cdot 8$. For the purpose of contrast, we will give two proofs—one using the First Principle of Mathematical Induction and one using the Second Principle.

Let S be the set of all integers greater than or equal to 28 of the form $a \cdot 5 + b \cdot 8$, where a and b are nonnegative. Obviously, $28 \in S$. Now assume that some integer $n \in S$, say, $n = a \cdot 5 + b \cdot 8$. We must show that $n + 1 \in S$. First, note that since $n \geq 28$, we cannot have both a and b less than 3. If $a \geq 3$, then

$$\begin{aligned} n + 1 &= (a \cdot 5 + b \cdot 8) + (-3 \cdot 5 + 2 \cdot 8) \\ &= (a - 3) \cdot 5 + (b + 2) \cdot 8. \end{aligned}$$

(Regarding chips, this last equation says that we may increase a bet from n to $n + 1$ by removing three blue chips from the pot and adding two red chips.) If $b \geq 3$, then

$$\begin{aligned} n + 1 &= (a \cdot 5 + b \cdot 8) + (5 \cdot 5 - 3 \cdot 8) \\ &= (a + 5) \cdot 5 + (b - 3) \cdot 8. \end{aligned}$$

(The bet can be increased by 1 by removing three red chips and adding five blue chips.) This completes the proof.

To prove the same statement by the Second Principle, we note that each of the integers 28, 29, 30, 31, and 32 is in S. Now assume that for some integer $n > 32$, S contains all integers k with $28 \leq k < n$. We must show that $n \in S$. Since $n - 5 \in S$, there are nonnegative integers a and b such that $n - 5 = a \cdot 5 + b \cdot 8$. But then $n = (a + 1) \cdot 5 + b \cdot 8$. Thus n is in S. ∎

Equivalence Relations

In mathematics, things that are considered different in one context may be viewed as equivalent in another context. We have already seen one such example. Indeed, the sums 2 + 1 and 4 + 4 are certainly different in ordinary arithmetic, but are the same under modulo 5 arithmetic. Congruent triangles that are situated differently in the plane are not the same, but they are often considered to be the same in plane geometry. In physics, vectors of the same magnitude and direction can produce different effects—a 10-pound weight placed 2 feet from a fulcrum produces a different effect than a 10-pound weight placed 1 foot from a fulcrum. But in linear algebra, vectors of the same magnitude and direction are considered to be the same. What is needed to make these distinctions precise is an appropriate generalization of the notion of equality; that is, we need a formal mechanism for specifying whether or not two quantities are the same in a given setting. This mechanism is an equivalence relation.

Definition Equivalence Relation

An *equivalence relation* on a set S is a set R of ordered pairs of elements of S such that

1. **$(a, a) \in R$ for all $a \in S$ (reflexive property).**
2. **$(a, b) \in R$ implies $(b, a) \in R$ (symmetric property).**
3. **$(a, b) \in R$ and $(b, c) \in R$ imply $(a, c) \in R$ (transitive property).**

When R is an equivalence relation on a set S, it is customary to write aRb instead of $(a, b) \in R$. Also, since an equivalence relation is just a generalization of equality, a suggestive symbol such as $\approx$, $\equiv$, or $\sim$ is usually used to denote the relation. Using this notation, the three conditions for an equivalence relation become $a \sim a$; $a \sim b$ implies $b \sim a$; and $a \sim b$ and $b \sim c$ imply $a \sim c$. If $\sim$ is an equivalence relation on a set S and $a \in S$, then the set $[a] = \{x \in S \mid x \sim a\}$ is called the *equivalence class of S containing a.*

EXAMPLE 13 Let S be the set of all triangles in a plane. If $a, b \in S$, define $a \sim b$ if a and b are similar—that is, if a and b have corresponding angles that are the same. Then $\sim$ is an equivalence relation on S. ∎

EXAMPLE 14 Let S be the set of all polynomials with real coefficients. If $f, g \in S$, define $f \sim g$ if $f' = g'$, where f' is the derivative of f. Then $\sim$ is an equivalence relation on S. Since two polynomials with equal derivatives differ by a constant, we see that for any f in S, $[f] = \{f + c \mid c \text{ is real}\}$. ∎

EXAMPLE 15 Let S be the set of integers and let n be a positive integer. If $a, b \in S$, define $a \equiv b$ if $a \bmod n = b \bmod n$ (that is, if $a - b$ is divisible by n). Then $\equiv$ is an equivalence relation on S and $[a] = \{a + kn \mid k \in S\}$. Since this particular relation is important in abstract algebra, we will take the trouble to verify that it is indeed an equivalence relation. Certainly, $a - a$ is divisible by n, so that $a \equiv a$ for all a in S. Next, assume that $a \equiv b$, say, $a - b = rn$. Then, $b - a = (-r)n$, and therefore $b \equiv a$. Finally, assume that $a \equiv b$ and $b \equiv c$, say, $a - b = rn$ and $b - c = sn$. Then, we have $a - c = (a - b) + (b - c) = rn + sn = (r + s)n$, so that $a \equiv c$. ∎

EXAMPLE 16 Let $\equiv$ be as in Example 15 and let $n = 7$. Then we have $16 \equiv 2$; $9 \equiv -5$; and $24 \equiv 3$. Also, $[1] = \{\ldots, -20, -13, -6, 1, 8, 15, \ldots\}$ and $[4] = \{\ldots, -17, -10, -3, 4, 11, 18, \ldots\}$. ∎

EXAMPLE 17 Let $S = \{(a, b) \mid a, b \text{ are integers}, b \neq 0\}$. If $(a, b), (c, d) \in S$, define $(a, b) \approx (c, d)$ if $ad = bc$. Then $\approx$ is an equivalence relation on S. [The motivation for this example comes from fractions. In fact, the pairs (a, b) and (c, d) are equivalent if the fractions a/b and c/d are equal.]

To verify that $\approx$ is an equivalence relation on S, note that $(a, b) \approx (a, b)$ requires that $ab = ba$, which is true. Next, we assume that $(a, b) \approx (c, d)$, so that $ad = bc$. We have $(c, d) \approx (a, b)$ provided that $cb = da$, which is true from commutativity of multiplication. Finally, we assume that $(a, b) \approx (c, d)$ and $(c, d) \approx (e, f)$ and prove that $(a, b) \approx (e, f)$. This amounts to using $ad = bc$ and $cf = de$ to show that $af = be$. Multiplying both sides of $ad = bc$ by f and replacing cf by de, we obtain $adf = bcf = bde$. Since $d \neq 0$, we can cancel d from the first and last terms. ∎

Definition Partition

A *partition* of a set S is a collection of nonempty disjoint subsets of S whose union is S. Figure 0.6 illustrates a partition of a set into four subsets.

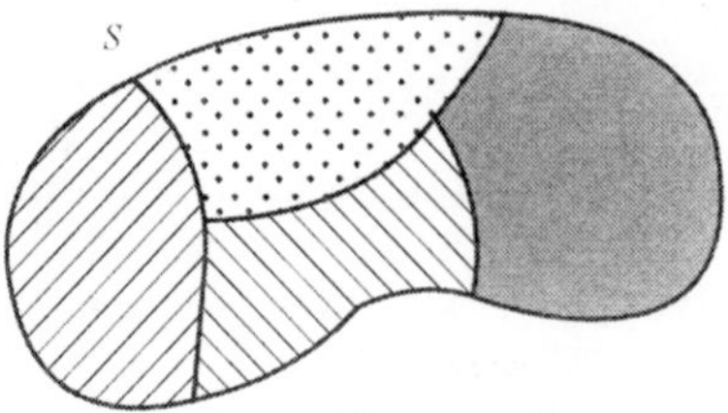

Figure 0.6 Partition of S into four subsets.

EXAMPLE 18 The sets $\{0\}$, $\{1, 2, 3, \ldots\}$, and $\{\ldots, -3, -2, -1\}$ constitute a partition of the set of integers. ∎

EXAMPLE 19 The set of nonnegative integers and the set of nonpositive integers do not partition the integers, since both contain 0. ∎

The next theorem reveals that equivalence relations and partitions are intimately intertwined.

Theorem 0.7 Equivalence Classes Partition

The equivalence classes of an equivalence relation on a set S constitute a partition of S. Conversely, for any partition P of S, there is an equivalence relation on S whose equivalence classes are the elements of P.

PROOF Let $\sim$ be an equivalence relation on a set S. For any $a \in S$, the reflexive property shows that $a \in [a]$. So, $[a]$ is nonempty and the union of all equivalence classes is S. Now, suppose that $[a]$ and $[b]$ are distinct equivalence classes. We must show that $[a] \cap [b] = \emptyset$. On the contrary, assume $c \in [a] \cap [b]$. We will show that $[a] \subseteq [b]$. To this end, let $x \in [a]$. We then have $c \sim a$, $c \sim b$, and $x \sim a$. By the symmetric property, we also have $a \sim c$. Thus, by transitivity, $x \sim c$, and by transitivity again, $x \sim b$. This proves $[a] \subseteq [b]$. Analogously, $[b] \subseteq [a]$. Thus, $[a] = [b]$, in contradiction to our assumption that $[a]$ and $[b]$ are distinct equivalence classes.

To prove the converse, let P be a collection of nonempty disjoint subsets of S whose union is S. Define $a \sim b$ if a and b belong to the same subset in the collection. We leave it to the reader to show that $\sim$ is an equivalence relation on S (Exercise 61). ∎

Functions (Mappings)

Although the concept of a function plays a central role in nearly every branch of mathematics, the terminology and notation associated with functions vary quite a bit. In this section, we establish ours.

Definition Function (Mapping)
A *function* (or *mapping*) ϕ *from a set A to a set B* is a rule that assigns to each element a of A exactly one element b of B. The set A is called the *domain of* ϕ, and B is called the *range of* ϕ. If ϕ assigns b to a, then b is called the *image of a under* ϕ. The subset of B comprising all the images of elements of A is called the *image of A under* ϕ.

We use the shorthand $\phi: A \to B$ to mean that ϕ is a mapping from A to B. We will write $\phi(a) = b$ or $\phi: a \to b$ to indicate that ϕ carries a to b.

There are often different ways to denote the same element of a set. In defining a function in such cases one must verify that the function values assigned to the elements depend not on the way the elements are expressed but on only the elements themselves. For example, the correspondence ϕ from the rational numbers to the integers given by $\phi(a/b) = a + b$ does not define a function since $1/2 = 2/4$ but $\phi(1/2) \neq \phi(2/4)$. To verify that a correspondence is a function, you assume that $x_1 = x_2$ and prove that $\phi(x_1) = \phi(x_2)$.

Definition Composition of Functions

Let $\phi: A \to B$ and $\psi: B \to C$. The *composition* $\psi\phi$ is the mapping from A to C defined by $(\psi\phi)(a) = \psi(\phi(a))$ for all a in A. The composition function $\psi\phi$ can be visualized as in Figure 0.7.

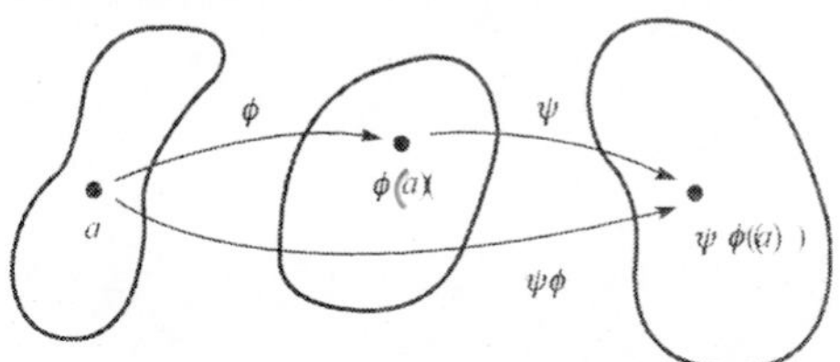

Figure 0.7 Composition of functions ϕ and ψ.

In calculus courses, the composition of f with g is written $(f \circ g)(x)$ and is defined by $(f \circ g)(x) = f(g(x))$. When we compose functions, we omit the "circle."

EXAMPLE 20 Let $f(x) = 2x + 3$ and $g(x) = x^2 + 1$. Then $(fg)(5) = f(g(5)) = f(26) = 55$; $(gf)(5) = g(f(5)) = g(13) = 170$. More generally, $(fg)(x) = f(g(x)) = f(x^2 + 1) = 2(x^2 + 1) + 3 = 2x^2 + 5$ and $(gf)(x) = g(f(x)) = g(2x + 3) = (2x + 3)^2 + 1 = 4x^2 + 12x + 9 + 1 = 4x^2 + 12x + 10$. Note that the function fg is not the same as the function gf. ∎

There are several kinds of functions that occur often enough to be given names.

Definition One-to-One Function

A function ϕ from a set A is called *one-to-one* if for every $a_1, a_2 \in A$, $\phi(a_1) = \phi(a_2)$ implies $a_1 = a_2$.

The term *one-to-one* is suggestive, since the definition ensures that one element of B can be the image of only one element of A. Alternatively, ϕ is one-to-one if $a_1 \neq a_2$ implies $\phi(a_1) \neq \phi(a_2)$. That is, different elements of A map to different elements of B. See Figure 0.8.

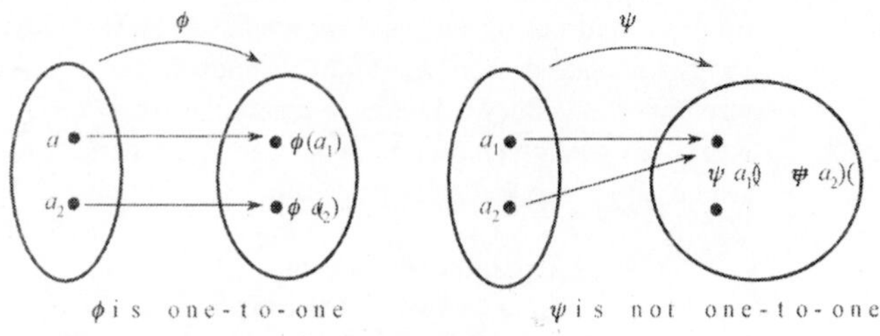

Figure 0.8

Definition Function from A onto B

A function ϕ from a set A to a set B is said to be *onto* B if each element of B is the image of at least one element of A. In symbols, $\phi: A \to B$ is onto if for each b in B there is at least one a in A such that $\phi(a) = b$. See Figure 0.9.

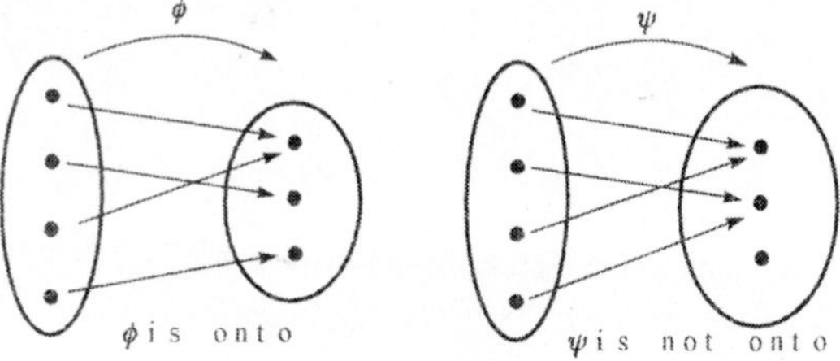

Figure 0.9

The next theorem summarizes the facts about functions we will need.

Theorem 0.8 Properties of Functions

Given functions $\alpha: A \to B$, $\beta: B \to C$, and $\gamma: C \to D$, then

1. *$\gamma(\beta\alpha) = (\gamma\beta)\alpha$ (associativity).*
2. *If α and β are one-to-one, then $\beta\alpha$ is one-to-one.*
3. *If α and β are onto, then $\beta\alpha$ is onto.*
4. *If α is one-to-one and onto, then there is a function α^{-1} from B onto A such that $(\alpha^{-1}\alpha)(a) = a$ for all a in A and $(\alpha\alpha^{-1})(b) = b$ for all b in B.*

PROOF We prove only part 1. The remaining parts are left as exercises (Exercise 57). Let $a \in A$. Then $(\gamma(\beta\alpha))(a) = \gamma((\beta\alpha)(a)) = \gamma(\beta(\alpha(a)))$. On the other hand, $((\gamma\beta)\alpha)(a) = (\gamma\beta)(\alpha(a)) = \gamma(\beta(\alpha(a)))$. So, $\gamma(\beta\alpha) = (\gamma\beta)\alpha$. ∎

It is useful to note that if α is one-to-one and onto, the function α^{-1} described in part 4 of Theorem 0.8 has the property that if $\alpha(s) = t$, then $\alpha^{-1}(t) = s$. That is, the image of t under α^{-1} is the unique element s that maps to t under α. In effect, α^{-1} "undoes" what α does.

EXAMPLE 21 Let **Z** denote the set of integers, **R** the set of real numbers, and **N** the set of nonnegative integers. The following table illustrates the properties of one-to-one and onto.

Domain	Range	Rule	One-to-One	Onto
Z	**Z**	$x \to x^3$	Yes	No
R	**R**	$x \to x^3$	Yes	Yes
Z	**N**	$x \to \|x\|$	No	Yes
Z	**Z**	$x \to x^2$	No	No

To verify that $x \to x^3$ is one-to-one in the first two cases, notice that if $x^3 = y^3$, we may take the cube roots of both sides of the equation to obtain $x = y$. Clearly, the mapping from **Z** to **Z** given by $x \to x^3$ is not onto, since 2 is the cube of no integer. However, $x \to x^3$ defines an onto function from **R** to **R**, since every real number is the cube of its cube root (that is, $\sqrt[3]{b} \to b$). The remaining verifications are left to the reader. ■

Exercises

I was interviewed in the Israeli Radio for five minutes and I said that more than 2000 years ago, Euclid proved that there are infinitely many primes. Immediately the host interrupted me and asked: "Are there still infinitely many primes?"

NOGA ALON

1. For $n = 5, 8, 12, 20$, and 25, find all positive integers less than n and relatively prime to n.
2. Determine $\gcd(2^4 \cdot 3^2 \cdot 5 \cdot 7^2, 2 \cdot 3^3 \cdot 7 \cdot 11)$ and $\text{lcm}(2^3 \cdot 3^2 \cdot 5, 2 \cdot 3^3 \cdot 7 \cdot 11)$.
3. Determine 51 mod 13, 342 mod 85, 62 mod 15, 10 mod 15, $(82 \cdot 73)$ mod 7, $(51 + 68)$ mod 7, $(35 \cdot 24)$ mod 11, and $(47 + 68)$ mod 11.
4. Find integers s and t such that $1 = 7 \cdot s + 11 \cdot t$. Show that s and t are not unique.
5. Show that if a and b are positive integers, then $ab = \text{lcm}(a, b) \cdot \gcd(a, b)$.
6. Suppose a and b are integers that divide the integer c. If a and b are relatively prime, show that ab divides c. Show, by example, that if a and b are not relatively prime, then ab need not divide c.

7. If a and b are integers and n is a positive integer, prove that $a \bmod n = b \bmod n$ if and only if n divides $a - b$.
8. Let $d = \gcd(a, b)$. If $a = da'$ and $b = db'$, show that $\gcd(a', b') = 1$.
9. Let n be a fixed positive integer greater than 1. If $a \bmod n = a'$ and $b \bmod n = b'$, prove that $(a + b) \bmod n = (a' + b') \bmod n$ and $(ab) \bmod n = (a'b') \bmod n$. (This exercise is referred to in Chapters 6, 8, 10, and 15.)
10. Let a and b be positive integers and let $d = \gcd(a, b)$ and $m = \operatorname{lcm}(a, b)$. If t divides both a and b, prove that t divides d. If s is a multiple of both a and b, prove that s is a multiple of m.
11. Let n and a be positive integers and let $d = \gcd(a, n)$. Show that the equation $ax \bmod n = 1$ has a solution if and only if $d = 1$. (This exercise is referred to in Chapter 2.)
12. Show that $5n + 3$ and $7n + 4$ are relatively prime for all n.
13. Suppose that m and n are relatively prime and r is any integer. Show that there are integers x and y such that $mx + ny = r$.
14. Let p, q, and r be primes other than 3. Show that 3 divides $p^2 + q^2 + r^2$.
15. Prove that every prime greater than 3 can be written in the form $6n + 1$ or $6n + 5$.
16. Determine $7^{1000} \bmod 6$ and $6^{1001} \bmod 7$.
17. Let a, b, s, and t be integers. If $a \bmod st = b \bmod st$, show that $a \bmod s = b \bmod s$ and $a \bmod t = b \bmod t$. What condition on s and t is needed to make the converse true? (This exercise is referred to in Chapter 8.)
18. Determine $8^{402} \bmod 5$.
19. Show that $\gcd(a, bc) = 1$ if and only if $\gcd(a, b) = 1$ and $\gcd(a, c) = 1$. (This exercise is referred to in Chapter 8.)
20. Let $p_1, p_2, \ldots, p_n$ be primes. Show that $p_1 p_2 \cdots p_n + 1$ is divisible by none of these primes.
21. Prove that there are infinitely many primes. (*Hint:* Use Exercise 20.)
22. Express $(-7 - 3i)^{-1}$ in standard form.
23. Express $\dfrac{-5 + 2i}{4 - 5i}$ in standard form.
24. Express $(\cos 360° + i \sin 360°)^{1/8}$ in standard form without trig expressions. (Note that $\cos 360° + i \sin 360° = 1$.)
25. Prove that for any positive integer n, $(\cos \theta + i \sin \theta)^{1/n} = \cos \frac{\theta}{n} + i \sin \frac{\theta}{n}$.
26. For every positive integer n, prove that $1 + 2 + \cdots + n = n(n + 1)/2$.

27. For every positive integer n, prove that a set with exactly n elements has exactly 2^n subsets (counting the empty set and the entire set).

28. Prove that $2^n 3^{2n} - 1$ is always divisible by 17.

29. Prove that there is some positive integer n such that $n, n + 1, n + 2, \ldots, n + 200$ are all composite.

30. (Generalized Euclid's Lemma) If p is a prime and p divides $a_1 a_2 \cdots a_n$, prove that p divides a_i for some i.

31. Use the Generalized Euclid's Lemma (see Exercise 30) to establish the uniqueness portion of the Fundamental Theorem of Arithmetic.

32. What is the largest bet that cannot be made with chips worth \$7.00 and \$9.00? Verify that your answer is correct with both forms of induction.

33. Prove that the First Principle of Mathematical Induction is a consequence of the Well Ordering Principle.

34. The Fibonacci numbers are 1, 1, 2, 3, 5, 8, 13, 21, 34, ... In general, the Fibonacci numbers are defined by $f_1 = 1$, $f_2 = 1$, and for $n \geq 3$, $f_n = f_{n-1} + f_{n-2}$. Prove that the nth Fibonacci number f_n satisfies $f_n < 2^n$.

35. Prove by induction on n that for all positive integers n, $n^3 + (n + 1)^3 + (n + 2)^3$ is a multiple of 9.

36. Suppose that there is a statement involving a positive integer parameter n and you have an argument that shows that whenever the statement is true for a particular n it is also true for $n + 2$. What remains to be done to prove the statement is true for every positive integer? Describe a situation in which this strategy would be applicable.

37. In the cut "As" from *Songs in the Key of Life*, Stevie Wonder mentions the equation $8 \times 8 \times 8 = 4$. Find all integers n for which this statement is true, modulo n.

38. Prove that for every integer n, $n^3 \bmod 6 = n \bmod 6$.

39. If it is 2:00 A.M. now, what time will it be 3736 hours from now?

40. Determine the check digit for a money order with identification number 7234541780.

41. Suppose that in one of the noncheck positions of a money order number, the digit 0 is substituted for the digit 9 or vice versa. Prove that this error will not be detected by the check digit. Prove that all other errors involving a single position are detected.

42. Suppose that a money order identification number and check digit of 21720421168 is erroneously copied as 27750421168. Will the check digit detect the error?

43. A transposition error involving distinct adjacent digits is one of the form . . . *ab* . . . → . . . *ba* . . . with $a \neq b$. Prove that the money order check-digit scheme will not detect such errors unless the check digit itself is transposed.

44. Determine the check digit for the Avis rental car with identification number 540047. (See Example 5.)

45. Show that a substitution of a digit a_i' for the digit a_i ($a_i' \neq a_i$) in a noncheck position of a UPS number is detected if and only if $|a_i - a_i'| \neq 7$.

46. Determine which transposition errors involving adjacent digits are detected by the UPS check digit.

47. Use the UPC scheme to determine the check digit for the number 07312400508.

48. Explain why the check digit for a money order for the number N is the repeated decimal digit in the real number $N \div 9$.

49. The 10-digit International Standard Book Number (ISBN-10) $a_1a_2a_3a_4a_5a_6a_7a_8\,a_9a_{10}$ has the property $(a_1, a_2, \ldots, a_{10}) \cdot (10, 9, 8, 7, 6, 5, 4, 3, 2, 1) \bmod 11 = 0$. The digit a_{10} is the check digit. When a_{10} is required to be 10 to make the dot product 0, the character X is used as the check digit. Verify the check digit for the ISBN-10 assigned to this book.

50. Suppose that an ISBN-10 has a smudged entry where the question mark appears in the number 0-716?-2841-9. Determine the missing digit.

51. Suppose three consecutive digits *abc* of an ISBN-10 are scrambled as *bca*. Which such errors will go undetected?

52. The ISBN-10 0-669-03925-4 is the result of a transposition of two adjacent digits not involving the first or last digit. Determine the correct ISBN-10.

53. Suppose the weighting vector for ISBN-10s were changed to (1, 2, 3, 4, 5, 6, 7, 8, 9, 10). Explain how this would affect the check digit.

54. Use the two-check-digit error-correction method described in this chapter to append two check digits to the number 73445860.

55. Suppose that an eight-digit number has two check digits appended using the error-correction method described in this chapter and it is incorrectly transcribed as 4302511568. If exactly one digit is incorrect, determine the correct number.

56. The state of Utah appends a ninth digit a_9 to an eight-digit driver's license number $a_1a_2 \ldots a_8$ so that $(9a_1 + 8a_2 + 7a_3 + 6a_4 + 5a_5 + 4a_6 + 3a_7 + 2a_8 + a_9) \bmod 10 = 0$. If you know that the license number 149105267 has exactly one digit incorrect, explain why the error cannot be in position 2, 4, 6, or 8.

57. Complete the proof of Theorem 0.8.

58. Let S be the set of real numbers. If $a, b \in S$, define $a \sim b$ if $a - b$ is an integer. Show that $\sim$ is an equivalence relation on S. Describe the equivalence classes of S.

59. Let S be the set of integers. If $a, b \in S$, define aRb if $ab \geq 0$. Is R an equivalence relation on S?

60. Let S be the set of integers. If $a, b \in S$, define aRb if $a + b$ is even. Prove that R is an equivalence relation and determine the equivalence classes of S.

61. Complete the proof of Theorem 0.7 by showing that $\sim$ is an equivalence relation on S.

62. Prove that 3, 5, and 7 are the only three consecutive odd integers that are prime.

63. What is the last digit of 3^{100}? What is the last digit of 2^{100}?

64. Prove that none of the integers 11, 111, 1111, 11111, . . . is a square of an integer.

65. (Cancellation Property) Suppose α, β, and γ are functions. If $\alpha\gamma = \beta\gamma$ and γ is one-to-one and onto, prove that $\alpha = \beta$.

Computer Exercises

Computer exercises for this chapter are available at the website:

http://www.d.umn.edu/~jgallian

Suggested Readings

Linda Deneen, "Secret Encryption with Public Keys," *The UMAP Journal* 8 (1987): 9–29.

This well-written article describes several ways in which modular arithmetic can be used to code secret messages. They range from a simple scheme used by Julius Caesar to a highly sophisticated scheme invented in 1978 and based on modular n arithmetic, where n has more than 200 digits.

J. A. Gallian, "Assigning Driver's License Numbers," *Mathematics Magazine* 64 (1991): 13–22.

This article describes various methods used by the states to assign driver's license numbers. Several include check digits for error detection. This article can be downloaded at **http://www.d.umn.edu/~jgallian/license.pdf**

J. A. Gallian, "The Mathematics of Identification Numbers," *The College Mathematics Journal* 22 (1991): 194–202.

This article is a comprehensive survey of check-digit schemes that are associated with identification numbers. This article can be downloaded at **http://www.d.umn.edu/~jgallian/ident.pdf**

J. A. Gallian and S. Winters, "Modular Arithmetic in the Marketplace," *The American Mathematical Monthly* 95 (1988): 548–551.

This article provides a more detailed analysis of the check-digit schemes presented in this chapter. In particular, the error detection rates for the various schemes are given. This article can be downloaded at **http://www.d.umn.edu/~jgallian/marketplace.pdf**

PART 2

Groups

For online student resources, visit this textbook's website at www.CengageBrain.com

1 Introduction to Groups

Symmetry is a vast subject, significant in art and nature. Mathematics lies at its root, and it would be hard to find a better one on which to demonstrate the working of the mathematical intellect.

HERMANN WEYL, *Symmetry*

Symmetries of a Square

Suppose we remove a square region from a plane, move it in some way, then put the square back into the space it originally occupied. Our goal in this chapter is to describe all possible ways in which this can be done. More specifically, we want to describe the possible relationships between the starting position of the square and its final position in terms of motions. However, we are interested in the net effect of a motion, rather than in the motion itself. Thus, for example, we consider a 90° rotation and a 450° rotation as equal, since they have the same net effect on every point. With this simplifying convention, it is an easy matter to achieve our goal.

To begin, we can think of the square region as being transparent (glass, say), with the corners marked on one side with the colors blue, white, pink, and green. This makes it easy to distinguish between motions that have different effects. With this marking scheme, we are now in a position to describe, in simple fashion, all possible ways in which a square object can be repositioned. See Figure 1.1. We now claim that any motion—no matter how complicated—is equivalent to one of these eight. To verify this claim, observe that the final position of the square is completely determined by the location and orientation (that is, face up or face down) of any particular corner. But, clearly, there are only four locations and two orientations for a given corner, so there are exactly eight distinct final positions for the corner.

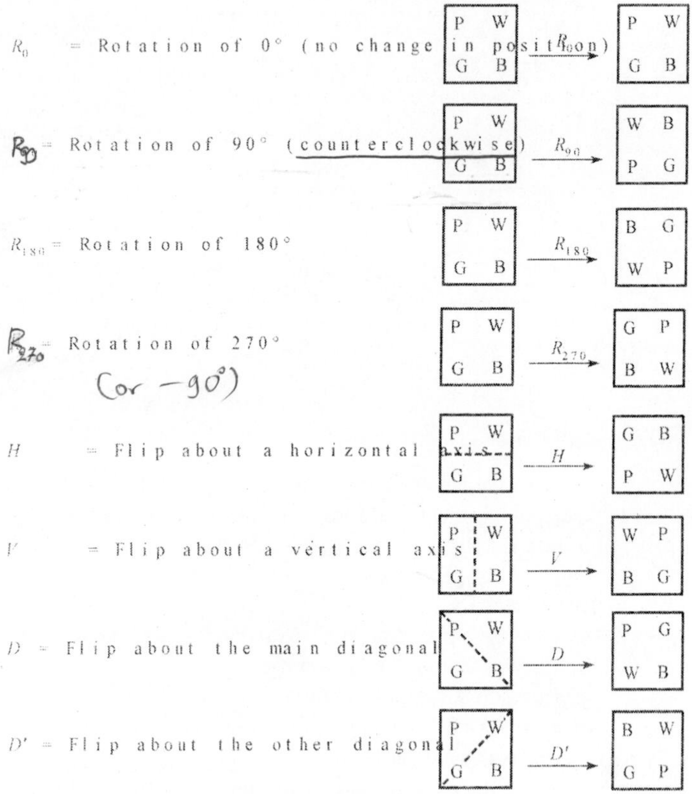

R_0 = Rotation of 0° (no change in position)

R_{90} = Rotation of 90° (counterclockwise)

R_{180} = Rotation of 180°

R_{270} = Rotation of 270°

H = Flip about a horizontal axis

V = Flip about a vertical axis

D = Flip about the main diagonal

D' = Flip about the other diagonal

Figure 1.1

Let's investigate some consequences of the fact that every motion is equal to one of the eight listed in Figure 1.1. Suppose a square is repositioned by a rotation of 90° followed by a flip about the horizontal axis of symmetry.

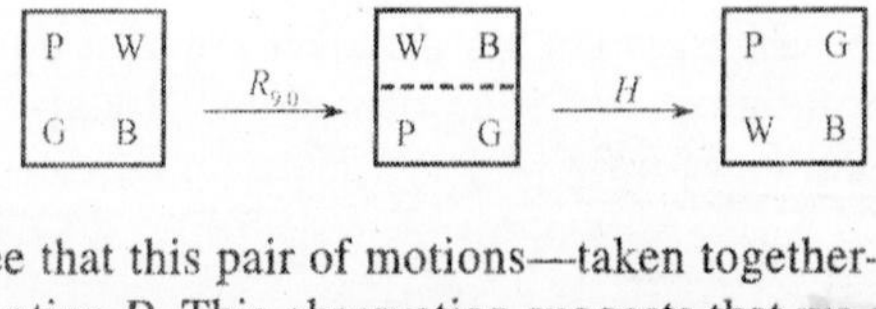

Thus, we see that this pair of motions—taken together—is equal to the single motion D. This observation suggests that we can compose two motions to obtain a single motion. And indeed we can, since the eight motions may be viewed as functions from the square region to itself, and as such we can combine them using function composition. With this in mind, we write $H R_{90} = D$ because in lower level math courses function composition $f \circ g$ means "g followed by f." The eight motions R_0, R_{90}, R_{180}, R_{270}, H, V, D, and D', together with the operation composition, form a mathematical system called the *dihedral group of order 8* (the order of a group is the number of elements it contains). It is denoted by D_4. Rather than introduce the formal definition of a group here, let's look at some properties of groups by way of the example D_4.

To facilitate future computations, we construct an *operation table* or *Cayley table* (so named in honor of the prolific English mathematician Arthur Cayley, who first introduced them in 1854) for D_4 below. The circled entry represents the fact that $D = HR_{90}$. (In general, ab denotes the entry at the intersection of the row with a at the left and the column with b at the top.)

	R_0	R_{90}	R_{180}	R_{270}	H	V	D	D'
R_0	R_0	R_{90}	R_{180}	R_{270}	H	V	D	D'
R_{90}	R_{90}	R_{180}	R_{270}	R_0	D'	D	H	V
R_{180}	R_{180}	R_{270}	R_0	R_{90}	V	H	D'	D
R_{270}	R_{270}	R_0	R_{90}	R_{180}	D	D'	V	H
H	H	D	V	D'	R_0	R_{180}	R_{90}	R_{270}
V	V	D'	H	D	R_{180}	R_0	R_{270}	R_{90}
D	D	V	D'	H	R_{270}	R_{90}	R_0	R_{180}
D'	D'	H	D	V	R_{90}	R_{270}	R_{180}	R_0

Notice how orderly this table looks! This is no accident. Perhaps the most important feature of this table is that it has been completely filled in without introducing any new motions. Of course, this is because, as we have already pointed out, any sequence of motions turns out to be the same as one of these eight. Algebraically, this says that if A and B are in D_4, then so is AB. This property is called *closure*, and it is one of the requirements for a mathematical system to be a group. Next, notice that if A is any element of D_4, then $AR_0 = R_0A = A$. Thus, combining any element A on either side with R_0 yields A back again. An element R_0 with this property is called an *identity*, and every group must have one. Moreover, we see that for each element A in D_4, there is exactly one element B in D_4 such that $AB = BA = R_0$. In this case, B is said to be the *inverse* of A and vice versa. For example, R_{90} and R_{270} are inverses of each other, and H is its own inverse. The term *inverse* is a descriptive one, for if A and B are inverses of each other, then B "undoes" whatever A "does," in the sense that A and B taken together in either order produce R_0, representing no change. Another striking feature of the table is that every element of D_4 appears exactly once in each row and column. This feature is something that all groups must have, and, indeed, it is quite useful to keep this fact in mind when constructing the table in the first place.

Another property of D_4 deserves special comment. Observe that $HD \neq DH$ but $R_{90}R_{180} = R_{180}R_{90}$. Thus, in a group, ab may or may not be the same as ba. If it happens that $ab = ba$ for all choices of group elements a and b, we say the group is *commutative* or—better yet—*Abelian* (in honor of the great Norwegian mathematician Niels Abel). Otherwise, we say the group is *non-Abelian*.

Thus far, we have illustrated, by way of D_4, three of the four conditions that define a group—namely, closure, existence of an identity, and existence of inverses. The remaining condition required for a group is *associativity*; that is, $(ab)c = a(bc)$ for all a, b, c in the set. To be sure that D_4 is indeed a group, we should check this equation for each of the $8^3 = 512$ possible choices of a, b, and c in D_4. In practice, however, this is rarely done! Here, for example, we simply observe that the eight motions are functions and the operation is function composition. Then, since function composition is associative, we do not have to check the equations.

The Dihedral Groups

The analysis carried out above for a square can similarly be done for an equilateral triangle or regular pentagon or, indeed, any regular n-gon ($n \geq 3$). The corresponding group is denoted by D_n and is called the *dihedral group of order 2n*.

The dihedral groups arise frequently in art and nature. Many of the decorative designs used on floor coverings, pottery, and buildings have one of the dihedral groups as a group of symmetry. Corporation logos are rich sources of dihedral symmetry [1]. Chrysler's logo has D_5 as a symmetry group, and that of Mercedes-Benz has D_3. The ubiquitous five-pointed star has symmetry group D_5. The phylum Echinodermata contains many sea animals (such as starfish, sea cucumbers, feather stars, and sand dollars) that exhibit patterns with D_5 symmetry.

Chemists classify molecules according to their symmetry. Moreover, symmetry considerations are applied in orbital calculations, in determining energy levels of atoms and molecules, and in the study of molecular vibrations. The symmetry group of a pyramidal molecule such as ammonia (NH_3), depicted in Figure 1.2, is D_3.

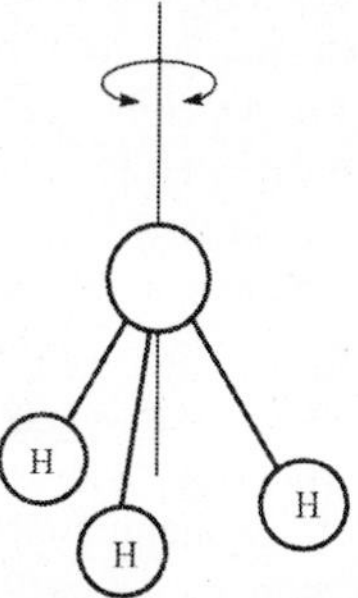

Figure 1.2 A pyramidal molecule with symmetry group D_3.

Mineralogists determine the internal structures of crystals (that is, rigid bodies in which the particles are arranged in three-dimensional repeating patterns—table salt and table sugar are two examples) by studying two-dimensional x-ray projections of the atomic makeup of the crystals. The symmetry present in the projections reveals the internal symmetry of the crystals themselves. Commonly occurring symmetry patterns are D_4 and D_6 (see Figure 1.3). Interestingly, it is mathematically impossible for a crystal to possess a D_n symmetry pattern with $n = 5$ or $n > 6$.

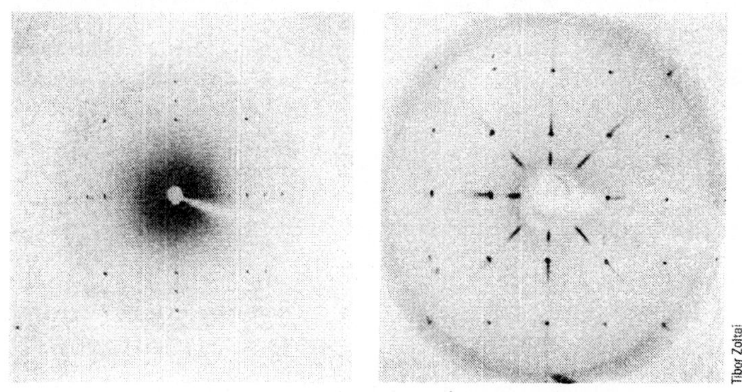

Figure 1.3 X-ray diffraction photos revealing D_4 symmetry patterns in crystals.

The dihedral group of order $2n$ is often called the *group of symmetries of a regular n-gon.* A *plane symmetry* of a figure F in a plane is a function from the plane to itself that carries F onto F and preserves distances; that is, for any points p and q in the plane, the distance from the image of p to the image of q is the same as the distance from p to q. (The term *symmetry* is from the Greek word *symmetros*, meaning "of like measure.") The *symmetry group* of a plane figure is the set of all symmetries of the figure. Symmetries in three dimensions are defined analogously. Obviously, a rotation of a plane about a point in the plane is a symmetry of the plane, and a rotation about a line in three dimensions is a symmetry in three-dimensional space. Similarly, any translation of a plane or of three-dimensional space is a symmetry. A *reflection across a line L* is that function that leaves every point of L fixed and takes any point q, not on L, to the point q' so that L is the perpendicular bisector of the line segment joining q and q' (see Figure 1.4). A reflection across a plane in three dimensions is defined analogously. Notice that the restriction of a 180° rotation about a line L in three dimensions to a plane containing L is a reflection across L in the plane. Thus, in the dihedral groups, the motions that we described as flips about axes of symmetry in three dimensions (for example, H, V, D, D') are reflections across lines in two dimensions. Just as a reflection across a line is a plane symmetry that cannot be achieved by a physical motion of the plane in two dimensions, a reflection across a plane is a three-dimensional symmetry that cannot be achieved by a physical motion of three-dimensional space. A cup, for instance, has reflective symmetry across the plane bisecting the cup, but this symmetry cannot be duplicated with a physical motion in three dimensions.

Figure 1.4

Many objects and figures have rotational symmetry but not reflective symmetry. A symmetry group consisting of the rotational symmetries of 0°, 360°/n, 2(360°)/n, . . . , $(n - 1)360°/n$, and no other symmetries, is called a *cyclic rotation group of order n* and is denoted by $R_{360/n}$. Cyclic rotation groups, along with dihedral groups, are favorites of artists, designers, and nature. Figure 1.5 illustrates with corporate logos the cyclic rotation groups of orders 2, 3, 4, 5, 6, 8, 16, and 20.

A study of symmetry in greater depth is given in Chapters 27 and 28.

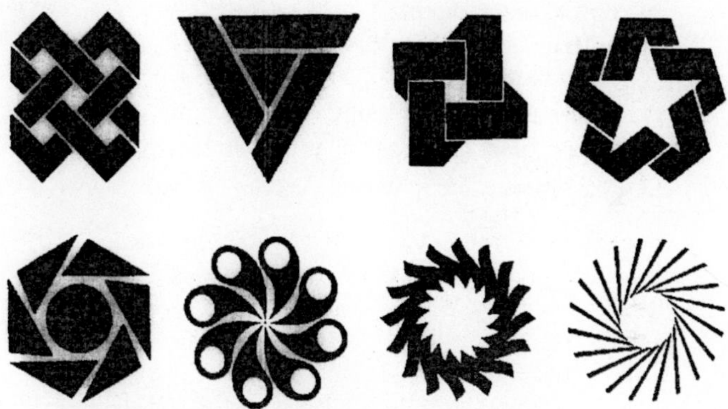

Figure 1.5 Logos with cyclic rotation symmetry groups.

Exercises

The only way to learn mathematics is to do mathematics.

PAUL R. HALMOS, *A Hilbert Space Problem Book*

1. With pictures and words, describe each symmetry in D_3 (the set of symmetries of an equilateral triangle).
2. Write out a complete Cayley table for D_3. Is D_3 Abelian?
3. In D_4, find all elements X such that
   a. $X^3 = V$;
   b. $X^3 = R_{90}$;
   c. $X^3 = R_0$;
   d. $X^2 = R_0$;
   e. $X^2 = H$.
4. Describe in pictures or words the elements of D_5 (symmetries of a regular pentagon).
5. For $n \geq 3$, describe the elements of D_n. (*Hint:* You will need to consider two cases—n even and n odd.) How many elements does D_n have?
6. In D_n, explain geometrically why a reflection followed by a reflection must be a rotation.
7. In D_n, explain geometrically why a rotation followed by a rotation must be a rotation.
8. In D_n, explain geometrically why a rotation and a reflection taken together in either order must be a reflection.
9. Associate the number 1 with a rotation and the number −1 with a reflection. Describe an analogy between multiplying these two numbers and multiplying elements of D_n.

10. If r_1, r_2, and r_3 represent rotations from D_n and f_1, f_2, and f_3 represent reflections from D_n, determine whether $r_1 r_2 f_1 r_3 f_2 f_3 r_3$ is a rotation or a reflection.

11. Find elements A, B, and C in D_4 such that $AB = BC$ but $A \neq C$. (Thus, "cross cancellation" is not valid.)

12. Explain what the following diagram proves about the group D_n.

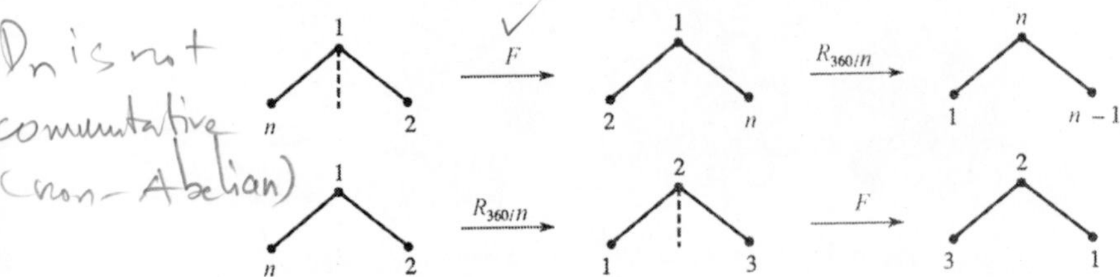

13. Describe the symmetries of a nonsquare rectangle. Construct the corresponding Cayley table.

14. Describe the symmetries of a parallelogram that is neither a rectangle nor a rhombus. Describe the symmetries of a rhombus that is not a rectangle.

15. Describe the symmetries of a noncircular ellipse. Do the same for a hyperbola.

16. Consider an infinitely long strip of equally spaced H's:

$$\cdots \text{H H H H} \cdots$$

Describe the symmetries of this strip. Is the group of symmetries of the strip Abelian?

17. For each of the snowflakes in the figure, find the symmetry group and locate the axes of reflective symmetry (disregard imperfections).

Photographs of snowflakes from the Bentley and Humphreys atlas.

18. Determine the symmetry group of the outer shell of the cross section of the human immunodeficiency virus (HIV) shown below.

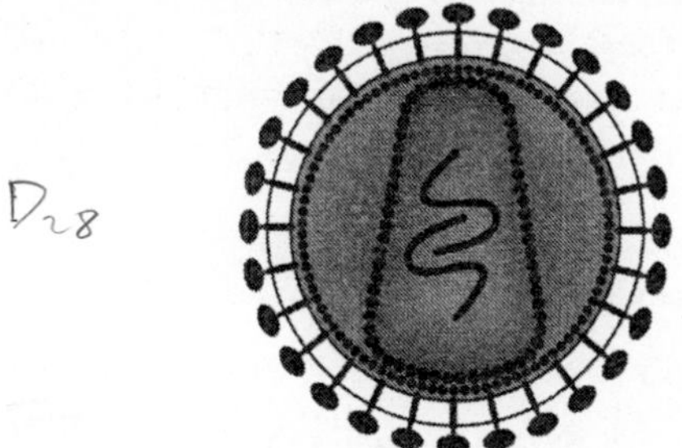

19. Does a fan blade have a cyclic symmetry group or a dihedral symmetry group?

20. Bottle caps that are pried off typically have 22 ridges around the rim. Find the symmetry group of such a cap.

21. What group theoretic property do uppercase letters F, G, J, L, P, Q, R have that is not shared by the remaining uppercase letters in the alphabet?

22. What symmetry property does the word "zoonosis" have when written in uppercase letters? (It means a disease of humans acquired from animals.)

23. What symmetry property do the words "mow," "sis," and "swims" have when written in uppercase letters?

24. For each design below, determine the symmetry group (ignore imperfections).

Suggested Reading

Michael Field and Martin Golubitsky, *Symmetry in Chaos*, Oxford University Press, 1992.

This book has many beautiful symmetric designs that arise in chaotic dynamic systems.

Suggested Website

http://britton.disted.camosun.bc.ca/jbsymteslk.htm

This spectacular website on symmetry and tessellations has numerous activities and links to many other sites on related topics. It is a wonderful website for K–12 teachers and students.

Niels Abel

Stock Montage

He [Abel] has left mathematicians something to keep them busy for five hundred years.

CHARLES HERMITE

A 500-kroner bank note first issued by Norway in 1948.

This stamp was issued in 1929 to commemorate the 100th anniversary of Abel's death.

NIELS HENRIK ABEL, one of the foremost mathematicians of the 19th century, was born in Norway on August 5, 1802. At the age of 16, he began reading the classic mathematical works of Newton, Euler, Lagrange, and Gauss. When Abel was 18 years old, his father died, and the burden of supporting the family fell upon him. He took in private pupils and did odd jobs, while continuing to do mathematical research. At the age of 19, Abel solved a problem that had vexed leading mathematicians for hundreds of years. He proved that, unlike the situation for equations of degree 4 or less, there is no finite (closed) formula for the solution of the general fifth-degree equation.

Although Abel died long before the advent of the subjects that now make up abstract algebra, his solution to the quintic problem laid the groundwork for many of these subjects. Just when his work was beginning to receive the attention it deserved, Abel contracted tuberculosis. He died on April 6, 1829, at the age of 26.

In recognition of the fact that there is no Nobel Prize for mathematics, in 2002 Norway established the Abel Prize as the "Nobel Prize in mathematics" in honor of its native son. At approximately the $1,000,000 level, the Abel Prize is now seen as an award equivalent to a Nobel Prize.

To find more information about Abel, visit:

http://www-groups.dcs.st-and .ac.uk/~history/

2 Groups

A good stock of examples, as large as possible, is indispensable for a thorough understanding of any concept, and when I want to learn something new, I make it my first job to build one.

PAUL R. HALMOS

Definition and Examples of Groups

The term *group* was used by Galois around 1830 to describe sets of one-to-one functions on finite sets that could be grouped together to form a set closed under composition. As is the case with most fundamental concepts in mathematics, the modern definition of a group that follows is the result of a long evolutionary process. Although this definition was given by both Heinrich Weber and Walther von Dyck in 1882, it did not gain universal acceptance until the 20th century.

Definition Binary Operation
Let G be a set. A *binary operation* on G is a function that assigns each ordered pair of elements of G an element of G.

A binary operation on a set G, then, is simply a method (or formula) by which the members of an ordered pair from G combine to yield a new member of G. This condition is called *closure*. The most familiar binary operations are ordinary addition, subtraction, and multiplication of integers. Division of integers is not a binary operation on the integers because an integer divided by an integer need not be an integer.

The binary operations addition modulo n and multiplication modulo n on the set $\{0, 1, 2, \ldots, n - 1\}$, which we denote by Z_n, play an extremely important role in abstract algebra. In certain situations we will want to combine the elements of Z_n by addition modulo n only; in other situations we will want to use both addition modulo n and multiplication modulo n to combine the elements. It will be clear from the context whether we are using addition only or addition and multiplication. For example, when multiplying matrices with entries from Z_n, we will need both addition modulo n and multiplication modulo n.

Definition Group

Let G be a set together with a binary operation (usually called multiplication) that assigns to each ordered pair (a, b) of elements of G an element in G denoted by ab. We say G is a *group* under this operation if the following three properties are satisfied.

1. ***Associativity.* The operation is associative; that is, $(ab)c = a(bc)$ for all a, b, c in G.**
2. ***Identity.* There is an element e (called the *identity*) in G such that $ae = ea = a$ for all a in G.**
3. ***Inverses.* For each element a in G, there is an element b in G (called an *inverse* of a) such that $ab = ba = e$.**

In words, then, a group is a set together with an associative operation such that there is an identity, every element has an inverse, and any pair of elements can be combined without going outside the set. Be sure to verify closure when testing for a group (see Example 5). Notice that if a is the inverse of b, then b is the inverse of a.

If a group has the property that $ab = ba$ for every pair of elements a and b, we say the group is *Abelian*. A group is *non-Abelian* if there is some pair of elements a and b for which $ab \neq ba$. When encountering a particular group for the first time, one should determine whether or not it is Abelian.

Now that we have the formal definition of a group, our first job is to build a good stock of examples. These examples will be used throughout the text to illustrate the theorems. (The best way to grasp the meat of a theorem is to see what it says in specific cases.) As we progress, the reader is bound to have hunches and conjectures that can be tested against the stock of examples. To develop a better understanding of the following examples, the reader should supply the missing details.

EXAMPLE 1 The set of integers Z (so denoted because the German word for numbers is *Zahlen*), the set of rational numbers Q (for quotient), and the set of real numbers $\mathbf{R}$ are all groups under ordinary addition. In each case, the identity is 0 and the inverse of a is $-a$. ∎

EXAMPLE 2 The set of integers under ordinary multiplication is not a group. Since the number 1 is the identity, property 3 fails. For example, there is no integer b such that $5b = 1$. ■

EXAMPLE 3 The subset $\{1, -1, i, -i\}$ of the complex numbers is a group under complex multiplication. Note that -1 is its own inverse, whereas the inverse of i is $-i$, and vice versa. ■

EXAMPLE 4 The set Q^+ of positive rationals is a group under ordinary multiplication. The inverse of any a is $1/a = a^{-1}$. ■

EXAMPLE 5 The set S of positive irrational numbers together with 1 under multiplication satisfies the three properties given in the definition of a group but is not a group. Indeed, $\sqrt{2} \cdot \sqrt{2} = 2$, so S is not closed under multiplication. ■

EXAMPLE 6 A rectangular array of the form $\begin{bmatrix} a & b \\ c & d \end{bmatrix}$ is called a 2×2 *matrix*. The set of all 2×2 matrices with real entries is a group under componentwise addition. That is,

$$\begin{bmatrix} a_1 & b_1 \\ c_1 & d_1 \end{bmatrix} + \begin{bmatrix} a_2 & b_2 \\ c_2 & d_2 \end{bmatrix} = \begin{bmatrix} a_1 + a_2 & b_1 + b_2 \\ c_1 + c_2 & d_1 + d_2 \end{bmatrix}$$

The identity is $\begin{bmatrix} 0 & 0 \\ 0 & 0 \end{bmatrix}$, and the inverse of $\begin{bmatrix} a & b \\ c & d \end{bmatrix}$ is $\begin{bmatrix} -a & -b \\ -c & -d \end{bmatrix}$. ■

EXAMPLE 7 The set $Z_n = \{0, 1, \ldots, n - 1\}$ for $n \geq 1$ is a group under addition modulo n. For any $j > 0$ in Z_n, the inverse of j is $n - j$. This group is usually referred to as the *group of integers modulo n*. ■

As we have seen, the real numbers, the 2×2 matrices with real entries, and the integers modulo n are all groups under the appropriate addition. But what about multiplication? In each case, the existence of some elements that do not have inverses prevents the set from being a group under the usual multiplication. However, we can form a group in each case by simply throwing out the rascals. Examples 8, 9, and 11 illustrate this.

EXAMPLE 8 The set $\mathbf{R}^*$ of nonzero real numbers is a group under ordinary multiplication. The identity is 1. The inverse of a is $1/a$. ■

EXAMPLE 9[†] The *determinant* of the 2×2 matrix $\begin{bmatrix} a & b \\ c & d \end{bmatrix}$ is the number $ad - bc$. If A is a 2×2 matrix, det A denotes the determinant of A. The set

$$GL(2, \mathbf{R}) = \left\{ \begin{bmatrix} a & b \\ c & d \end{bmatrix} \middle| \, a, b, c, d \in \mathbf{R}, ad - bc \neq 0 \right\}$$

of 2×2 matrices with real entries and nonzero determinants is a non-Abelian group under the operation

$$\begin{bmatrix} a_1 & b_1 \\ c_1 & d_1 \end{bmatrix} \begin{bmatrix} a_2 & b_2 \\ c_2 & d_2 \end{bmatrix} = \begin{bmatrix} a_1a_2 + b_1c_2 & a_1b_2 + b_1d_2 \\ c_1a_2 + d_1c_2 & c_1b_2 + d_1d_2 \end{bmatrix}.$$

The first step in verifying that this set is a group is to show that the product of two matrices with nonzero determinants also has a nonzero determinant. This follows from the fact that for any pair of 2×2 matrices A and B, $\det(AB) = (\det A)(\det B)$.

Associativity can be verified by direct (but cumbersome) calculations. The identity is $\begin{bmatrix} 1 & 0 \\ 0 & 1 \end{bmatrix}$; the inverse of $\begin{bmatrix} a & b \\ c & d \end{bmatrix}$ is

$$\begin{bmatrix} \dfrac{d}{ad - bc} & \dfrac{-b}{ad - bc} \\ \dfrac{-c}{ad - bc} & \dfrac{a}{ad - bc} \end{bmatrix}$$

(explaining the requirement that $ad - bc \neq 0$). This very important non-Abelian group is called the *general linear group* of 2×2 matrices over **R**. ∎

EXAMPLE 10 The set of all 2×2 matrices with real entries is not a group under the operation defined in Example 9. Inverses do not exist when the determinant is 0. ∎

Now that we have shown how to make subsets of the real numbers and subsets of the set of 2×2 matrices into multiplicative groups, we next consider the integers under multiplication modulo n.

[†]For simplicity, we have restricted our matrix examples to the 2×2 case. However, readers who have had linear algebra can readily generalize to $n \times n$ matrices.

▮ EXAMPLE 11 (L. EULER, 1761) By Exercise 11 in Chapter 0, an integer a has a multiplicative inverse modulo n if and only if a and n are relatively prime. So, for each $n > 1$, we define $U(n)$ to be the set of all positive integers less than n and relatively prime to n. Then $U(n)$ is a group under multiplication modulo n. (We leave it to the reader to check that this set is closed under this operation.)

For $n = 10$, we have $U(10) = \{1, 3, 7, 9\}$. The Cayley table for $U(10)$ is

mod 10	1	3	7	9
1	1	3	7	9
3	3	9	1	7
7	7	1	9	3
9	9	7	3	1

(Recall that ab mod n is the unique integer r with the property $a \cdot b = nq + r$, where $0 \leq r < n$ and $a \cdot b$ is ordinary multiplication.) In the case that n is a prime, $U(n) = \{1, 2, \ldots, n - 1\}$. ▮

In his classic book *Lehrbuch der Algebra*, published in 1895, Heinrich Weber gave an extensive treatment of the groups $U(n)$ and described them as the most important examples of finite Abelian groups.

▮ EXAMPLE 12 The set $\{0, 1, 2, 3\}$ is not a group under multiplication modulo 4. Although 1 and 3 have inverses, the elements 0 and 2 do not. ▮

▮ EXAMPLE 13 The set of integers under subtraction is not a group, since the operation is not associative. ▮

With the examples given thus far as a guide, it is wise for the reader to pause here and think of his or her own examples. Study actively! Don't just read along and be spoon-fed by the book.

▮ EXAMPLE 14 The complex numbers $\mathbf{C} + \{a + bi \mid a, b \in \mathbf{R}, i^2 = -1\}$ are a group under the operation $(a + bi) + (c + di) = (a + c) + (b + d)i$. The inverse of $a + bi$ is $-a - bi$. The nonzero complex numbers $\mathbf{C}^*$ are a group under the operation $(a + bi)(c + di) = (ac - bd) + (ad + bc)i$. The inverse of $a + bi$ is $\frac{1}{a + bi} = \frac{1}{a + bi}\frac{a - bi}{a - bi} = \frac{1}{a^2 + b^2}a - \frac{1}{a^2 + b^2}bi$. ▮

EXAMPLE 15 For all integers $n \geq 1$, the set of complex nth roots of unity

$$\left\{ \cos \frac{k \cdot 360^\circ}{n} + i \sin \frac{k \cdot 360^\circ}{n} \;\middle|\; k = 0, 1, 2, \ldots, n - 1 \right\}$$

(i.e., complex zeros of $x^n - 1$) is a group under multiplication. (See DeMoivre's Theorem—Example 10 in Chapter 0.) Compare this group with the one in Example 3. ∎

Recall from Chapter 0 that the complex number $\cos \theta + i \sin \theta$ can be represented geometrically as the point $(\cos \theta, \sin \theta)$ in a plane coordinatized by a real horizontal axis and a vertical imaginary axis, where θ is the angle formed by the line segment joining the origin and the point $(\cos \theta, \sin \theta)$ and the positive real axis. Thus, the six complex zeros of $x^6 = 1$ are located at points around the circle of radius 1, 60° apart, as shown in Figure 2.1.

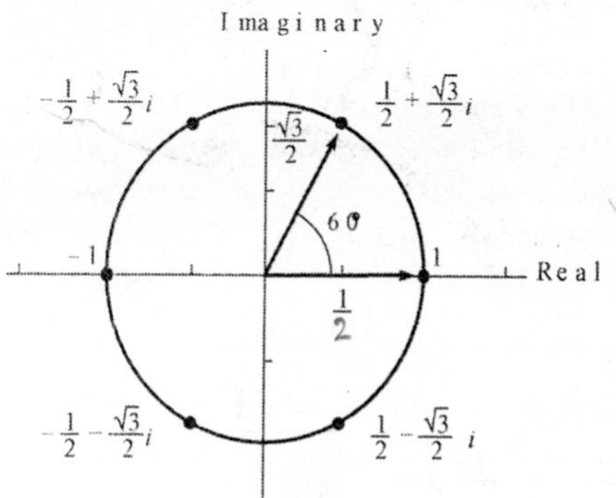

Figure 2.1

EXAMPLE 16 The set $\mathbf{R}^n = \{(a_1, a_2, \ldots, a_n) \mid a_1, a_2, \ldots, a_n \in \mathbf{R}\}$ is a group under componentwise addition [i.e., $(a_1, a_2, \ldots, a_n) + (b_1, b_2, \ldots, b_n) = (a_1 + b_1, a_2 + b_2, \ldots, a_n + b_n)$]. ∎

EXAMPLE 17 For a fixed point (a, b) in $\mathbf{R}^2$, define $T_{a,b}: \mathbf{R}^2 \to \mathbf{R}^2$ by $(x, y) \to (x + a, y + b)$. Then $G = \{T_{a,b} \mid a, b \in \mathbf{R}\}$ is a group under function composition. Straightforward calculations show that $T_{a,b}T_{c,d} = T_{a+c,b+d}$. From this formula we may observe that G is closed, $T_{0,0}$ is the identity, the inverse of $T_{a,b}$ is $T_{-a,-b}$, and G is Abelian. Function composition is always associative. The elements of G are called *translations*. ∎

EXAMPLE 18 The set of all 2×2 matrices with determinant 1 with entries from Q (rationals), **R** (reals), **C** (complex numbers), or Z_p (p a prime) is a non-Abelian group under matrix multiplication. This group is called the *special linear group* of 2×2 matrices over Q, **R**, **C**, or Z_p, respectively. If the entries are from F, where F is any of the above, we denote this group by $SL(2, F)$. For the group $SL(2, F)$, the formula given in Example 9 for the inverse of $\begin{bmatrix} a & b \\ c & d \end{bmatrix}$ simplifies to $\begin{bmatrix} d & -b \\ -c & a \end{bmatrix}$. When the matrix entries are from Z_p, we use modulo p arithmetic to compute determinants, matrix products, and inverses. To illustrate the case $SL(2, Z_5)$, consider the element $A = \begin{bmatrix} 3 & 4 \\ 4 & 4 \end{bmatrix}$. Then $\det A = (3 \cdot 4 - 4 \cdot 4) \bmod 5 = -4 \bmod 5 = 1$, and the inverse of A is $\begin{bmatrix} 4 & -4 \\ -4 & 3 \end{bmatrix} = \begin{bmatrix} 4 & 1 \\ 1 & 3 \end{bmatrix}$. Note that $\begin{bmatrix} 3 & 4 \\ 4 & 4 \end{bmatrix}\begin{bmatrix} 4 & 1 \\ 1 & 3 \end{bmatrix} = \begin{bmatrix} 1 & 0 \\ 0 & 1 \end{bmatrix}$ when the arithmetic is done modulo 5. ∎

Example 9 is a special case of the following general construction.

EXAMPLE 19 Let F be any of Q, **R**, **C**, or Z_p (p a prime). The set $GL(2, F)$ of all 2×2 matrices with nonzero determinants and entries from F is a non-Abelian group under matrix multiplication. As in Example 18, when F is Z_p, modulo p arithmetic is used to calculate determinants, matrix products, and inverses. The formula given in Example 9 for the inverse of $\begin{bmatrix} a & b \\ c & d \end{bmatrix}$ remains valid for elements from $GL(2, Z_p)$, provided we interpret division by $ad - bc$ as multiplication by the inverse of $(ad - bc)$ modulo p. For example, in $GL(2, Z_7)$, consider $\begin{bmatrix} 4 & 5 \\ 6 & 3 \end{bmatrix}$. Then the determinant $(ad - bc) \bmod 7$ is $(12 - 30) \bmod 7 = -18 \bmod 7 = 3$ and the inverse of 3 is 5 [since $(3 \cdot 5) \bmod 7 = 1$]. So, the inverse of $\begin{bmatrix} 4 & 5 \\ 6 & 3 \end{bmatrix}$ is $\begin{bmatrix} 3 \cdot 5 & 2 \cdot 5 \\ 1 \cdot 5 & 4 \cdot 5 \end{bmatrix} = \begin{bmatrix} 1 & 3 \\ 5 & 6 \end{bmatrix}$. [The reader should check that $\begin{bmatrix} 4 & 5 \\ 6 & 3 \end{bmatrix}\begin{bmatrix} 1 & 3 \\ 5 & 6 \end{bmatrix} = \begin{bmatrix} 1 & 0 \\ 0 & 1 \end{bmatrix}$ in $GL(2, Z_7)$]. ∎

The group $GL(n, F)$ is called the *general linear group* of $n \times n$ matrices over F.

EXAMPLE 20 The set $\{1, 2, \ldots, n - 1\}$ is a group under multiplication modulo n if and only if n is prime. ∎

EXAMPLE 21 The set of all symmetries of the infinite ornamental pattern in which arrowheads are spaced uniformly a unit apart along a line is an Abelian group under composition. Let T denote a translation to the right by one unit, T^{-1} a translation to the left by one unit, and H a reflection across the horizontal line of the figure. Then, every member of the group is of the form $x_1x_2 \cdots x_n$, where each $x_i \in \{T, T^{-1}, H\}$. In this case, we say that T, T^{-1}, and H *generate* the group. ■

Table 2.1 summarizes many of the specific groups that we have presented thus far.

As the previous examples demonstrate, the notion of a group is a very broad one indeed. The goal of the axiomatic approach is to find properties general enough to permit many diverse examples having these properties and specific enough to allow one to deduce many interesting consequences.

The goal of abstract algebra is to discover truths about algebraic systems (that is, sets with one or more binary operations) that are independent of the specific nature of the operations. All one knows or needs to know is that these operations, whatever they may be, have

Table 2.1 Summary of Group Examples (F can be any of Q, **R**, **C**, or Z_p; L is a reflection)

Group	Operation	Identity	Form of Element	Inverse	Abelian
Z	Addition	0	k	$-k$	Yes
Q^+	Multiplication	1	m/n, $m, n > 0$	n/m	Yes
Z_n	Addition mod n	0	k	$n - k$	Yes
$\mathbf{R}^*$	Multiplication	1	x	$1/x$	Yes
$\mathbf{C}^*$	Multiplication	1	$a + bi$	$\frac{1}{a^2 + b^2}a - \frac{1}{a^2 - b^2}bi$	Yes
$GL(2, F)$	Matrix multiplication	$\begin{bmatrix} 1 & 0 \\ 0 & 1 \end{bmatrix}$	$\begin{bmatrix} a & b \\ c & d \end{bmatrix}$, $ad - bc \neq 0$	$\begin{bmatrix} \frac{d}{ad - bc} & \frac{-b}{ad - bc} \\ \frac{-c}{ad - bc} & \frac{a}{ad - bc} \end{bmatrix}$	No
$U(n)$	Multiplication mod n	1	k, $\gcd(k, n) = 1$	Solution to $kx \bmod n = 1$	Yes
$\mathbf{R}^n$	Componentwise addition	$(0, 0, \ldots, 0)$	$(a_1, a_2, \ldots, a_n)$	$(-a_1, -a_2, \ldots, -a_n)$	Yes
$SL(2, F)$	Matrix multiplication	$\begin{bmatrix} 1 & 0 \\ 0 & 1 \end{bmatrix}$	$\begin{bmatrix} a & b \\ c & d \end{bmatrix}$, $ad - bc = 1$	$\begin{bmatrix} d & -b \\ -c & a \end{bmatrix}$	No
D_n	Composition	R_0	R_α, L	$R_{360 - \alpha}$, L	No

certain properties. We then seek to deduce consequences of these properties. This is why this branch of mathematics is called *abstract* algebra. It must be remembered, however, that when a specific group is being discussed, a specific operation must be given (at least implicitly).

Elementary Properties of Groups

Now that we have seen many diverse examples of groups, we wish to deduce some properties that they share. The definition itself raises some fundamental questions. Every group has *an* identity. Could a group have more than one? Every group element has *an* inverse. Could an element have more than one? The examples suggest not. But examples can only suggest. One cannot prove that every group has a unique identity by looking at examples, because each example inherently has properties that may not be shared by all groups. We are forced to restrict ourselves to the properties that all groups have; that is, we must view groups as abstract entities rather than argue by example. The next three theorems illustrate the abstract approach.

Theorem 2.1 Uniqueness of the Identity

In a group G, there is only one identity element.

PROOF Suppose both e and e' are identities of G. Then,

1. $ae = a$ for all a in G, and
2. $e'a = a$ for all a in G.

The choices of $a = e'$ in (part 1) and $a = e$ in (part 2) yield $e'e = e'$ and $e'e = e$. Thus, e and e' are both equal to $e'e$ and so are equal to each other. ∎

Because of this theorem, we may unambiguously speak of "the identity" of a group and denote it by 'e' (because the German word for identity is *Einheit*).

Theorem 2.2 Cancellation

In a group G, the right and left cancellation laws hold; that is, ba = ca implies b = c, and ab = ac implies b = c.

PROOF Suppose $ba = ca$. Let a' be an inverse of a. Then multiplying on the right by a' yields $(ba)a' = (ca)a'$. Associativity yields $b(aa') = c(aa')$. Then $be = ce$ and, therefore, $b = c$ as desired. Similarly, one can prove that $ab = ac$ implies $b = c$ by multiplying by a' on the left. ■

A consequence of the cancellation property is the fact that in a Cayley table for a group, each group element occurs exactly once in each row and column (see Exercise 31). Another consequence of the cancellation property is the uniqueness of inverses.

Theorem 2.3 Uniqueness of Inverses

For each element a in a group G, there is a unique element b in G such that $ab = ba = e$.

PROOF Suppose b and c are both inverses of a. Then $ab = e$ and $ac = e$, so that $ab = ac$. Canceling the a on both sides gives $b = c$, as desired. ■

As was the case with the identity element, it is reasonable, in view of Theorem 2.3, to speak of "the inverse" of an element g of a group; in fact, we may unambiguously denote it by g^{-1}. This notation is suggested by that used for ordinary real numbers under multiplication. Similarly, when n is a positive integer, the associative law allows us to use g^n to denote the unambiguous product

$$\underbrace{gg \cdots g}_{n \text{ factors}}.$$

We define $g^0 = e$. When n is negative, we define $g^n = (g^{-1})^{|n|}$ [for example, $g^{-3} = (g^{-1})^3$]. Unlike for real numbers, in an abstract group we do not permit noninteger exponents such as $g^{1/2}$. With this notation, the familiar laws of exponents hold for groups; that is, for all integers m and n and any group element g, we have $g^m g^n = g^{m+n}$ and $(g^m)^n = g^{mn}$. Although the way one manipulates the group expressions $g^m g^n$ and $(g^m)^n$ coincides with the laws of exponents for real numbers, the laws of exponents fail to hold for expressions involving two group elements. Thus, for groups in general, $(ab)^n \neq a^n b^n$ (see Exercise 23).

The important thing about the existence of a unique inverse for each group element a is that for every element b in the group there is a unique solution in the group of the equations $ax = b$ and $xa = b$. Namely, $x = a^{-1}b$ in the first case and $x = ba^{-1}$ in the second case. In contrast, in the set $\{0, 1, 2, 3, 4, 5\}$, the equation $2x = 4$ has the solutions $x = 2$ and $x = 5$ under the operation multiplication mod 6. However, this set is not a group under multiplication mod 6.

Also, one must be careful with this notation when dealing with a specific group whose binary operation is addition and is denoted by "+." In this case, the definitions and group properties expressed in multiplicative notation must be translated to additive notation. For example, the inverse of g is written as $-g$. Likewise, for example, g^3

Table **2.2**

	Multiplicative Group		Additive Group
$a \cdot b$ or ab	Multiplication	$a + b$	Addition
e or 1	Identity or one	0	Zero
a^{-1}	Multiplicative inverse of a	$-a$	Additive inverse of a
a^n	Power of a	na	Multiple of a
ab^{-1}	Quotient	$a - b$	Difference

means $g + g + g$ and is usually written as $3g$, whereas g^{-3} means $(-g) + (-g) + (-g)$ and is written as $-3g$. When additive notation is used, do not interpret "ng" as combining n and g under the group operation; n may not even be an element of the group! Table 2.2 shows the common notation and corresponding terminology for groups under multiplication and groups under addition. As is the case for real numbers, we use $a - b$ as an abbreviation for $a + (-b)$.

Because of the associative property, we may unambiguously write the expression abc, for this can be reasonably interpreted as only $(ab)c$ or $a(bc)$, which are equal. In fact, by using induction and repeated application of the associative property, one can prove a general associative property that essentially means that parentheses can be inserted or deleted at will without affecting the value of a product involving any number of group elements. Thus,

$$a^2(bcdb^2) = a^2b(cd)b^2 = (a^2b)(cd)b^2 = a(abcdb)b,$$

and so on.

Although groups do not have the property that $(ab)^n = a^nb^n$, there is a simple relationship between $(ab)^{-1}$ and a^{-1} and b^{-1}.

Theorem 2.4 Socks–Shoes Property

For group elements a and b, $(ab)^{-1} = b^{-1}a^{-1}$.

PROOF Since $(ab)(ab)^{-1} = e$ and $(ab)(b^{-1}a^{-1}) = a(bb^{-1})a^{-1} = aea^{-1} = aa^{-1} = e$, we have by Theorem 2.3 that $(ab)^{-1} = b^{-1}a^{-1}$. ■

Historical Note

We conclude this chapter with a bit of history concerning the noncommutativity of matrix multiplication. In 1925, quantum theory was replete with annoying and puzzling ambiguities. It was Werner Heisenberg who recognized the cause. He observed that the product of the quantum-theoretical analogs of the classical Fourier series did not necessarily commute. For all his boldness, this shook Heisenberg. As he later recalled [2, p. 94]:

> In my paper the fact that *XY* was not equal to *YX* was very disagreeable to me. I felt this was the only point of difficulty in the whole scheme, otherwise I would be perfectly happy. But this difficulty had worried me and I was not able to solve it.

Heisenberg asked his teacher, Max Born, if his ideas were worth publishing. Born was fascinated and deeply impressed by Heisenberg's new approach. Born wrote [1, p. 217]:

> After having sent off Heisenberg's paper to the *Zeitschrift für Physik* for publication, I began to ponder over his symbolic multiplication, and was soon so involved in it that I thought about it for the whole day and could hardly sleep at night. For I felt there was something fundamental behind it, the consummation of our endeavors of many years. And one morning, about the 10 July 1925, I suddenly saw light: Heisenberg's symbolic multiplication was nothing but the matrix calculus, well-known to me since my student days from Rosanes' lectures in Breslau.

Born and his student, Pascual Jordan, reformulated Heisenberg's ideas in terms of matrices, but it was Heisenberg who was credited with the formulation. In his autobiography, Born lamented [1, p. 219]:

> Nowadays the textbooks speak without exception of Heisenberg's matrices, Heisenberg's commutation law, and Dirac's field quantization.
>
> In fact, Heisenberg knew at that time very little of matrices and had to study them.

Upon learning in 1933 that he was to receive the Nobel Prize with Dirac and Schrödinger for this work, Heisenberg wrote to Born [1, p. 220]:

> If I have not written to you for such a long time, and have not thanked you for your congratulations, it was partly because of my rather bad conscience with respect to you. The fact that I am to receive the Nobel Prize alone, for work done in Göttingen in collaboration—you, Jordan, and I—this fact depresses me and I hardly know what to write to you. I am, of course, glad that our common efforts are now appreciated, and I enjoy the recollection of the beautiful time of collaboration. I also believe that all good physicists know how great was your and Jordan's contribution to the structure of quantum mechanics—and this remains unchanged by a wrong decision from outside. Yet I myself can do nothing but thank you again for all the fine collaboration, and feel a little ashamed.

The story has a happy ending, however, because Born received the Nobel Prize in 1954 for his fundamental work in quantum mechanics.

Exercises

"For example" is not proof.

JEWISH PROVERB

1. Which of the following binary operations are closed?
   a. subtraction of positive integers
   b. division of nonzero integers
   c. function composition of polynomials with real coefficients
   d. multiplication of 2×2 matrices with integer entries
2. Which of the following binary operations are associative?
   a. multiplication mod n
   b. division of nonzero rationals
   c. function composition of polynomials with real coefficients
   d. multiplication of 2×2 matrices with integer entries
3. Which of the following binary operations are commutative?
   a. substraction of integers
   b. division of nonzero real numbers
   c. function composition of polynomials with real coefficients
   d. multiplication of 2×2 matrices with real entries
4. Which of the following sets are closed under the given operation?
   a. $\{0, 4, 8, 12\}$ addition mod 16
   b. $\{0, 4, 8, 12\}$ addition mod 15
   c. $\{1, 4, 7, 13\}$ multiplication mod 15
   d. $\{1, 4, 5, 7\}$ multiplication mod 9
5. In each case, find the inverse of the element under the given operation.
   a. 13 in Z_{20}
   b. 13 in $U(14)$
   c. $n-1$ in $U(n)$ $(n > 2)$
   d. $3-2i$ in $\mathbf{C}^*$, the group of nonzero complex numbers under multiplication
6. In each case, perform the indicated operation.
   a. In $\mathbf{C}^*$, $(7 + 5i)(-3 + 2i)$
   b. In $GL(2, Z_{13})$, det $\begin{bmatrix} 7 & 4 \\ 1 & 5 \end{bmatrix}$
   c. In $GL(2, \mathrm{R})$, $\begin{bmatrix} 6 & 3 \\ 8 & 2 \end{bmatrix}^{-1}$
   d. In $GL(2, Z_{13})$, $\begin{bmatrix} 6 & 3 \\ 8 & 2 \end{bmatrix}^{-1}$

7. Give two reasons why the set of odd integers under addition is not a group.
8. Referring to Example 13, verify the assertion that subtraction is not associative.
9. Show that $\{1, 2, 3\}$ under multiplication modulo 4 is not a group but that $\{1, 2, 3, 4\}$ under multiplication modulo 5 is a group.
10. Show that the group $GL(2, \mathbf{R})$ of Example 9 is non-Abelian by exhibiting a pair of matrices A and B in $GL(2, \mathbf{R})$ such that $AB \neq BA$.
11. Find the inverse of the element $\begin{bmatrix} 2 & 6 \\ 3 & 5 \end{bmatrix}$ in $GL(2, Z_{11})$.
12. Give an example of group elements a and b with the property that $a^{-1}ba \neq b$.
13. Translate each of the following multiplicative expressions into its additive counterpart. Assume that the operation is commutative.
    a. a^2b^3
    b. $a^{-2}(b^{-1}c)^2$
    c. $(ab^2)^{-3}c^2 = e$
14. For group elements a, b, and c, express $(ab)^3$ and $(ab^{-2}c)^{-2}$ without parentheses.
15. Let G be a group and let $H = \{x^{-1} \mid x \in G\}$. Show that $G = H$ as sets.
16. Show that the set $\{5, 15, 25, 35\}$ is a group under multiplication modulo 40. What is the identity element of this group? Can you see any relationship between this group and $U(8)$?
17. (From the GRE Practice Exam)* Let p and q be distinct primes. Suppose that H is a proper subset of the integers that is a group under addition that contains exactly three elements of the set $\{p, p + q, pq, p^q, q^p\}$. Determine which of the following are the three elements in H.
    a. pq, p^q, q^p
    b. $p + q, pq, p^q$
    c. $p, p + q, pq$
    d. p, p^q, q^p
    e. p, pq, p^q
18. List the members of $H = \{x^2 \mid x \in D_4\}$ and $K = \{x \in D_4 \mid x^2 = e\}$.
19. Prove that the set of all 2×2 matrices with entries from $\mathbf{R}$ and determinant $+1$ is a group under matrix multiplication.
20. For any integer $n > 2$, show that there are at least two elements in $U(n)$ that satisfy $x^2 = 1$.
21. An abstract algebra teacher intended to give a typist a list of nine integers that form a group under multiplication modulo 91. Instead,

*GRE materials selected from the GRE Practice Exam, Question 9 by Educational Testing Service. Reprinted by permission of Educational Testing Service, the copyright owner.

one of the nine integers was inadvertently left out, so that the list appeared as 1, 9, 16, 22, 53, 74, 79, 81. Which integer was left out? (This really happened!)

22. Let G be a group with the property that for any x, y, z in the group, $xy = zx$ implies $y = z$. Prove that G is Abelian. ("Left-right cancellation" implies commutativity.)

23. (Law of Exponents for Abelian Groups) Let a and b be elements of an Abelian group and let n be any integer. Show that $(ab)^n = a^n b^n$. Is this also true for non-Abelian groups?

24. (Socks–Shoes Property) Draw an analogy between the statement $(ab)^{-1} = b^{-1}a^{-1}$ and the act of putting on and taking off your socks and shoes. Find distinct nonidentity elements a and b from a non-Abelian group such that $(ab)^{-1} = a^{-1}b^{-1}$. Find an example that shows that in a group, it is possible to have $(ab)^{-2} \neq b^{-2}a^{-2}$. What would be an appropriate name for the group property $(abc)^{-1} = c^{-1}b^{-1}a^{-1}$?

25. Prove that a group G is Abelian if and only if $(ab)^{-1} = a^{-1}b^{-1}$ for all a and b in G.

26. Prove that in a group, $(a^{-1})^{-1} = a$ for all a.

27. For any elements a and b from a group and any integer n, prove that $(a^{-1}ba)^n = a^{-1}b^n a$.

28. If $a_1, a_2, \ldots, a_n$ belong to a group, what is the inverse of $a_1a_2 \cdots a_n$?

29. The integers 5 and 15 are among a collection of 12 integers that form a group under multiplication modulo 56. List all 12.

30. Give an example of a group with 105 elements. Give two examples of groups with 44 elements.

31. Prove that every group table is a *Latin square*[†]; that is, each element of the group appears exactly once in each row and each column.

32. Construct a Cayley table for $U(12)$.

33. Suppose the table below is a group table. Fill in the blank entries.

	e	a	b	c	d
e	e				
a		b			e
b		c	d	e	
c		d		a	b
d					

[†]Latin squares are useful in designing statistical experiments. There is also a close connection between Latin squares and finite geometries.

34. Prove that in a group, $(ab)^2 = a^2b^2$ if and only if $ab = ba$.
35. Let a, b, and c be elements of a group. Solve the equation $axb = c$ for x. Solve $a^{-1}xa = c$ for x.
36. Let a and b belong to a group G. Find an x in G such that $xabx^{-1} = ba$.
37. Let G be a finite group. Show that the number of elements x of G such that $x^3 = e$ is odd. Show that the number of elements x of G such that $x^2 \neq e$ is even.
38. Give an example of a group with elements a, b, c, d, and x such that $axb = cxd$ but $ab \neq cd$. (Hence "middle cancellation" is not valid in groups.)
39. Suppose that G is a group with the property that for every choice of elements in G, $axb = cxd$ implies $ab = cd$. Prove that G is Abelian. ("Middle cancellation" implies commutativity.)
40. Find an element X in D_4 such that $R_{90}VXH = D'$.
41. Suppose F_1 and F_2 are distinct reflections in a dihedral group D_n. Prove that $F_1F_2 \neq R_0$.
42. Suppose F_1 and F_2 are distinct reflections in a dihedral group D_n such that $F_1F_2 = F_2F_1$. Prove that $F_1F_2 = R_{180}$.
43. Let R be any fixed rotation and F any fixed reflection in a dihedral group. Prove that $R^kFR^k = F$.
44. Let R be any fixed rotation and F any fixed reflection in a dihedral group. Prove that $FR^kF = R^{-k}$. Why does this imply that D_n is non-Abelian?
45. In the dihedral group D_n, let $R = R_{360/n}$ and let F be any reflection. Write each of the following products in the form R^i or R^iF, where $0 \leq i < n$.
    a. In D_4, $FR^{-2}FR^5$
    b. In D_5, $R^{-3}FR^4FR^{-2}$
    c. In D_6, $FR^5FR^{-2}F$
46. Prove that the set of all rational numbers of the form 3^m6^n, where m and n are integers, is a group under multiplication.
47. Prove that if G is a group with the property that the square of every element is the identity, then G is Abelian. (This exercise is referred to in Chapter 26.)
48. Prove that the set of all 3×3 matrices with real entries of the form

$$\begin{bmatrix} 1 & a & b \\ 0 & 1 & c \\ 0 & 0 & 1 \end{bmatrix}$$

is a group. (Multiplication is defined by

$$\begin{bmatrix} 1 & a & b \\ 0 & 1 & c \\ 0 & 0 & 1 \end{bmatrix} \begin{bmatrix} 1 & a' & b' \\ 0 & 1 & c' \\ 0 & 0 & 1 \end{bmatrix} = \begin{bmatrix} 1 & a + a' & b' + ac' + b \\ 0 & 1 & c' + c \\ 0 & 0 & 1 \end{bmatrix}.$$

This group, sometimes called the *Heisenberg group* after the Nobel Prize–winning physicist Werner Heisenberg, is intimately related to the Heisenberg Uncertainty Principle of quantum physics.)

49. Prove the assertion made in Example 20 that the set $\{1, 2, \ldots, n - 1\}$ is a group under multiplication modulo n if and only if n is prime.

50. In a finite group, show that the number of nonidentity elements that satisfy the equation $x^5 = e$ is a multiple of 5. If the stipulation that the group be finite is omitted, what can you say about the number of nonidentity elements that satisfy the equation $x^5 = e$?

51. List the six elements of GL(2, Z_2). Show that this group is non-Abelian by finding two elements that do not commute. (This exercise is referred to in Chapter 7.)

52. Let $G = \left\{ \begin{bmatrix} a & a \\ a & a \end{bmatrix} \middle| a \in \mathbf{R}, a \neq 0 \right\}$. Show that G is a group under matrix multiplication. Explain why each element of G has an inverse even though the matrices have 0 determinants. (Compare with Example 10.)

53. Suppose that in the definition of a group G, the condition that there exists an element e with the property $ae = ea = a$ for all a in G is replaced by $ae = a$ for all a in G. Show that $ea = a$ for all a in G. (Thus, a one-sided identity is a two-sided identity.)

54. Suppose that in the definition of a group G, the condition that for each element a in G there exists an element b in G with the property $ab = ba = e$ is replaced by the condition $ab = e$. Show that $ba = e$. (Thus, a one-sided inverse is a two-sided inverse.)

Computer Exercises

Software for the computer exercises in this chapter is available at the website:

http://www.d.umn.edu/~jgallian

References

1. Max Born, *My Life: Recollections of a Nobel Laureate*, New York: Charles Scribner's Sons, 1978.
2. J. Mehra and H. Rechenberg, *The Historical Development of Quantum Theory*, Vol. 3, New York: Springer-Verlag, 1982.

Suggested Readings

Marcia Ascher, *Ethnomathematics*, Pacific Grove, CA: Brooks/Cole, 1991.

Chapter 3 of this book describes how the dihedral group of order 8 can be used to encode the social structure of the kin system of family relationships among a tribe of native people of Australia.

Arie Bialostocki, "An Application of Elementary Group Theory to Central Solitaire," *The College Mathematics Journal*, May 1998: 208–212.

The author uses properties of groups to analyze the peg board game central solitaire (which also goes by the name peg solitaire).

J. E. White, "Introduction to Group Theory for Chemists," *Journal of Chemical Education* 44 (1967): 128–135.

Students interested in the physical sciences may find this article worthwhile. It begins with easy examples of groups and builds up to applications of group theory concepts and terminology to chemistry.

---

(52) I_2 is not the identity of $\begin{bmatrix} a & a \\ a & a \end{bmatrix}$

since $\begin{bmatrix} 1 & 0 \\ 0 & 1 \end{bmatrix} \neq \begin{bmatrix} a & a \\ a & a \end{bmatrix}$

$I_G = \begin{bmatrix} 1/2 & 1/2 \\ 1/2 & 1/2 \end{bmatrix}$

$\Rightarrow Inv(G) = \begin{bmatrix} 1/4a & 1/4a \\ 1/4a & 1/4a \end{bmatrix}$

3 Finite Groups; Subgroups

In our own time, in the period 1960–1980, we have seen particle physics emerge as the playground of group theory.

FREEMAN DYSON

Terminology and Notation

As we will soon discover, finite groups—that is, groups with finitely many elements—have interesting arithmetic properties. To facilitate the study of finite groups, it is convenient to introduce some terminology and notation.

Definition Order of a Group

The number of elements of a group (finite or infinite) is called its *order*. We will use $|G|$ to denote the order of G.

Thus, the group Z of integers under addition has infinite order, whereas the group $U(10) = \{1, 3, 7, 9\}$ under multiplication modulo 10 has order 4.

Definition Order of an Element

The *order* of an element g in a group G is the smallest positive integer n such that $g^n = e$. (In additive notation, this would be $ng = 0$.) If no such integer exists, we say that g has *infinite order*. The order of an element g is denoted by $|g|$.

So, to find the order of a group element g, you need only compute the sequence of products $g, g^2, g^3, \ldots,$ until you reach the identity for the first time. The exponent of this product (or coefficient if the operation is addition) is the order of g. If the identity never appears in the sequence, then g has infinite order.

EXAMPLE 1 Consider $U(15) = \{1, 2, 4, 7, 8, 11, 13, 14\}$ under multiplication modulo 15. This group has order 8. To find the order of the element 7, say, we compute the sequence $7^1 = 7$, $7^2 = 4$, $7^3 = 13$, $7^4 = 1$, so $|7| = 4$. To find the order of 11, we compute $11^1 = 11$, $11^2 = 1$, so $|11| = 2$. Similar computations show that $|1| = 1$, $|2| = 4$, $|4| = 2$, $|8| = 4$, $|13| = 4$, $|14| = 2$. [Here is a trick that makes these calculations easier. Rather than compute the sequence 13^1, 13^2, 13^3, 13^4, we may observe that $13 = -2 \bmod 15$, so that $13^2 = (-2)^2 = 4$, $13^3 = -2 \cdot 4 = -8$, $13^4 = (-2)(-8) = 1$.]† ■

■ **EXAMPLE 2** Consider Z_{10} under addition modulo 10. Since $1 \cdot 2 = 2$, $2 \cdot 2 = 4$, $3 \cdot 2 = 6$, $4 \cdot 2 = 8$, $5 \cdot 2 = 0$, we know that $|2| = 5$. Similar computations show that $|0| = 1$, $|7| = 10$, $|5| = 2$, $|6| = 5$. (Here $2 \cdot 2$ is an abbreviation for $2 + 2$, $3 \cdot 2$ is an abbreviation for $2 + 2 + 2$, etc.) ■

■ **EXAMPLE 3** Consider Z under ordinary addition. Here every nonzero element has infinite order, since the sequence $a, 2a, 3a, \ldots$ never includes 0 when $a \neq 0$. ■

The perceptive reader may have noticed among our examples of groups in Chapter 2 that some are subsets of others with the same binary operation. The group $SL(2, \mathbf{R})$ in Example 18, for instance, is a subset of the group $GL(2, \mathbf{R})$ in Example 9. Similarly, the group of complex numbers $\{1, -1, i, -i\}$ under multiplication is a subset of the group described in Example 15 for n equal to any multiple of 4. This situation arises so often that we introduce a special term to describe it.

Definition Subgroup
If a subset H of a group G is itself a group under the operation of G, we say that H is a *subgroup* of G.

We use the notation $H \leq G$ to mean that H is a subgroup of G. If we want to indicate that H is a subgroup of G but is not equal to G itself, we write $H < G$. Such a subgroup is called a *proper subgroup*. The subgroup $\{e\}$ is called the *trivial subgroup* of G; a subgroup that is not $\{e\}$ is called a *nontrivial subgroup* of G.

Notice that Z_n under addition modulo n is *not* a subgroup of Z under addition, since addition modulo n is not the operation of Z.

Subgroup Tests

When determining whether or not a subset H of a group G is a subgroup of G, one need not directly verify the group axioms. The next

† The website **http://www.google.com** provides a convenient way to do modular arithmetic. For example, to compute 13^4 mod 15, just type "13^4 mod 15" in the search box.

three results provide simple tests that suffice to show that a subset of a group is a subgroup.

Theorem 3.1 One-Step Subgroup Test

Let G be a group and H a nonempty subset of G. If ab^{-1} is in H whenever a and b are in H, then H is a subgroup of G. (In additive notation, if $a - b$ is in H whenever a and b are in H, then H is a subgroup of G.)

PROOF Since the operation of H is the same as that of G, it is clear that this operation is associative. Next, we show that e is in H. Since H is nonempty, we may pick some x in H. Then, letting $a = x$ and $b = x$ in the hypothesis, we have $e = xx^{-1} = ab^{-1}$ is in H. To verify that x^{-1} is in H whenever x is in H, all we need to do is to choose $a = e$ and $b = x$ in the statement of the theorem. Finally, the proof will be complete when we show that H is closed; that is, if x, y belong to H, we must show that xy is in H also. Well, we have already shown that y^{-1} is in H whenever y is; so, letting $a = x$ and $b = y^{-1}$, we have $xy = x(y^{-1})^{-1} = ab^{-1}$ is in H. ■

Although we have dubbed Theorem 3.1 the One-Step Subgroup Test, there are actually four steps involved in applying the theorem. (After you gain some experience, the first three steps will be routine.) Notice the similarity between the last three steps listed below and the three steps involved in the Second Principle of Mathematical Induction.

1. Identify the property P that distinguishes the elements of H; that is, identify a defining condition.
2. Prove that the identity has property P. (This verifies that H is nonempty.)
3. *Assume* that two elements a and b have property P.
4. Use the assumption that a and b have property P to show that ab^{-1} has property P.

The procedure is illustrated in Examples 4 and 5.

EXAMPLE 4 Let G be an Abelian group with identity e. Then $H = \{x \in G \mid x^2 = e\}$ is a subgroup of G. Here, the defining property of H is the condition $x^2 = e$. So, we first note that $e^2 = e$, so that H is nonempty. Now we assume that a and b belong to H. This means that $a^2 = e$ and $b^2 = e$. Finally, we must show that $(ab^{-1})^2 = e$. Since G is Abelian, $(ab^{-1})^2 = ab^{-1}ab^{-1} = a^2(b^{-1})^2 = a^2(b^2)^{-1} = ee^{-1} = e$. Therefore, ab^{-1} belongs to H and, by the One-Step Subgroup Test, H is a subgroup of G. ■

In many instances, a subgroup will consist of all elements that have a particular form. Then the property P is that the elements have that particular form. This is illustrated in the following example.

EXAMPLE 5 Let G be an Abelian group under multiplication with identity e. Then $H = \{x^2 \mid x \in G\}$ is a subgroup of G. (In words, H is the set of all "squares.") Since $e^2 = e$, the identity has the correct form. Next, we write two elements of H in the correct form, say, a^2 and b^2. We must show that $a^2(b^2)^{-1}$ also has the correct form; that is, $a^2(b^2)^{-1}$ is the square of some element. Since G is Abelian, we may write $a^2(b^2)^{-1}$ as $(ab^{-1})^2$, which is the correct form. Thus, H is a subgroup of G. ∎

Beginning students often prefer to use the next theorem instead of Theorem 3.1.

Theorem 3.2 Two-Step Subgroup Test

Let G be a group and let H be a nonempty subset of G. If ab is in H whenever a and b are in H (H is closed under the operation), and a^{-1} is in H whenever a is in H (H is closed under taking inverses), then H is a subgroup of G.

PROOF By Theorem 3.1, it suffices to show that $a, b \in H$ implies $ab^{-1} \in H$. So, we suppose that $a, b \in H$. Since H is closed under taking inverses, we also have $b^{-1} \in H$. Thus, $ab^{-1} \in H$ by closure under multiplication. ∎

When applying the Two-Step Subgroup Test, we proceed exactly as in the case of the One-Step Subgroup Test, except we use the assumption that a and b have property P to prove that ab has property P and that a^{-1} has property P.

EXAMPLE 6 Let G be an Abelian group. Then $H = \{x \in G \mid |x| \text{ is finite}\}$ is a subgroup of G. Since $e^1 = e$, $H \neq \theta$. To apply the Two-Step Subgroup Test we assume that a and b belong to H and prove that ab and a^{-1} belong to H. Let $|a| = m$ and $|b| = n$. Then, because G is Abelian, we have $(ab)^{mn} = (a^m)^n(b^n)^m = e^ne^m = e$. Thus, ab has finite order (this does not show that $|ab| = mn$). Moveover, $(a^{-1})^m = (a^m)^{-1} = e^{-1} = e$ shows that a^{-1} has finite order. ∎

We next illustrate how to use the Two-Step Subgroup Test by introducing an important technique for creating new subgroups of Abelian groups from existing ones. The method will be extended to some subgroups of non-Abelian groups in later chapters.

EXAMPLE 7 Let G be an Abelian group and H and K be subgroups of G. Then $HK = \{hk \mid h \in H, k \in K\}$ is a subgroup of G. First note that $e = ee$ belongs to HK because e is in both H and K. Now suppose that a and b are in HK. Then by definition of H there are elements h_1, $h_2 \in H$ and $k_1, k_2 \in K$ such that $a = h_1k_1$ and $b = h_2k_2$. We must prove that $ab \in HK$ and $a^{-1} \in HK$. Observe that because G is Abelian and H and K are subgroups of G, we have $ab = h_1k_1h_2k_2 = (h_1h_2)(k_1k_2) \in HK$. Likewise, $a^{-1} = (h_1k_1)^{-1} = k_1^{-1}h_1^{-1} = h_1^{-1}k_1^{-1} \in HK$. ■

How do you prove that a subset of a group is *not* a subgroup? Here are three possible ways, any one of which guarantees that the subset is not a subgroup:

1. Show that the identity is not in the set.
2. Exhibit an element of the set whose inverse is not in the set.
3. Exhibit two elements of the set whose product is not in the set.

EXAMPLE 8 Let G be the group of nonzero real numbers under multiplication, $H = \{x \in G \mid x = 1 \text{ or } x \text{ is irrational}\}$ and $K = \{x \in G \mid x \geq 1\}$. Then H is not a subgroup of G, since $\sqrt{2} \in H$ but $\sqrt{2} \cdot \sqrt{2} = 2 \notin H$. Also, K is not a subgroup, since $2 \in K$ but $2^{-1} \notin K$. ■

When dealing with finite groups, it is easier to use the following subgroup test.

Theorem 3.3 Finite Subgroup Test

Let H be a nonempty finite subset of a group G. If H is closed under the operation of G, then H is a subgroup of G.

PROOF In view of Theorem 3.2, we need only prove that $a^{-1} \in H$ whenever $a \in H$. If $a = e$, then $a^{-1} = a$ and we are done. If $a \neq e$, consider the sequence $a, a^2, \ldots$. By closure, all of these elements belong to H. Since H is finite, not all of these elements are distinct. Say $a^i = a^j$ and $i > j$. Then, $a^{i-j} = e$; and since $a \neq e$, $i - j > 1$. Thus, $aa^{i-j-1} = a^{i-j} = e$ and, therefore, $a^{i-j-1} = a^{-1}$. But $i - j - 1 \geq 1$ implies $a^{i-j-1} \in H$ and we are done. ■

Examples of Subgroups

The proofs of the next few theorems show how our subgroup tests work. We first introduce an important notation. For any element a from a group, we let $\langle a \rangle$ denote the set $\{a^n \mid n \in Z\}$. In particular, observe that the exponents of a include all negative integers as well as 0 and the positive integers (a^0 is defined to be the identity).

Theorem 3.4 $\langle a \rangle$ Is a Subgroup

Let G be a group, and let a be any element of G. Then, $\langle a \rangle$ is a subgroup of G.

PROOF Since $a \in \langle a \rangle$, $\langle a \rangle$ is not empty. Let $a^n, a^m \in \langle a \rangle$. Then, $a^n(a^m)^{-1} = a^{n-m} \in \langle a \rangle$; so, by Theorem 3.1, $\langle a \rangle$ is a subgroup of G. ∎

The subgroup $\langle a \rangle$ is called the *cyclic subgroup of G generated by a*. In the case that $G = \langle a \rangle$, we say that G is *cyclic* and a is a *generator* of G. (A cyclic group may have many generators.) Notice that although the list $\ldots, a^{-2}, a^{-1}, a^0, a^1, a^2, \ldots$ has infinitely many entries, the set $\{a^n \mid n \in Z\}$ might have only finitely many elements. Also note that, since $a^i a^j = a^{i+j} = a^{j+i} = a^j a^i$, every cyclic group is Abelian.

EXAMPLE 9 In $U(10)$, $\langle 3 \rangle = \{3, 9, 7, 1\} = U(10)$, for $3^1 = 3$, $3^2 = 9$, $3^3 = 7$, $3^4 = 1$, $3^5 = 3^4 \cdot 3 = 1 \cdot 3$, $3^6 = 3^4 \cdot 3^2 = 9, \ldots$; $3^{-1} = 7$ (since $3 \cdot 7 = 1$), $3^{-2} = 9$, $3^{-3} = 3$, $3^{-4} = 1$, $3^{-5} = 3^{-4} \cdot 3^{-1} = 1 \cdot 7$, $3^{-6} = 3^{-4} \cdot 3^{-2} = 1 \cdot 9 = 9, \ldots$. ∎

EXAMPLE 10 In Z_{10}, $\langle 2 \rangle = \{2, 4, 6, 8, 0\}$. Remember, a^n means na when the operation is addition. ∎

EXAMPLE 11 In Z, $\langle -1 \rangle = Z$. Here each entry in the list $\ldots, -2(-1), -1(-1), 0(-1), 1(-1), 2(-1), \ldots$ represents a distinct group element. ∎

EXAMPLE 12 In D_n, the dihedral group of order $2n$, let R denote a rotation of $360/n$ degrees. Then,

$$R^n = R_{360^\circ} = e, \qquad R^{n+1} = R, \qquad R^{n+2} = R^2, \ldots.$$

Similarly, $R^{-1} = R^{n-1}, R^{-2} = R^{n-2}, \ldots$, so that $\langle R \rangle = \{e, R, \ldots, R^{n-1}\}$. We see, then, that the powers of R "cycle back" periodically with period n. Visually, raising R to successive positive powers is the same as moving counterclockwise around the following circle one node at a time, whereas raising R to successive negative powers is the same as moving around the circle clockwise one node at a time.

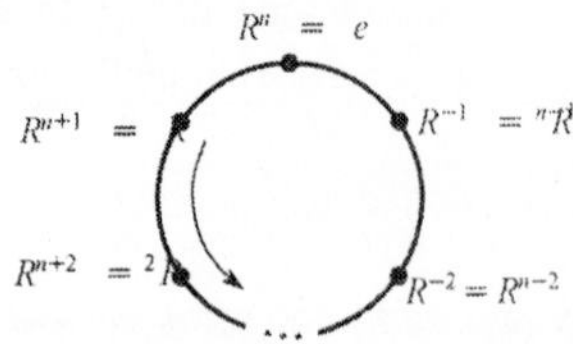

In Chapter 4 we will show that $|\langle a\rangle| = |a|$; that is, the order of the subgroup generated by a is the order of a itself. (Actually, the definition of $|a|$ was chosen to ensure the validity of this equation.)

For any element a of a group G, it is useful to think of $\langle a\rangle$ as the smallest subgroup of G containing a. This notion can be extended to any collection S of elements from a group G by defining $\langle S\rangle$ as the subgroup of G with the property that $\langle S\rangle$ contains S and if H is any subgroup of G containing S, then H also contains $\langle S\rangle$. Thus, $\langle S\rangle$ is the smallest subgroup of G that contains S. The set $\langle S\rangle$ is called *the subgroup generated by S*. We illustrate this concept in the next example. The verifications are left to the reader (Exercise 40).

EXAMPLE 13 In Z_{20}, $\langle 8, 14\rangle = \{0, 2, 4, \ldots, 18\} = \langle 2\rangle$; in Z, $\langle 8, 13\rangle = Z$; in D_4, $\langle H, V\rangle = \{H, H^2, V, HV\} = \{R_0, R_{180}, H, V\}$; in D_4, $\langle R_{90}, V\rangle = \{R_{90}, R_{90}^2, R_{90}^3, R_{90}^4, V, R_{90}V, R_{90}^2V, R_{90}^3V\} = D_4$; in $\mathbf{C}^*$, the group of nonzero complex numbers under multiplication, $\langle 1, i\rangle = \{1, -1, i, -i\} = \langle i\rangle$; in $\mathbf{C}$, the group of complex numbers under addition, $\langle 1, i\rangle = \{a + bi \mid a, b \in Z\}$ (This group is called the "Gaussian integers"); in $\mathbf{R}$, the group of real numbers under addition, $\langle 2, \pi, \sqrt{2}\rangle = \{2a + b\pi + c\sqrt{2} \mid a, b, c \in Z\}$; in a group in which a, b, c, and d commute, $\langle a, b, c, d\rangle = \{a^q b^r c^s d^t \mid q, r, s, t \in Z\}$. ■

We next consider one of the most important subgroups.

Definition Center of a Group

The *center*, $Z(G)$, of a group G is the subset of elements in G that commute with every element of G. In symbols,

$$Z(G) = \{a \in G \mid ax = xa \text{ for all } x \text{ in } G\}.$$

[The notation $Z(G)$ comes from the fact that the German word for center is *Zentrum*. The term was coined by J. A. de Séguier in 1904.]

Theorem 3.5 Center Is a Subgroup

The center of a group G is a subgroup of G.

PROOF For variety, we shall use Theorem 3.2 to prove this result. Clearly, $e \in Z(G)$, so $Z(G)$ is nonempty. Now, suppose $a, b \in Z(G)$. Then $(ab)x = a(bx) = a(xb) = (ax)b = (xa)b = x(ab)$ for all x in G; and, therefore, $ab \in Z(G)$.

Next, assume that $a \in Z(G)$. Then we have $ax = xa$ for all x in G. What we want is $a^{-1}x = xa^{-1}$ for all x in G. Informally, all we need do to obtain the second equation from the first one is simultaneously to bring the a's across the equals sign:

$$ax = xa$$

becomes $xa^{-1} = a^{-1}x$. (Be careful here; groups need not be commutative. The a on the left comes across as a^{-1} on the left, and the a on the right comes across as a^{-1} on the right.) Formally, the desired equation can be obtained from the original one by multiplying it on the left and right by a^{-1}, like so:

$$\begin{aligned} a^{-1}(ax)a^{-1} &= a^{-1}(xa)a^{-1}, \\ (a^{-1}a)xa^{-1} &= a^{-1}x(aa^{-1}), \\ exa^{-1} &= a^{-1}xe, \\ xa^{-1} &= a^{-1}x. \end{aligned}$$

This shows that $a^{-1} \in Z(G)$ whenever a is. ∎

For practice, let's determine the centers of the dihedral groups.

EXAMPLE 14 For $n \geq 3$,

$$Z(D_n) = \begin{cases} \{R_0, R_{180}\} & \text{when } n \text{ is even}, \\ \{R_0\} & \text{when } n \text{ is odd}. \end{cases}$$

To verify this, first observe that since every rotation in D_n is a power of $R_{360/n}$, rotations commute with rotations. We now investigate when a rotation commutes with a reflection. Let R be any rotation in D_n and let F be any reflection in D_n. Observe that since RF is a reflection we have $RF = (RF)^{-1} = F^{-1}R^{-1} = FR^{-1}$. Thus, it follows that R and F commute if and only if $FR = RF = FR^{-1}$. By cancellation, this holds if and only if $R = R^{-1}$. But $R = R^{-1}$ only when $R = R_0$ or $R = R_{180}$, and R_{180} is in D_n only when n is even. So, we have proved that $Z(D_n) = \{R_0\}$ when n is odd and $Z(D_n) = \{R_0, R_{180}\}$ when n is even. ∎

Although an element from a non-Abelian group does not necessarily commute with every element of the group, there are always some elements with which it will commute. For example, every element a commutes with all powers of a. This observation prompts the next definition and theorem.

Definition Centralizer of a in G

Let a be a fixed element of a group G. The *centralizer of a in G*, $C(a)$, is the set of all elements in G that commute with a. In symbols, $C(a) = \{g \in G \mid ga = ag\}$.

EXAMPLE 15 In D_4, we have the following centralizers:

$$\begin{aligned} C(R_0) &= D_4 = C(R_{180}), \\ C(R_{90}) &= \{R_0, R_{90}, R_{180}, R_{270}\} = C(R_{270}), \\ C(H) &= \{R_0, H, R_{180}, V\} = C(V), \\ C(D) &= \{R_0, D, R_{180}, D'\} = C(D'). \end{aligned}$$

Notice that each of the centralizers in Example 15 is actually a subgroup of D_4. The next theorem shows that this was not a coincidence.

Theorem 3.6 $C(a)$ Is a Subgroup

For each a in a group G, the centralizer of a is a subgroup of G.

PROOF A proof similar to that of Theorem 3.5 is left to the reader to supply (Exercise 41).

Notice that for every element a of a group G, $Z(G) \subseteq C(a)$. Also, observe that G is Abelian if and only if $C(a) = G$ for all a in G.

Exercises

The purpose of proof is to understand, not to verify.

ARNOLD ROSS

1. For each group in the following list, find the order of the group and the order of each element in the group. What relation do you see between the orders of the elements of a group and the order of the group?

$$Z_{12}, \quad U(10), \quad U(12), \quad U(20), \quad D_4$$

2. Let Q be the group of rational numbers under addition and let Q^* be the group of nonzero rational numbers under multiplication. In Q, list the elements in $\langle \frac{1}{2} \rangle$. In Q^*, list the elements in $\langle \frac{1}{2} \rangle$.

3. Let Q and Q^* be as in Exercise 2. Find the order of each element in Q and in Q^*.

4. Prove that in any group, an element and its inverse have the same order.

5. Without actually computing the orders, explain why the two elements in each of the following pairs of elements from Z_{30} must have the same order: $\{2, 28\}$, $\{8, 22\}$. Do the same for the following pairs of elements from $U(15)$: $\{2, 8\}$, $\{7, 13\}$.

6. In the group Z_{12}, find $|a|$, $|b|$, and $|a + b|$ for each case.

a. $a = 6, b = 2$

b. $a = 3, b = 8$

c. $a = 5, b = 4$

Do you see any relationship between $|a|$, $|b|$, and $|a + b|$?

7. If a, b, and c are group elements and $|a| = 6$, $|b| = 7$, express $(a^4c^{-2}b^4)^{-1}$ without using negative exponents.

8. What can you say about a subgroup of D_3 that contains R_{240} and a reflection F? What can you say about a subgroup of D_3 that contains two reflections?

9. What can you say about a subgroup of D_4 that contains R_{270} and a reflection? What can you say about a subgroup of D_4 that contains H and D? What can you say about a subgroup of D_4 that contains H and V?

10. How many subgroups of order 4 does D_4 have?

11. Determine all elements of finite order in R^*, the group of nonzero real numbers under multiplication.

12. If a and b are group elements and $ab \neq ba$, prove that $aba \neq e$.

13. Suppose that H is a nonempty subset of a group G that is closed under the group operation and has the property that if a is not in H then a^{-1} is not in H. Is H a subgroup?

14. Let G be the group of polynomials under addition with coefficients from Z_{10}. Find the orders of $f(x) = 7x^2 + 5x + 4$, $g(x) = 4x^2 + 8x + 6$, and $f(x) + g(x) = x^2 + 3x$. If $h(x) = a_nx^n + a_{n-1}x^{n-1} + \cdots + a_0$ belongs to G, determine $|h(x)|$ given that $\gcd(a_1, a_2, \ldots, a_n) = 1$; $\gcd(a_1, a_2, \ldots, a_n) = 2$; $\gcd(a_1, a_2, \ldots, a_n) = 5$; and $\gcd(a_1, a_2, \ldots, a_n) = 10$.

15. If a is an element of a group G and $|a| = 7$, show that a is the cube of some element of G.

16. Suppose that H is a nonempty subset of a group G with the property that if a and b belong to H then $a^{-1}b^{-1}$ belongs to H. Prove or disprove that this is enough to guarantee that H is a subgroup of G.
17. Prove that if an Abelian group has more than three elements of order 2, then it has at least 7 elements of order 2. Find an example that shows this is not true for non-Abelian groups.
18. Suppose that a is a group element and $a^6 = e$. What are the possibilities for $|a|$? Provide reasons for your answer.
19. If a is a group element and a has infinite order, prove that $a^m \neq a^n$ when $m \neq n$.
20. Let x belong to a group. If $x^2 \neq e$ and $x^6 = e$, prove that $x^4 \neq e$ and $x^5 \neq e$. What can we say about the order of x?
21. Show that if a is an element of a group G, then $|a| \leq |G|$.
22. Show that $U(14) = \langle 3 \rangle = \langle 5 \rangle$. [Hence, $U(14)$ is cyclic.] Is $U(14) = \langle 11 \rangle$?
23. Show that $U(20) \neq \langle k \rangle$ for any k in $U(20)$. [Hence, $U(20)$ is not cyclic.]
24. Suppose n is an even positive integer and H is a subgroup of Z_n. Prove that either every member of H is even or exactly half of the members of H are even.
25. Prove that for every subgroup of D_n, either every member of the subgroup is a rotation or exactly half of the members are rotations.
26. Prove that a group with two elements of order 2 that commute must have a subgroup of order 4.
27. For every even integer n, show that D_n has a subgroup of order 4.
28. Suppose that H is a proper subgroup of Z under addition and H contains 18, 30, and 40. Determine H.
29. Suppose that H is a proper subgroup of Z under addition and that H contains 12, 30, and 54. What are the possibilities for H?
30. Prove that the dihedral group of order 6 does not have a subgroup of order 4.
31. For each divisor $k > 1$ of n, let $U_k(n) = \{x \in U(n) \mid x \bmod k = 1\}$. [For example, $U_3(21) = \{1, 4, 10, 13, 16, 19\}$ and $U_7(21) = \{1, 8\}$.] List the elements of $U_4(20)$, $U_5(20)$, $U_5(30)$, and $U_{10}(30)$. Prove that $U_k(n)$ is a subgroup of $U(n)$. Let $H = \{x \in U(10) \mid x \bmod 3 = 1\}$. Is H a subgroup of $U(10)$? (This exercise is referred to in Chapter 8.)
32. If H and K are subgroups of G, show that $H \cap K$ is a subgroup of G. (Can you see that the same proof shows that the intersection of any number of subgroups of G, finite or infinite, is again a subgroup of G?)

33. Let G be a group. Show that $Z(G) = \cap_{a \in G} C(a)$. [This means the intersection of *all* subgroups of the form $C(a)$.]

34. Let G be a group, and let $a \in G$. Prove that $C(a) = C(a^{-1})$.

35. For any group element a and any integer k, show that $C(a) \subseteq C(a^k)$. Use this fact to complete the following statement: "In a group, if x commutes with a, then . . . ." Is the converse true?

36. Complete the partial Cayley group table given below.

	1	2	3	4	5	6	7	8
1	1	2	3	4	5	6	7	8
2	2	1	4	3	6	5	8	7
3	3	4	2	1	7	8	6	5
4	4	3	1	2	8	7	5	6
5	5	6	8	7	1			
6	6	5	7	8		1		
7	7	8	5	6			1	
8	8	7	6	5				1

37. Suppose G is the group defined by the following Cayley table.

	1	2	3	4	5	6	7	8
1	1	2	3	4	5	6	7	8
2	2	1	8	7	6	5	4	3
3	3	4	5	6	7	8	1	2
4	4	3	2	1	8	7	6	5
5	5	6	7	8	1	2	3	4
6	6	5	4	3	2	1	8	7
7	7	8	1	2	3	4	5	6
8	8	7	6	5	4	3	2	1

a. Find the centralizer of each member of G.

b. Find $Z(G)$.

c. Find the order of each element of G. How are these orders arithmetically related to the order of the group?

38. If a and b are distinct group elements, prove that either $a^2 \neq b^2$ or $a^3 \neq b^3$.

39. Let S be a subset of a group and let H be the intersection of all subgroups of G that contain S.

a. Prove that $\langle S \rangle = H$.

b. If S is nonempty, prove that $\langle S \rangle = \{s_1^{n_1} s_2^{n_2} \ldots s_m^{n_m} \mid m \geq 1, s_i \in S, n_i \in Z\}$. (The s_i terms need not be distinct.)

40. In the group Z, find

a. $\langle 8, 14 \rangle$;

b. $\langle 8, 13 \rangle$;

c. $\langle 6, 15 \rangle$;

d. $\langle m, n \rangle$;

e. $\langle 12, 18, 45 \rangle$.

In each part, find an integer k such that the subgroup is $\langle k \rangle$.

41. Prove Theorem 3.6.

42. If H is a subgroup of G, then by the *centralizer* $C(H)$ of H we mean the set $\{x \in G \mid xh = hx \text{ for all } h \in H\}$. Prove that $C(H)$ is a subgroup of G.

43. Must the centralizer of an element of a group be Abelian?

44. Must the center of a group be Abelian?

45. Let G be an Abelian group with identity e and let n be some fixed integer. Prove that the set of all elements of G that satisfy the equation $x^n = e$ is a subgroup of G. Give an example of a group G in which the set of all elements of G that satisfy the equation $x^2 = e$ does not form a subgroup of G. (This exercise is referred to in Chapter 11.)

46. Suppose a belongs to a group and $|a| = 5$. Prove that $C(a) = C(a^3)$. Find an element a from some group such that $|a| = 6$ and $C(a) \neq C(a^3)$.

47. Let G be the set of all polynomials with coefficients from the set $\{0, 1, 2, 3\}$. We can make G a group under addition by adding the polynomials in the usual way, except that we use modulo 4 to combine the coefficients. With this group operation, determine the orders of the elements of G. Determine a necessary and sufficient condition for an element of G to have order 2.

48. In each case, find elements a and b from a group such that $|a| = |b| = 2$.

a. $|ab| = 3$ **b.** $|ab| = 4$ **c.** $|ab| = 5$

Can you see any relationship among $|a|$, $|b|$, and $|ab|$?

49. Suppose a group contains elements a and b such that $|a| = 4$, $|b| = 2$, and $a^3b = ba$. Find $|ab|$.

50. Suppose a and b are group elements such that $|a| = 2$, $b \neq e$, and $aba = b^2$. Determine $|b|$.

51. Let a be a group element of order n, and suppose that d is a positive divisor of n. Prove that $|a^d| = n/d$.

52. Consider the elements $A = \begin{bmatrix} 0 & -1 \\ 1 & 0 \end{bmatrix}$ and $B = \begin{bmatrix} 0 & 1 \\ -1 & -1 \end{bmatrix}$ from $SL(2, \mathbf{R})$. Find $|A|$, $|B|$, and $|AB|$. Does your answer surprise you?

53. Consider the element $A = \begin{bmatrix} 1 & 1 \\ 0 & 1 \end{bmatrix}$ in $SL(2, \mathbf{R})$. What is the order of A? If we view $A = \begin{bmatrix} 1 & 1 \\ 0 & 1 \end{bmatrix}$ as a member of $SL(2, Z_p)$ (p is a prime), what is the order of A?

54. For any positive integer n and any angle θ, show that in the group $SL(2, \mathbf{R})$,

$$\begin{bmatrix} \cos\theta & -\sin\theta \\ \sin\theta & \cos\theta \end{bmatrix}^n = \begin{bmatrix} \cos n\theta & -\sin n\theta \\ \sin n\theta & \cos n\theta \end{bmatrix}.$$

Use this formula to find the order of

$$\begin{bmatrix} \cos 60° & -\sin 60° \\ \sin 60° & \cos 60° \end{bmatrix} \text{ and } \begin{bmatrix} \cos\sqrt{2}° & -\sin\sqrt{2}° \\ \sin\sqrt{2}° & \cos\sqrt{2}° \end{bmatrix}.$$

(Geometrically, $\begin{bmatrix} \cos\theta & -\sin\theta \\ \sin\theta & \cos\theta \end{bmatrix}$ represents a rotation of the plane θ degrees.)

55. Let G be the symmetry group of a circle. Show that G has elements of every finite order as well as elements of infinite order.

56. Let x belong to a group and $|x| = 6$. Find $|x^2|$, $|x^3|$, $|x^4|$, and $|x^5|$. Let y belong to a group and $|y| = 9$. Find $|y^i|$ for $i = 2, 3, \ldots, 8$. Do these examples suggest any relationship between the order of the power of an element and the order of the element?

57. D_4 has seven cyclic subgroups. List them.

58. $U(15)$ has six cyclic subgroups. List them.

59. Prove that a group of even order must have an element of order 2.

60. Suppose G is a group that has exactly eight elements of order 3. How many subgroups of order 3 does G have?

61. Let H be a subgroup of a finite group G. Suppose that g belongs to G and n is the smallest positive integer such that $g^n \in H$. Prove that n divides $|g|$.

62. Compute the orders of the following groups.

a. $U(3)$, $U(4)$, $U(12)$

b. $U(5)$, $U(7)$, $U(35)$

c. $U(4)$, $U(5)$, $U(20)$

d. $U(3)$, $U(5)$, $U(15)$

On the basis of your answers, make a conjecture about the relationship among $|U(r)|$, $|U(s)|$, and $|U(rs)|$.

63. Let $\mathbf{R}^*$ be the group of nonzero real numbers under multiplication and let $H = \{x \in \mathbf{R}^* \mid x^2 \text{ is rational}\}$. Prove that H is a subgroup of $\mathbf{R}^*$. Can the exponent 2 be replaced by any positive integer and still have H be a subgroup?

64. Compute $|U(4)|$, $|U(10)|$, and $|U(40)|$. Do these groups provide a counterexample to your answer to Exercise 62? If so, revise your conjecture.

65. Find a cyclic subgroup of order 4 in $U(40)$.

66. Find a noncyclic subgroup of order 4 in $U(40)$.

67. Let $G = \left\{ \begin{bmatrix} a & b \\ c & d \end{bmatrix} \middle| a, b, c, d \in Z \right\}$ under addition. Let $H = \left\{ \begin{bmatrix} a & b \\ c & d \end{bmatrix} \in G \mid a+b+c+d = 0 \right\}$. Prove that H is a subgroup of G. What if 0 is replaced by 1?

68. Let $H = \{A \in GL(2, \mathbf{R}) \mid \det A \text{ is an integer power of } 2\}$. Show that H is a subgroup of $GL(2, \mathbf{R})$.

69. Let H be a subgroup of $\mathbf{R}$ under addition. Let $K = \{2^a \mid a \in H\}$. Prove that K is a subgroup of $\mathbf{R}^*$ under multiplication.

70. Let G be a group of functions from $\mathbf{R}$ to $\mathbf{R}^*$, where the operation of G is multiplication of functions. Let $H = \{f \in G \mid f(2) = 1\}$. Prove that H is a subgroup of G. Can 2 be replaced by any real number?

71. Let $G = GL(2, \mathbf{R})$ and $H = \left\{ \begin{bmatrix} a & 0 \\ 0 & b \end{bmatrix} \middle| a \text{ and } b \text{ are nonzero integers} \right\}$ under the operation of matrix multiplication. Prove or disprove that H is a subgroup of $GL(2, \mathbf{R})$.

72. Let $H = \{a + bi \mid a, b \in \mathbf{R}, ab \geq 0\}$. Prove or disprove that H is a subgroup of $\mathbf{C}$ under addition.

73. Let $H = \{a + bi \mid a, b \in \mathbf{R}, a^2 + b^2 = 1\}$. Prove or disprove that H is a subgroup of $\mathbf{C}^*$ under multiplication. Describe the elements of H geometrically.

74. Let G be a finite Abelian group and let a and b belong to G. Prove that the set $\langle a, b \rangle = \{a^i b^j \mid i, j \in Z\}$ is a subgroup of G. What can you say about $|\langle a, b \rangle|$ in terms of $|a|$ and $|b|$?

75. Let H be a subgroup of a group G. Prove that the set $HZ(G) = \{hz \mid h \in H, z \in Z(G)\}$ is a subgroup of G. This exercise is referred to in this chapter.

76. Let G be a group and H a subgroup. For any element g of G, define $gH = \{gh \mid h \in H\}$. If G is Abelian and g has order 2, show that the set $K = H \cup gH$ is a subgroup of G. Is your proof valid if we drop the assumption that G is Abelian and let $K = Z(G) \cup gZ(G)$?

77. Let a belong to a group and $|a| = m$. If n is relatively prime to m, show that a can be written as the nth power of some element in the group.

78. Let F be a reflection in the dihedral group D_n and R a rotation in D_n. Determine $C(F)$ when n is odd. Determine $C(F)$ when n is even. Determine $C(R)$.

79. Let $G = GL(2, \mathbf{R})$.

a. Find $C\left(\begin{bmatrix} 1 & 1 \\ 1 & 0 \end{bmatrix}\right)$.

b. Find $C\left(\begin{bmatrix} 0 & 1 \\ 1 & 0 \end{bmatrix}\right)$.

c. Find $Z(G)$.

80. Let G be a finite group with more than one element. Show that G has an element of prime order.

Computer Exercises

Computer exercises for this chapter are available at the website:

http://www.d.umn.edu/~jgallian

Suggested Readings

Ruth Berger, "Hidden Group Structure," *Mathematics Magazine* 78 (2005): 45–48.

In this note, the author investigates groups obtained from $U(n)$ by multiplying each element by some k in $U(n)$. Such groups have identities that are not obvious.

J. Gallian and M. Reid, "Abelian Forcing Sets," *American Mathematical Monthly* 100 (1993): 580–582.

A set S is called *Abelian forcing* if the only groups that satisfy $(ab)^n = a^nb^n$ for all a and b in the group and all n in S are the Abelian ones. This paper characterizes the Abelian forcing sets. It can be downloaded at **http://www.d.umn.edu/~jgallian/forcing.pdf**

Gina Kolata, "Perfect Shuffles and Their Relation to Math," *Science* 216 (1982): 505–506.

This is a delightful nontechnical article that discusses how group theory and computers were used to solve a difficult problem about shuffling a deck of cards. Serious work on the problem was begun by an undergraduate student as part of a programming course.

Suggested Software

Allen Hibbard and Kenneth Levasseur, *Exploring Abstract Algebra with Mathematica*, New York: Springer-Verlag, 1999.

This book, intended as a supplement for a course in abstract algebra, consists of 14 group labs, 13 ring labs, and documentation for the *Abstract Algebra* software on which the labs are based. The software uses the Mathematica language, and only a basic familiarity with the program is required. The software can be freely downloaded at **http://www.central.edu/eaam/** and can be used independently of the book.

4 Cyclic Groups

The notion of a "group," viewed only 30 years ago as the epitome of sophistication, is today one of the mathematical concepts most widely used in physics, chemistry, biochemistry, and mathematics itself.

ALEXEY SOSINSKY, 1991

Properties of Cyclic Groups

Recall from Chapter 3 that a group G is called *cyclic* if there is an element a in G such that $G = \{a^n \mid n \in Z\}$. Such an element a is called a *generator* of G. In view of the notation introduced in the preceding chapter, we may indicate that G is a cyclic group generated by a by writing $G = \langle a \rangle$.

In this chapter, we examine cyclic groups in detail and determine their important characteristics. We begin with a few examples.

EXAMPLE 1 The set of integers Z under ordinary addition is cyclic. Both 1 and -1 are generators. (Recall that, when the operation is addition, 1^n is interpreted as

$$\underbrace{1 + 1 + \cdots + 1}_{n \text{ terms}}$$

when n is positive and as

$$\underbrace{(-1) + (-1) + \cdots + (-1)}_{|n| \text{ terms}}$$

when n is negative.) ∎

EXAMPLE 2 The set $Z_n = \{0, 1, \ldots, n - 1\}$ for $n \geq 1$ is a cyclic group under addition modulo n. Again, 1 and $-1 = n - 1$ are generators. ∎

Unlike Z, which has only two generators, Z_n may have many generators (depending on which n we are given).

EXAMPLE 3 $Z_8 = \langle 1 \rangle = \langle 3 \rangle = \langle 5 \rangle = \langle 7 \rangle$. To verify, for instance, that $Z_8 = \langle 3 \rangle$, we note that $\langle 3 \rangle = \{3, 3 + 3, 3 + 3 + 3, \ldots\}$ is the set $\{3, 6, 1, 4, 7, 2, 5, 0\} = Z_8$. Thus, 3 is a generator of Z_8. On the other hand, 2 is not a generator, since $\langle 2 \rangle = \{0, 2, 4, 6\} \neq Z_8$. ∎

EXAMPLE 4 (See Example 11 in Chapter 2.) $U(10) = \{1, 3, 7, 9\} = \{3^0, 3^1, 3^3, 3^2\} = \langle 3 \rangle$. Also, $\{1, 3, 7, 9\} = \{7^0, 7^3, 7^1, 7^2\} = \langle 7 \rangle$. So both 3 and 7 are generators for $U(10)$. ∎

Quite often in mathematics, a "nonexample" is as helpful in understanding a concept as an example. With regard to cyclic groups, $U(8)$ serves this purpose; that is, $U(8)$ is not a cyclic group. How can we verify this? Well, note that $U(8) = \{1, 3, 5, 7\}$. But

$$\begin{aligned} \langle 1 \rangle &= \{1\}, \\ \langle 3 \rangle &= \{3, 1\}, \\ \langle 5 \rangle &= \{5, 1\}, \\ \langle 7 \rangle &= \{7, 1\}, \end{aligned}$$

so $U(8) \neq \langle a \rangle$ for any a in $U(8)$.

With these examples under our belts, we are now ready to tackle cyclic groups in an abstract way and state their key properties.

Theorem 4.1 Criterion for $a^i = a^j$

Let G be a group, and let a belong to G. If a has infinite order, then $a^i = a^j$ if and only if $i = j$. If a has finite order, say, n, then $\langle a \rangle = \{e, a, a^2, \ldots, a^{n-1}\}$ and $a^i = a^j$ if and only if n divides $i - j$.

PROOF If a has infinite order, there is no nonzero n such that a^n is the identity. Since $a^i = a^j$ implies $a^{i-j} = e$, we must have $i - j = 0$, and the first statement of the theorem is proved.

Now assume that $|a| = n$. We will prove that $\langle a \rangle = \{e, a, \ldots, a^{n-1}\}$. Certainly, the elements $e, a, \ldots, a^{n-1}$ are in $\langle a \rangle$.

Now, suppose that a^k is an arbitrary member of $\langle a \rangle$. By the division algorithm, there exist integers q and r such that

$$k = qn + r \quad \text{with} \quad 0 \leq r < n.$$

Then $a^k = a^{qn+r} = a^{qn}a^r = (a^n)^q a^r = ea^r = a^r$, so that $a^k \in \{e, a, a^2, \ldots, a^{n-1}\}$. This proves that $\langle a \rangle = \{e, a, a^2, \ldots, a^{n-1}\}$.

Next, we assume that $a^i = a^j$ and prove that n divides $i - j$. We begin by observing that $a^i = a^j$ implies $a^{i-j} = e$. Again, by the division algorithm, there are integers q and r such that

$$i - j = qn + r \qquad \text{with} \qquad 0 \leq r < n.$$

Then $a^{i-j} = a^{qn+r}$, and therefore $e = a^{i-j} = a^{qn+r} = (a^n)^q a^r = e^q a^r = ea^r = a^r$. Since n is the least positive integer such that a^n is the identity, we must have $r = 0$, so that n divides $i - j$.

Conversely, if $i - j = nq$, then $a^{i-j} = a^{nq} = e^q = e$, so that $a^i = a^j$. ∎

Theorem 4.1 reveals the reason for the dual use of the notation and terminology for the order of an element and the order of a group.

Corollary 1 $|a| = |\langle a \rangle|$

For any group element a, $|a| = |\langle a \rangle|$.

One special case of Theorem 4.1 occurs so often that it deserves singling out.

Corollary 2 $a^k = e$ Implies That $|a|$ Divides k

Let G be a group and let a be an element of order n in G. If $a^k = e$, then n divides k.

PROOF Since $a^k = e = a^0$, we know by Theorem 4.1 that n divides $k - 0$. ∎

Theorem 4.1 and its corollaries for the case $|a| = 6$ are illustrated in Figure 4.1.

What is important about Theorem 4.1 in the finite case is that it says that multiplication in $\langle a \rangle$ is essentially done by *addition* modulo n. That is, if $(i + j) \bmod n = k$, then $a^i a^j = a^k$. Thus, no matter what group G is, or how the element a is chosen, multiplication in $\langle a \rangle$ works the same as addition in Z_n whenever $|a| = n$. Similarly, if a has infinite order,

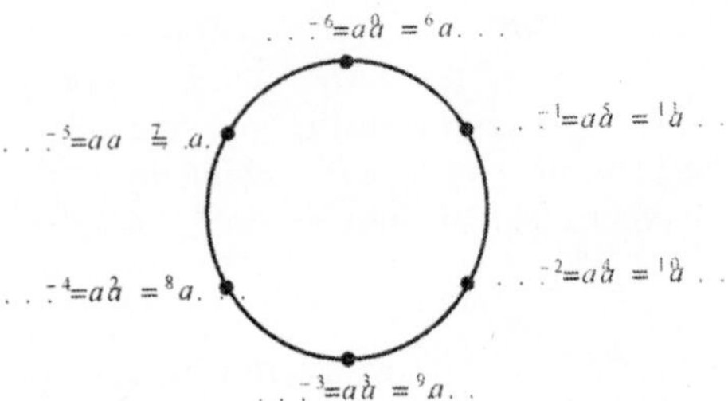

Figure 4.1

then multiplication in $\langle a \rangle$ works the same as addition in Z, since $a^i a^j = a^{i+j}$ and no modular arithmetic is done.

For these reasons, the cyclic groups Z_n and Z serve as prototypes for all cyclic groups, and algebraists say that there is essentially only one cyclic group of each order. What is meant by this is that, although there may be many different sets of the form $\{a^n \mid n \in Z\}$, there is essentially only one way to operate on these sets. Algebraists do not really care what the elements of a set are; they care only about the algebraic properties of the set—that is, the ways in which the elements of a set can be combined. We will return to this theme in the chapter on isomorphisms (Chapter 6).

The next theorem provides a simple method for computing $|a^k|$ knowing only $|a|$, and its first corollary provides a simple way to tell when $\langle a^i \rangle = \langle a^j \rangle$.

Theorem 4.2 $\langle a^k \rangle = \langle a^{\gcd(n,k)} \rangle$ **and** $|a^k| = n/\gcd(n, k)$

Let a be an element of order n in a group and let k be a positive integer. Then $\langle a^k \rangle = \langle a^{\gcd(n,k)} \rangle$ ***and*** $|a^k| = n/\gcd(n, k)$.

PROOF To simplify the notation, let $d = \gcd(n, k)$ and let $k = dr$. Since $a^k = (a^d)^r$, we have by closure that $\langle a^k \rangle \subseteq \langle a^d \rangle$. By Theorem 0.2 (the gcd theorem), there are integers s and t such that $d = ns + kt$. So, $a^d = a^{ns+kt} = a^{ns}a^{kt} = (a^n)^s(a^k)^t = e(a^k)^t = (a^k)^t \in \langle a^k \rangle$. This proves $\langle a^d \rangle \subseteq \langle a^k \rangle$. So, we have verified that $\langle a^k \rangle = \langle a^{\gcd(n,k)} \rangle$.

We prove the second part of the theorem by showing first that $|a^d| = n/d$ for any divisor d of n. Clearly, $(a^d)^{n/d} = a^n = e$, so that $|a^d| \leq n/d$. On the other hand, if i is a positive integer less than n/d, then $(a^d)^i \neq e$ by definition of $|a|$. We now apply this fact with $d = \gcd(n, k)$ to obtain $|a^k| = |\langle a^k \rangle| = |\langle a^{\gcd(n,k)} \rangle| = |a^{\gcd(n,k)}| = n/\gcd(n, k)$. ∎

The advantage of Theorem 4.2 is that it allows us to replace one generator of a cyclic subgroup with a more convenient one. For example, if $|a| = 30$, we have $\langle a^{26}\rangle = \langle a^2\rangle$, $\langle a^{23}\rangle = \langle a\rangle$, $\langle a^{22}\rangle = \langle a^2\rangle$, $\langle a^{21}\rangle = \langle a^3\rangle$. From this we can easily see that $|a^{23}| = 30$ and $|a^{22}| = 15$. Moreover, if one wants to list the elements of, say, $\langle a^{21}\rangle$, it is easier to list the elements of $\langle a^3\rangle$ instead. (Try it doing it both ways!).

Theorem 4.2 establishes an important relationship between the order of an element in a finite cyclic group and the order of the group.

Corollary 1 Orders of Elements in Finite Cyclic Groups

In a finite cyclic group, the order of an element divides the order of the group.

Corollary 2 Criterion for $\langle a^i\rangle = \langle a^j\rangle$ and $|a^i| = |a^j|$

Let $|a| = n$. Then $\langle a^i\rangle = \langle a^j\rangle$ if and only if $\gcd(n, i) = \gcd(n, j)$, and $|a^i| = |a^j|$ if and only if $\gcd(n, i) = \gcd(n, j)$.

PROOF Theorem 4.2 shows that $\langle a^i\rangle = \langle a^{\gcd(n,i)}\rangle$ and $\langle a^j\rangle = \langle a^{\gcd(n,j)}\rangle$, so that the proof reduces to proving that $\langle a^{\gcd(n,i)}\rangle = \langle a^{\gcd(n,j)}\rangle$ if and only if $\gcd(n, i) = \gcd(n, j)$. Certainly, $\gcd(n, i) = \gcd(n, j)$ implies that $\langle a^{\gcd(n,i)}\rangle = \langle a^{\gcd(n,j)}\rangle$. On the other hand, $\langle a^{\gcd(n,i)}\rangle = \langle a^{\gcd(n,j)}\rangle$ implies that $|a^{\gcd(n,i)}| = |a^{\gcd(n,j)}|$, so that by the second conclusion of Theorem 4.2, we have $n/\gcd(n, i) = n/\gcd(n, j)$, and therefore $\gcd(n, i) = \gcd(n, j)$. ∎

The second part of the corollary follows from the first part and Corollary 1 of Theorem 4.1.

The next two corollaries are important special cases of the preceding corollary.

Corollary 3 Generators of Finite Cyclic Groups

Let $|a| = n$. Then $\langle a\rangle = \langle a^j\rangle$ if and only if $\gcd(n, j) = 1$, and $|a| = |\langle a^j\rangle|$ if and only if $\gcd(n, j) = 1$.

Corollary 4 Generators of Z_n

An integer k in Z_n is a generator of Z_n if and only if $\gcd(n, k) = 1$.

The value of Corollary 3 is that once one generator of a cyclic group has been found, all generators of the cyclic group can easily be determined.

For example, consider the subgroup of all rotations in D_6. Clearly, one generator is R_{60}. And, since $|R_{60}| = 6$, we see by Corollary 3 that the only other generator is $(R_{60})^5 = R_{300}$. Of course, we could have readily deduced this information without the aid of Corollary 3 by direct calculations. So, to illustrate the real power of Corollary 3, let us use it to find all generators of the cyclic group $U(50)$. First, note that direct computations show that $|U(50)| = 20$ and that 3 is one of its generators. Thus, in view of Corollary 3, the complete list of generators for $U(50)$ is

$$\begin{aligned} 3 \bmod 50 &= 3, & 3^{11} \bmod 50 &= 47, \\ 3^3 \bmod 50 &= 27, & 3^{13} \bmod 50 &= 23, \\ 3^7 \bmod 50 &= 37, & 3^{17} \bmod 50 &= 13, \\ 3^9 \bmod 50 &= 33, & 3^{19} \bmod 50 &= 17. \end{aligned}$$

Admittedly, we had to do some arithmetic here, but it certainly entailed much less work than finding all the generators by simply determining the order of each element of $U(50)$ one by one.

The reader should keep in mind that Theorem 4.2 and its corollaries apply only to elements of finite order.

Classification of Subgroups of Cyclic Groups

The next theorem tells us how many subgroups a finite cyclic group has and how to find them.

Theorem 4.3 Fundamental Theorem of Cyclic Groups

Every subgroup of a cyclic group is cyclic. Moreover, if $|\langle a\rangle| = n$, then the order of any subgroup of $\langle a\rangle$ is a divisor of n; and, for each positive divisor k of n, the group $\langle a\rangle$ has exactly one subgroup of order k—namely, $\langle a^{n/k}\rangle$.

Before we prove this theorem, let's see what it means. Understanding what a theorem means is a prerequisite to understanding its proof. Suppose $G = \langle a\rangle$ and G has order 30. The first and second parts of the theorem say that if H is any subgroup of G, then H has the form $\langle a^{30/k}\rangle$ for some k that is a divisor of 30. The third part of the theorem says that G has one subgroup of each of the orders 1, 2, 3, 5, 6, 10, 15, and 30—and no others. The proof will also show how to find these subgroups.

PROOF Let $G = \langle a\rangle$ and suppose that H is a subgroup of G. We must show that H is cyclic. If it consists of the identity alone, then clearly H is cyclic. So we may assume that $H \neq \{e\}$. We now claim that H contains an element of the form a^t, where t is positive. Since $G = \langle a \rangle$, every element of H has the form a^t; and when a^t belongs to H with $t < 0$, then a^{-t} belongs to H also and $-t$ is positive. Thus, our claim is verified. Now let m be the least positive integer such that $a^m \in H$. By closure, $\langle a^m \rangle \subseteq H$. We next claim that $H = \langle a^m \rangle$. To prove this claim, it suffices to let b be an arbitrary member of H and show that b is in $\langle a^m \rangle$. Since $b \in G = \langle a \rangle$, we have $b = a^k$ for some k. Now, apply the division algorithm to k and m to obtain integers q and r such that $k = mq + r$ where $0 \leq r < m$. Then $a^k = a^{mq+r} = a^{mq}a^r$, so that $a^r = a^{-mq}a^k$. Since $a^k = b \in H$ and $a^{-mq} = (a^m)^{-q}$ is in H also, $a^r \in H$. But, m is the *least* positive integer such that $a^m \in H$, and $0 \leq r < m$, so r must be 0. Therefore, $b = a^k = a^{mq} = (a^m)^q \in \langle a^m \rangle$. This proves the assertion of the theorem that every subgroup of a cyclic group is cyclic.

To prove the next portion of the theorem, suppose that $|\langle a \rangle| = n$ and H is any subgroup of $\langle a \rangle$. We have already shown that $H = \langle a^m \rangle$, where m is the least positive integer such that $a^m \in H$. Using $e = b = a^n$ as in the preceding paragraph, we have $n = mq$.

Finally, let k be any positive divisor of n. We will show that $\langle a^{n/k} \rangle$ is the one and only subgroup of $\langle a \rangle$ of order k. From Theorem 4.2, we see that $\langle a^{n/k} \rangle$ has order $n/\gcd(n, n/k) = n/(n/k) = k$. Now let H be any subgroup of $\langle a \rangle$ of order k. We have already shown above that $H = \langle a^m \rangle$, where m is a divisor of n. Then $m = \gcd(n, m)$ and $k = |a^m| = |a^{\gcd(n,m)}| = n/\gcd(n, m) = n/m$. Thus, $m = n/k$ and $H = \langle a^{n/k} \rangle$. ■

Returning for a moment to our discussion of the cyclic group $\langle a \rangle$, where a has order 30, we may conclude from Theorem 4.3 that the subgroups of $\langle a \rangle$ are precisely those of the form $\langle a^m \rangle$, where m is a divisor of 30. Moreover, if k is a divisor of 30, the subgroup of order k is $\langle a^{30/k} \rangle$. So the list of subgroups of $\langle a \rangle$ is:

$$
\begin{aligned}
\langle a \rangle &= \{e, a, a^2, \ldots, a^{29}\} && \text{order 30,}\\
\langle a^2 \rangle &= \{e, a^2, a^4, \ldots, a^{28}\} && \text{order 15,}\\
\langle a^3 \rangle &= \{e, a^3, a^6, \ldots, a^{27}\} && \text{order 10,}\\
\langle a^5 \rangle &= \{e, a^5, a^{10}, a^{15}, a^{20}, a^{25}\} && \text{order 6,}\\
\langle a^6 \rangle &= \{e, a^6, a^{12}, a^{18}, a^{24}\} && \text{order 5,}\\
\langle a^{10} \rangle &= \{e, a^{10}, a^{20}\} && \text{order 3,}\\
\langle a^{15} \rangle &= \{e, a^{15}\} && \text{order 2,}\\
\langle a^{30} \rangle &= \{e\} && \text{order 1.}
\end{aligned}
$$

In general, if $\langle a \rangle$ has order n and k divides n, then $\langle a^{n/k} \rangle$ is the unique subgroup of order k.

Taking the group in Theorem 4.3 to be Z_n and a to be 1, we obtain the following important special case.

Corollary Subgroups of Z_n

For each positive divisor k of n, the set $\langle n/k \rangle$ is the unique subgroup of Z_n of order k; moreover, these are the only subgroups of Z_n.

EXAMPLE 5 The list of subgroups of Z_{30} is

$$\begin{array}{rll} \langle 1 \rangle = \{0, 1, 2, \ldots, 29\} & & \text{order } 30, \\ \langle 2 \rangle = \{0, 2, 4, \ldots, 28\} & & \text{order } 15, \\ \langle 3 \rangle = \{0, 3, 6, \ldots, 27\} & & \text{order } 10, \\ \langle 5 \rangle = \{0, 5, 10, 15, 20, 25\} & & \text{order } 6, \\ \langle 6 \rangle = \{0, 6, 12, 18, 24\} & & \text{order } 5, \\ \langle 10 \rangle = \{0, 10, 20\} & & \text{order } 3, \\ \langle 15 \rangle = \{0, 15\} & & \text{order } 2, \\ \langle 30 \rangle = \{0\} & & \text{order } 1. \end{array}$$

Theorems 4.2 and 4.3 provide a simple way to find all the generators of the subgroups of a finite cyclic group.

EXAMPLE 6 To find the generators of the subgroup of order 9 in Z_{36}, we observe that $36/9 = 4$ is one generator. To find the others, we have from Corollary 3 of Theorem 4.2 that they are all elements of Z_{36} of the form $4j$, where $\gcd(9, j) = 1$. Thus,

$$\langle 4 \cdot 1 \rangle = \langle 4 \cdot 2 \rangle = \langle 4 \cdot 4 \rangle = \langle 4 \cdot 5 \rangle = \langle 4 \cdot 7 \rangle = \langle 4 \cdot 8 \rangle.$$

In the generic case, to find all the subgroups of $\langle a \rangle$ of order 9 where $|a| = 36$, we have

$$\langle (a^4)^1 \rangle = \langle (a^4)^2 \rangle = \langle (a^4)^4 \rangle = \langle (a^4)^5 \rangle = \langle (a^4)^7 \rangle = \langle (a^4)^8 \rangle.$$

In particular, note that once you have the generator $a^{n/d}$ for the subgroup of order d where d is a divisor of $|a| = n$, all the generators of $\langle a^d \rangle$ have the form $(a^d)^j$ where $j \in U(d)$.

By combining Theorems 4.2 and 4.3, we can easily count the number of elements of each order in a finite cyclic group. For convenience, we introduce an important number-theoretic function called the *Euler phi function*. Let $\phi(1) = 1$, and for any integer $n > 1$, let $\phi(n)$ denote the number of positive integers less than n and relatively prime to n. Notice that by definition of the group $U(n)$, $|U(n)| = \phi(n)$. The first 12 values of $\phi(n)$ are given in Table 4.1.

Table 4.1 Values of $\phi(n)$

n	1	2	3	4	5	6	7	8	9	10	11	12
$\phi(n)$	1	1	2	2	4	2	6	4	6	4	10	4

Theorem 4.4 Number of Elements of Each Order in a Cyclic Group

If d is a positive divisor of n, the number of elements of order d in a cyclic group of order n is $\phi(d)$.

PROOF By Theorem 4.3, the group has exactly one subgroup of order d—call it $\langle a \rangle$. Then every element of order d also generates the subgroup $\langle a \rangle$ and, by Corollary 3 of Theorem 4.2, an element a^k generates $\langle a \rangle$ if and only if $\gcd(k, d) = 1$. The number of such elements is precisely $\phi(d)$. ■

Notice that for a finite cyclic group of order n, the number of elements of order d for any divisor d of n depends only on d. Thus, Z_8, Z_{640}, and Z_{80000} each have $\phi(8) = 4$ elements of order 8.

Although there is no formula for the number of elements of each order for arbitrary finite groups, we still can say something important in this regard.

Corollary Number of Elements of Order d in a Finite Group

In a finite group, the number of elements of order d is a multiple of $\phi(d)$.

PROOF If a finite group has no elements of order d, the statement is true, since $\phi(d)$ divides 0. Now suppose that $a \in G$ and $|a| = d$. By Theorem 4.4, we know that $\langle a \rangle$ has $\phi(d)$ elements of order d. If all elements of order d in G are in $\langle a \rangle$, we are done. So, suppose that there is an element b in G of order d that is not in $\langle a \rangle$. Then, $\langle b \rangle$ also has $\phi(d)$ elements of order d. This means that we have found $2\phi(d)$ elements of order d in G provided that $\langle a \rangle$ and $\langle b \rangle$ have no elements of order d in common. If there is an element c of order d that belongs to both $\langle a \rangle$ and $\langle b \rangle$, then we have $\langle a \rangle = \langle c \rangle = \langle b \rangle$, so that $b \in \langle a \rangle$, which is a contradiction. Continuing in this fashion, we see that the number of elements of order d in a finite group is a multiple of $\phi(d)$. ■

On its face, the value of Theorem 4.4 and its corollary seem limited for large values of n, because it is tedious to determine the number of positive integers less than or equal to n and relatively prime to n by examining them one by one. However, the following properties of the ϕ function make computing $\phi(n)$ simple: For any prime p, $\phi(p^n) = p^n - p^{n-1}$ (see Exercise 85) and for relatively prime m and n, $\phi(mn) = \phi(m)\phi(n)$. Thus, $\phi(40) = \phi(8)\phi(5) = 4 \cdot 4 = 16$; $\phi(75) = \phi(5^2)\phi(3) = (25 - 5) \cdot 2 = 40$.

The relationships among the various subgroups of a group can be illustrated with a *subgroup lattice* of the group. This is a diagram that includes all the subgroups of the group and connects a subgroup H at one level to a subgroup K at a higher level with a sequence of line segments if and only if H is a proper subgroup of K. Although there are many ways to draw such a diagram, the connections between the subgroups must be the same. Typically, one attempts to present the diagram in an eye-pleasing fashion. The lattice diagram for Z_{30} is shown in Figure 4.2. Notice that $\langle 10\rangle$ is a subgroup of both $\langle 2\rangle$ and $\langle 5\rangle$, but $\langle 6\rangle$ is not a subgroup of $\langle 10\rangle$.

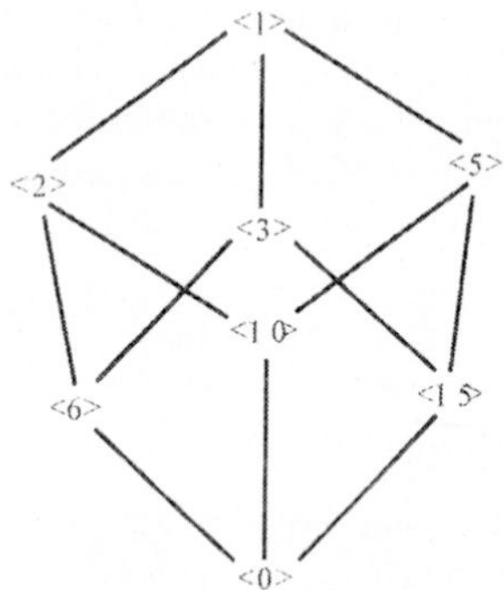

Figure 4.2 Subgroup lattice of Z_{30}.

The precision of Theorem 4.3 can be appreciated by comparing the ease with which we are able to identify the subgroups of Z_{30} with that of doing the same for, say, $U(30)$ or D_{30}. And these groups have relatively simple structures among noncyclic groups.

We will prove in Chapter 7 that a certain portion of Theorem 4.3 extends to arbitrary finite groups; namely, the order of a subgroup divides the order of the group itself. We will also see, however, that a finite group need not have exactly one subgroup corresponding to each divisor of the order of the group. For some divisors, there may be none at all, whereas for other divisors, there may be many. Indeed, D_4, the dihedral group of order 8, has five subgroups of order 2 and three of order 4.

One final remark about the importance of cyclic groups is appropriate. Although cyclic groups constitute a very narrow class of finite groups, we will see in Chapter 11 that they play the role of building blocks for all finite Abelian groups in much the same way that primes are the building blocks for the integers and that chemical elements are the building blocks for the chemical compounds.

Exercises

It is not unreasonable to use the hypothesis.

ARNOLD ROSS

1. Find all generators of Z_6, Z_8, and Z_{20}.
2. Suppose that $\langle a\rangle$, $\langle b\rangle$, and $\langle c\rangle$ are cyclic groups of orders 6, 8, and 20, respectively. Find all generators of $\langle a\rangle$, $\langle b\rangle$, and $\langle c\rangle$.
3. List the elements of the subgroups $\langle 20\rangle$ and $\langle 10\rangle$ in Z_{30}. Let a be a group element of order 30. List the elements of the subgroups $\langle a^{20}\rangle$ and $\langle a^{10}\rangle$.
4. List the elements of the subgroups $\langle 3\rangle$ and $\langle 15\rangle$ in Z_{18}. Let a be a group element of order 18. List the elements of the subgroups $\langle a^3\rangle$ and $\langle a^{15}\rangle$.
5. List the elements of the subgroups $\langle 3\rangle$ and $\langle 7\rangle$ in $U(20)$.
6. What do Exercises 3, 4, and 5 have in common? Try to make a generalization that includes these three cases.
7. Find an example of a noncyclic group, all of whose proper subgroups are cyclic.
8. Let a be an element of a group and let $|a| = 15$. Compute the orders of the following elements of G.
   a. a^3, a^6, a^9, a^{12}
   b. a^5, a^{10}
   c. a^2, a^4, a^8, a^{14}
9. How many subgroups does Z_{20} have? List a generator for each of these subgroups. Suppose that $G = \langle a\rangle$ and $|a| = 20$. How many subgroups does G have? List a generator for each of these subgroups.
10. In Z_{24}, list all generators for the subgroup of order 8. Let $G = \langle a\rangle$ and let $|a| = 24$. List all generators for the subgroup of order 8.
11. Let G be a group and let $a \in G$. Prove that $\langle a^{-1}\rangle = \langle a\rangle$.
12. In Z, find all generators of the subgroup $\langle 3\rangle$. If a has infinite order, find all generators of the subgroup $\langle a^3\rangle$.
13. In Z_{24}, find a generator for $\langle 21\rangle \cap \langle 10\rangle$. Suppose that $|a| = 24$. Find a generator for $\langle a^{21}\rangle \cap \langle a^{10}\rangle$. In general, what is a generator for the subgroup $\langle a^m\rangle \cap \langle a^n\rangle$?
14. Suppose that a cyclic group G has exactly three subgroups: G itself, $\{e\}$, and a subgroup of order 7. What is $|G|$? What can you say if 7 is replaced with p where p is a prime?

15. Let G be an Abelian group and let $H = \{g \in G \mid |g| \text{ divides } 12\}$. Prove that H is a subgroup of G. Is there anything special about 12 here? Would your proof be valid if 12 were replaced by some other positive integer? State the general result.
16. Find a collection of distinct subgroups $\langle a_1\rangle, \langle a_2\rangle, \ldots, \langle a_n\rangle$ of Z_{240} with the property that $\langle a_1\rangle \subset \langle a_2\rangle \subset \cdots \subset \langle a_n\rangle$ with n as large as possible.
17. Complete the following statement: $|a| = |a^2|$ if and only if $|a| \ldots$.
18. If a cyclic group has an element of infinite order, how many elements of finite order does it have?
19. List the cyclic subgroups of $U(30)$.
20. Suppose that G is an Abelian group of order 35 and every element of G satisfies the equation $x^{35} = e$. Prove that G is cyclic. Does your argument work if 35 is replaced with 33?
21. Let G be a group and let a be an element of G.
    a. If $a^{12} = e$, what can we say about the order of a?
    b. If $a^m = e$, what can we say about the order of a?
    c. Suppose that $|G| = 24$ and that G is cyclic. If $a^8 \neq e$ and $a^{12} \neq e$, show that $\langle a\rangle = G$.
22. Prove that a group of order 3 must be cyclic.
23. Let Z denote the group of integers under addition. Is every subgroup of Z cyclic? Why? Describe all the subgroups of Z. Let a be a group element with infinite order. Describe all subgroups of $\langle a\rangle$.
24. For any element a in any group G, prove that $\langle a\rangle$ is a subgroup of $C(a)$ (the centralizer of a).
25. If d is a positive integer, $d \neq 2$, and d divides n, show that the number of elements of order d in D_n is $\phi(d)$. How many elements of order 2 does D_n have?
26. Find all generators of Z. Let a be a group element that has infinite order. Find all generators of $\langle a\rangle$.
27. Prove that C^*, the group of nonzero complex numbers under multiplication, has a cyclic subgroup of order n for every positive integer n.
28. Let a be a group element that has infinite order. Prove that $\langle a^i\rangle = \langle a^j\rangle$ if and only if $i = \pm j$.
29. List all the elements of order 8 in $Z_{8000000}$. How do you know your list is complete? Let a be a group element such that $|a| = 8000000$. List all elements of order 8 in $\langle a\rangle$. How do you know your list is complete?
30. Suppose a and b belong to a group, a has odd order, and $aba^{-1} = b^{-1}$. Show that $b^2 = e$.

31. Let G be a finite group. Show that there exists a fixed positive integer n such that $a^n = e$ for all a in G. (Note that n is independent of a.)

32. Determine the subgroup lattice for Z_{12}.

33. Determine the subgroup lattice for Z_{p^2q}, where p and q are distinct primes.

34. Determine the subgroup lattice for Z_8.

35. Determine the subgroup lattice for Z_{p^n}, where p is a prime and n is some positive integer.

36. Prove that a finite group is the union of proper subgroups if and only if the group is not cyclic.

37. Show that the group of positive rational numbers under multiplication is not cyclic.

38. Consider the set $\{4, 8, 12, 16\}$. Show that this set is a group under multiplication modulo 20 by constructing its Cayley table. What is the identity element? Is the group cyclic? If so, find all of its generators.

39. Give an example of a group that has exactly 6 subgroups (including the trivial subgroup and the group itself). Generalize to exactly n subgroups for any positive integer n.

40. Let m and n be elements of the group Z. Find a generator for the group $\langle m \rangle \cap \langle n \rangle$.

41. Suppose that a and b are group elements that commute and have orders m and n. If $\langle a \rangle \cap \langle b \rangle = \{e\}$, prove that the group contains an element whose order is the least common multiple of m and n. Show that this need not be true if a and b do not commute.

42. Suppose that a and b belong to a group G, a and b commute, and $|a|$ and $|b|$ are finite. What are the possibilities for $|ab|$?

43. Suppose that a and b belong to a group G, a and b commute, and $|a|$ and $|b|$ are finite. Prove that G has an element of order $\text{lcm}(|a|, |b|)$.

44. Let F and F' be distinct reflections in D_{21}. What are the possibilities for $|FF'|$?

45. Suppose that H is a subgroup of a group G and $|H| = 10$. If a belongs to G and a^6 belongs to H, what are the possibilities for $|a|$?

46. Which of the following numbers could be the exact number of elements of order 21 in a group: 21600, 21602, 21604?

47. If G is an infinite group, what can you say about the number of elements of order 8 in the group? Generalize.

48. Suppose that K is a proper subgroup of D_{35} and K contains at least two reflections. What are the possible orders of K? Explain your reasoning.

49. For each positive integer n, prove that C^*, the group of nonzero complex numbers under multiplication, has exactly $\phi(n)$ elements of order n.
50. Prove or disprove that $H = \{n \in Z \mid n \text{ is divisible by both 8 and 10}\}$ is a subgroup of Z.
51. Suppose that G is a finite group with the property that every nonidentity element has prime order (for example, D_3 and D_5). If $Z(G)$ is not trivial, prove that every nonidentity element of G has the same order.
52. Prove that an infinite group must have an infinite number of subgroups.
53. Let p be a prime. If a group has more than $p - 1$ elements of order p, why can't the group be cyclic?
54. Suppose that G is a cyclic group and that 6 divides $|G|$. How many elements of order 6 does G have? If 8 divides $|G|$, how many elements of order 8 does G have? If a is one element of order 8, list the other elements of order 8.
55. List all the elements of Z_{40} that have order 10. Let $|x| = 40$. List all the elements of $\langle x \rangle$ that have order 10.
56. Reformulate the corollary of Theorem 4.4 to include the case when the group has infinite order.
57. Determine the orders of the elements of D_{33} and how many there are of each.
58. If G is a cyclic group and 15 divides the order of G, determine the number of solutions in G of the equation $x^{15} = e$. If 20 divides the order of G, determine the number of solutions of $x^{20} = e$. Generalize.
59. If G is an Abelian group and contains cyclic subgroups of orders 4 and 5, what other sizes of cyclic subgroups must G contain? Generalize.
60. If G is an Abelian group and contains cyclic subgroups of orders 4 and 6, what other sizes of cyclic subgroups must G contain? Generalize.
61. Prove that no group can have exactly two elements of order 2.
62. Given the fact that $U(49)$ is cyclic and has 42 elements, deduce the number of generators that $U(49)$ has without actually finding any of the generators.
63. Let a and b be elements of a group. If $|a| = 10$ and $|b| = 21$, show that $\langle a \rangle \cap \langle b \rangle = \{e\}$.
64. Let a and b belong to a group. If $|a|$ and $|b|$ are relatively prime, show that $\langle a \rangle \cap \langle b \rangle = \{e\}$.

65. Let a and b belong to a group. If $|a| = 24$ and $|b| = 10$, what are the possibilities for $|\langle a\rangle \cap \langle b\rangle|$?

66. Prove that $U(2^n)$ $(n \geq 3)$ is not cyclic.

67. Suppose that G is a group of order 16 and that, by direct computation, you know that G has at least nine elements x such that $x^8 = e$. Can you conclude that G is not cyclic? What if G has at least five elements x such that $x^4 = e$? Generalize.

68. Prove that Z_n has an even number of generators if $n > 2$. What does this tell you about $\phi(n)$?

69. If $|a^5| = 12$, what are the possibilities for $|a|$? If $|a^4| = 12$, what are the possibilities for $|a|$?

70. Suppose that $|x| = n$. Find a necessary and sufficient condition on r and s such that $\langle x^r\rangle \subseteq \langle x^s\rangle$.

71. Suppose a is a group element such that $|a^{28}| = 10$ and $|a^{22}| = 20$. Determine $|a|$.

72. Let a be a group element such that $|a| = 48$. For each part, find a divisor k of 48 such that

a. $\langle a^{21}\rangle = \langle a^k\rangle$;
b. $\langle a^{14}\rangle = \langle a^k\rangle$;
c. $\langle a^{18}\rangle = \langle a^k\rangle$.

73. Let p be a prime. Show that in a cyclic group of order $p^n - 1$, every element is a pth power (that is, every element can be written in the form a^p for some a).

74. Prove that $H = \left\{ \begin{bmatrix} 1 & n \\ 0 & 1 \end{bmatrix} \middle|\, n \in Z \right\}$ is a cyclic subgroup of $GL(2, \mathbf{R})$.

75. Let a and b belong to a group. If $|a| = 12$, $|b| = 22$, and $\langle a\rangle \cap \langle b\rangle \neq \{e\}$, prove that $a^6 = b^{11}$.

76. (2008 GRE Practice Exam) If x is an element of a cyclic group of order 15 and exactly two of x^3, x^5, and x^9 are equal, determine $|x^{13}|$.

77. Determine the number of cyclic subgroups of order 4 in D_n.

78. If n is odd, prove that D_n has no subgroup of order 4.

79. If $n \geq 4$ and is even, show that D_n has exactly $n/2$ noncyclic subgroups of order 4.

80. If $n \geq 4$ and n is divisible by 2 but not by 4, prove that D_n has exactly $n/2$ subgroups of order 4.

81. How many subgroups of order n does D_n have?

82. Let G be the set of all polynomials of the form $ax^2 + bx + c$ with coefficients from the set $\{0, 1, 2\}$. We can make G a group under addition by adding the polynomials in the usual way, except that we use modulo 3 to combine the coefficients. With this operation, prove that G is a group of order 27 that is not cyclic.

83. Let a and b belong to some group. Suppose that $|a| = m$, $|b| = n$, and m and n are relatively prime. If $a^k = b^k$ for some integer k, prove that mn divides k.

84. For every integer n greater than 2, prove that the group $U(n^2 - 1)$ is not cyclic.

85. Prove that for any prime p and positive integer n, $\phi(p^n) = p^n - p^{n-1}$.

86. Give an example of an infinite group that has exactly two elements of order 4.

Computer Exercises

Computer exercises for this chapter are available at the website:

http://www.d.umn.edu/~jgallian

Suggested Reading

Deborah L. Massari, "The Probability of Generating a Cyclic Group," *Pi Mu Epsilon Journal* 7 (1979): 3–6.

In this easy-to-read paper, it is shown that the probability of a randomly chosen element from a cyclic group being a generator of the group depends only on the set of prime divisors of the order of the group, and not on the order itself. This article, written by an undergraduate student, received first prize in a Pi Mu Epsilon paper contest.

James Joseph Sylvester

Stock Montage

I really love my subject.

J. J. SYLVESTER

JAMES JOSEPH SYLVESTER was the most influential mathematician in America in the 19th century. Sylvester was born on September 3, 1814, in London and showed his mathematical genius early. At the age of 14, he studied under De Morgan and won several prizes for his mathematics, and at the unusually young age of 25, he was elected a fellow of the Royal Society.

After receiving B.A. and M.A. degrees from Trinity College in Dublin in 1841, Sylvester began a professional life that was to include academics, law, and actuarial careers. In 1876, at the age of 62, he was appointed to a prestigious position at the newly founded Johns Hopkins University. During his seven years at Johns Hopkins, Sylvester pursued research in pure mathematics with tremendous vigor and enthusiasm. He also founded the *American Journal of Mathematics*, the first journal in America devoted to mathematical research. Sylvester returned to England in 1884 to a professorship at Oxford, a position he held until his death on March 15, 1897.

Sylvester's major contributions to mathematics were in the theory of equations, matrix theory, determinant theory, and invariant theory (which he founded with Cayley). His writings and lectures—flowery and eloquent, pervaded with poetic flights, emotional expressions, bizarre utterances, and paradoxes—reflected the personality of this sensitive, excitable, and enthusiastic man. We quote three of his students.[†] E. W. Davis commented on Sylvester's teaching methods.

> Sylvester's methods! He had none. "Three lectures will be delivered on a New Universal Algebra," he would say; then, "The course must be extended to twelve." It did last all the rest of that year. The following year the course was to be *Substitutions-Theorie*, by Netto. We all got the text. He lectured about three times, following the text closely and stopping sharp at the end of the hour. Then he began to think about matrices again. "I must give one lecture a week on those," he said. He could not confine himself to the hour, nor to the one lecture a week. Two weeks were passed, and Netto was forgotten entirely and never mentioned again. Statements like the following were not infrequent in his lectures: "I haven't proved this, but I am as sure as I can be of anything that it must be so. From this it will follow, etc." At the next lecture it turned out that what he was so sure of was false. Never mind, he kept on forever guessing and trying, and presently a wonderful discovery followed, then another and another. Afterward he would go back and work it all over again, and surprise us with all sorts of side lights. He then made another leap in the dark, more treasures were discovered, and so on forever.

[†]F. Cajori, *Teaching and History of Mathematics in the United States*, Washington: Government Printing Office, 1890, 265–266.

Sylvester's enthusiasm for teaching and his influence on his students are captured in the following passage written by Sylvester's first student at Johns Hopkins, G. B. Halsted.

> A short, broad man of tremendous vitality, . . . Sylvester's capacious head was ever lost in the highest cloud-lands of pure mathematics. Often in the dead of night he would get his favorite pupil, that he might communicate the very last product of his creative thought. Everything he saw suggested to him something new in the higher algebra. This transmutation of everything into new mathematics was a revelation to those who knew him intimately. They began to do it themselves.

Another characteristic of Sylvester, which is very unusual among mathematicians, was his apparent inability to remember mathematics! W. P. Durfee had the following to say.

> Sylvester had one remarkable peculiarity. He seldom remembered theorems, propositions, etc., but had always to deduce them when he wished to use them. In this he was the very antithesis of Cayley, who was thoroughly conversant with everything that had been done in every branch of mathematics.
>
> I remember once submitting to Sylvester some investigations that I had been engaged on, and he immediately denied my first statement, saying that such a proposition had never been heard of, let alone proved. To his astonishment, I showed him a paper of his own in which he had proved the proposition; in fact, I believe the object of his paper had been the very proof which was so strange to him.

For more information about Sylvester, visit:

http://www-groups.dcs.st-and .ac.uk/~history/

Supplementary Exercises for Chapters 1–4

If you really want something in this life, you have to work for it. Now quiet, they're about to announce the lottery numbers!

HOMER SIMPSON

True/false questions for Chapters 1–4 are available on the Web at:

http://www.d.umn.edu/~jgallian/TF

1. Let G be a group and let H be a subgroup of G. For any fixed x in G, define $xHx^{-1} = \{xhx^{-1} \mid h \in H\}$. Prove the following.
   a. xHx^{-1} is a subgroup of G.
   b. If H is cyclic, then xHx^{-1} is cyclic.
   c. If H is Abelian, then xHx^{-1} is Abelian.
   The group xHx^{-1} is called a *conjugate* of H. (Note that conjugation preserves structure.)

2. Let G be a group and let H be a subgroup of G. Define $N(H) = \{x \in G \mid xHx^{-1} = H\}$. Prove that $N(H)$ (called the *normalizer* of H) is a subgroup of G.†

3. Let G be a group. For each $a \in G$, define $\text{cl}(a) = \{xax^{-1} \mid x \in G\}$. Prove that these subsets of G partition G. [cl(a) is called the *conjugacy class* of a.]

4. The group defined by the following table is called the *group of quaternions*. Use the table to determine each of the following.
   a. The center
   b. cl(a)
   c. cl(b)
   d. All cyclic subgroups

	e	a	a^2	a^3	b	ba	ba^2	ba^3
e	e	a	a^2	a^3	b	ba	ba^2	ba^3
a	a	a^2	a^3	e	ba^3	b	ba	ba^2
a^2	a^2	a^3	e	a	ba^2	ba^3	b	ba
a^3	a^3	e	a	a^2	ba	ba^2	ba^3	b
b	b	ba	ba^2	ba^3	a^2	a^3	e	a
ba	ba	ba^2	ba^3	b	a	a^2	a^3	e
ba^2	ba^2	ba^3	b	ba	e	a	a^2	a^3
ba^3	ba^3	b	ba	ba^2	a^3	e	a	a^2

†This very important subgroup was first used by L. Sylow in 1872 to prove the existence of certain kinds of subgroups in a group. His work is discussed in Chapter 24.

5. (Conjugation preserves order.) Prove that, in any group, $|xax^{-1}| = |a|$. (This exercise is referred to in Chapter 24.)
6. Prove that, in any group, $|ab| = |ba|$.
7. If a and b are group elements, prove that $|ab| = |a^{-1}b^{-1}|$.
8. Prove that a group of order 4 cannot have a subgroup of order 3.
9. If a, b, and c are elements of a group, give an example to show that it need not be the case that $|abc| = |cba|$.
10. Let a and b belong to a group G. Prove that there is an element x in G such that $xax = b$ if and only if $ab = c^2$ for some element c in G.
11. Prove that if a is the only element of order 2 in a group, then a lies in the center of the group.
12. Let G be the plane symmetry group of the infinite strip of equally spaced H's shown below.

Let x be the reflection about Axis 1 and let y be the reflection about Axis 2. Calculate $|x|$, $|y|$, and $|xy|$. Must the product of elements of finite order have finite order? (This exercise is referred to in Chapter 27.)

13. What are the orders of the elements of D_{15}? How many elements have each of these orders?
14. Prove that a group of order 4 is Abelian.
15. Prove that a group of order 5 must be cyclic.
16. Prove that an Abelian group of order 6 must be cyclic.
17. Let G be an Abelian group and let n be a fixed positive integer. Let $G^n = \{g^n \mid g \in G\}$. Prove that G^n is a subgroup of G. Give an example showing that G^n need not be a subgroup of G when G is non-Abelian. (This exercise is referred to in Chapter 11.)
18. Let $G = \{a + b\sqrt{2}\}$, where a and b are rational numbers not both 0. Prove that G is a group under ordinary multiplication.
19. (1969 Putnam Competition) Prove that no group is the union of two proper subgroups. Does the statement remain true if "two" is replaced by "three"?
20. Prove that the subset of elements of finite order in an Abelian group forms a subgroup. (This subgroup is called the *torsion subgroup*.) Is the same thing true for non-Abelian groups?
21. Let p be a prime and let G be an Abelian group. Show that the set of all elements whose orders are powers of p is a subgroup of G.

22. Suppose that a and b are group elements. If $|b| = 2$ and $bab = a^4$, determine the possibilities for $|a|$.
23. Suppose that a finite group is generated by two elements a and b (that is, every element of the group can be expressed as some product of a's and b's). Given that $a^3 = b^2 = e$ and $ba^2 = ab$, construct the Cayley table for the group. We have already seen an example of a group that satisfies these conditions. Name it.
24. If a is an element from a group and $|a| = n$, prove that $C(a) = C(a^k)$ when k is relatively prime to n.
25. Let x and y belong to a group G. If $xy \in Z(G)$, prove that $xy = yx$.
26. Suppose that H and K are nontrivial subgroups of Q under addition. Show that $H \cap K$ is a nontrivial subgroup of Q. Is this true if Q is replaced by $\mathbf{R}$?
27. Let H be a subgroup of G and let g be an element of G. Prove that $N(gHg^{-1}) = gN(H)g^{-1}$. See Exercise 2 for the notation.
28. Let H be a subgroup of a group G and let $|g| = n$. If g^m belongs to H, and m and n are relatively prime, prove that g belongs to H.
29. Find a group that contains elements a and b such that $|a| = 2$, $|b| = 11$, and $|ab| = 2$.
30. Suppose that G is a group with exactly eight elements of order 10. How many cyclic subgroups of order 10 does G have?
31. (1989 Putnam Competition) Let S be a nonempty set with an associative operation that is left and right cancellative ($xy = xz$ implies $y = z$, and $yx = zx$ implies $y = z$). Assume that for every a in S the set $\{a^n \mid n = 1, 2, 3, \ldots\}$ is finite. Must S be a group?
32. Let $H_1, H_2, H_3, \ldots$ be a sequence of subgroups of a group with the property that $H_1 \subseteq H_2 \subseteq H_3 \ldots$. Prove that the union of the sequence is a subgroup.
33. Let n be an integer greater than 1. Find a noncyclic subgroup of $U(4n)$ of order 4 that contains the element $2n - 1$.
34. Let G be an Abelian group and $H = \{x \in G \mid x^n = e$ for some odd integer n (n may vary with x)$\}$. Prove that H is a subgroup of G. Is H a subgroup if "odd" is replaced by "even"?
35. Let $H = \{A \in GL(2, \mathbf{R}) \mid \det A$ is rational$\}$. Prove or disprove that H is a subgroup of $GL(2, \mathbf{R})$. What if "rational" is replaced by "an integer"?
36. Suppose that G is a group that has exactly one nontrivial proper subgroup. Prove that G is cyclic and $|G| = p^2$, where p is prime.
37. Suppose that G is a group and G has exactly two nontrivial proper subgroups. Prove that G is cyclic and $|G| = pq$, where p and q are distinct primes, or that G is cyclic and $|G| = p^3$, where p is prime.

38. If $|a^2| = |b^2|$, prove or disprove that $|a| = |b|$.

39. (1995 Putnam Competition) Let S be a set of real numbers that is closed under multiplication. Let T and U be disjoint subsets of S whose union is S. Given that the product of any three (not necessarily distinct) elements of T is in T and that the product of any three elements of U is in U, show that at least one of the two subsets T and U is closed under multiplication.

40. If p is an odd prime, prove that there is no group that has exactly p elements of order p.

41. Give an example of a group G with infinitely many distinct subgroups $H_1, H_2, H_3, \ldots$ such that $H_1 \subset H_2 \subset H_3 \ldots$.

42. Suppose a and b are group elements and $b \neq e$. If $a^{-1}ba = b^2$ and $|a| = 3$, find $|b|$. What is $|b|$, if $|a| = 5$? What can you say about $|b|$ in the case where $|a| = k$?

43. Let a and b belong to a group G. Show that there is an element g in G such that $g^{-1}abg = ba$.

44. Suppose G is a group and $x^3y^3 = y^3x^3$ for every x and y in G. Let $H = \{x \in G \mid |x| \text{ is relatively prime to } 3\}$. Prove that elements of H commute with each other and that H is a subgroup of G. Is your argument valid if 3 is replaced by an arbitrary positive integer n? Explain why or why not.

45. Let G be a finite group and let S be a subset of G that contains more than half of the elements of G. Show that every element of G can be expressed in the form s_1s_2 where s_1 and s_2 belong to S.

46. Let G be a group and let f be a function from G to some set. Show that $H = \{g \in G \mid f(xg) = f(x) \text{ for all } x \in G\}$ is a subgroup of G. In the case that G is the group of real numbers under addition and $f(x) = \sin x$, describe H.

47. Let G be a cyclic group of order n and let H be the subgroup of order d. Show that $H = \{x \in G \mid |x| \text{ divides } d\}$.

48. Let a be an element of maximum order from a finite Abelian group G. Prove that for any element b in G, $|b|$ divides $|a|$. Show by example that this need not be true for finite non-Abelian groups.

49. Define an operation $*$ on the set of integers by $a * b = a + b - 1$. Show that the set of integers under this operation is a cyclic group.

50. Let n be an integer greater than 1. Find a noncyclic subgroup of $U(4n)$ of order 4 that contains the element $2n - 1$.

5 Permutation Groups

Wigner's discovery about the electron permutation group was just the beginning. He and others found many similar applications and nowadays group theoretical methods—especially those involving characters and representations—pervade all branches of quantum mechanics.

GEORGE MACKEY, *Proceedings of the American Philosophical Society*

Definition and Notation

In this chapter, we study certain groups of functions, called permutation groups, from a set A to itself. In the early and mid-19th century, groups of permutations were the only groups investigated by mathematicians. It was not until around 1850 that the notion of an abstract group was introduced by Cayley, and it took another quarter century before the idea firmly took hold.

Definitions Permutation of A, Permutation Group of A

A *permutation* of a set A is a function from A to A that is both one-to-one and onto. A *permutation group* of a set A is a set of permutations of A that forms a group under function composition.

Although groups of permutations of any nonempty set A of objects exist, we will focus on the case where A is finite. Furthermore, it is customary, as well as convenient, to take A to be a set of the form $\{1, 2, 3, \ldots, n\}$ for some positive integer n. Unlike in calculus, where most functions are defined on infinite sets and are given by formulas, in algebra, permutations of finite sets are usually given by an explicit listing of each element of the domain and its corresponding functional value. For example, we define a permutation α of the set $\{1, 2, 3, 4\}$ by specifying

$$\alpha(1) = 2, \qquad \alpha(2) = 3, \qquad \alpha(3) = 1, \qquad \alpha(4) = 4.$$

A more convenient way to express this correspondence is to write α in array form as

$$\alpha = \begin{bmatrix} 1 & 2 & 3 & 4 \\ 2 & 3 & 1 & 4 \end{bmatrix}.$$

Here $\alpha(j)$ is placed directly below j for each j. Similarly, the permutation β of the set $\{1, 2, 3, 4, 5, 6\}$ given by

$$\beta(1) = 5, \quad \beta(2) = 3, \quad \beta(3) = 1, \quad \beta(4) = 6, \quad \beta(5) = 2, \quad \beta(6) = 4$$

is expressed in array form as

$$\beta = \begin{bmatrix} 1 & 2 & 3 & 4 & 5 & 6 \\ 5 & 3 & 1 & 6 & 2 & 4 \end{bmatrix}.$$

Composition of permutations expressed in array notation is carried out from right to left by going from top to bottom, then again from top to bottom. For example, let

$$\sigma = \begin{bmatrix} 1 & 2 & 3 & 4 & 5 \\ 2 & 4 & 3 & 5 & 1 \end{bmatrix}$$

and

$$\gamma = \begin{bmatrix} 1 & 2 & 3 & 4 & 5 \\ 5 & 4 & 1 & 2 & 3 \end{bmatrix};$$

then

$$\gamma\sigma = \begin{bmatrix} 1 & 2 & 3 & 4 & 5 \\ & \downarrow & & & \\ 5 & 4 & 1 & 2 & 3 \end{bmatrix} \begin{bmatrix} 1 & 2 & 3 & 4 & 5 \\ \downarrow & & & & \\ 2 & 4 & 3 & 5 & 1 \end{bmatrix} = \begin{bmatrix} 1 & 2 & 3 & 4 & 5 \\ 4 & 2 & 1 & 3 & 5 \end{bmatrix}$$

On the right we have 4 under 1, since $(\gamma\sigma)(1) = \gamma(\sigma(1)) = \gamma(2) = 4$, so $\gamma\sigma$ sends 1 to 4. The remainder of the bottom row $\gamma\sigma$ is obtained in a similar fashion.

We are now ready to give some examples of permutation groups.

EXAMPLE 1 Symmetric Group S_3 Let S_3 denote the set of all one-to-one functions from $\{1, 2, 3\}$ to itself. Then S_3, under function composition, is a group with six elements. The six elements are

$$\varepsilon = \begin{bmatrix} 1 & 2 & 3 \\ 1 & 2 & 3 \end{bmatrix}, \quad \alpha = \begin{bmatrix} 1 & 2 & 3 \\ 2 & 3 & 1 \end{bmatrix}, \quad \alpha^2 = \begin{bmatrix} 1 & 2 & 3 \\ 3 & 1 & 2 \end{bmatrix},$$

$$\beta = \begin{bmatrix} 1 & 2 & 3 \\ 1 & 3 & 2 \end{bmatrix}, \quad \alpha\beta = \begin{bmatrix} 1 & 2 & 3 \\ 2 & 1 & 3 \end{bmatrix}, \quad \alpha^2\beta = \begin{bmatrix} 1 & 2 & 3 \\ 3 & 2 & 1 \end{bmatrix}.$$

Note that $\beta\alpha = \begin{bmatrix} 1 & 2 & 3 \\ 3 & 2 & 1 \end{bmatrix} = \alpha^2\beta \neq \alpha\beta$, so that S_3 is non-Abelian. ∎

The relation $\beta\alpha = \alpha^2\beta$ can be used to compute other products in S_3 without resorting to the arrays. For example, $\beta\alpha^2 = (\beta\alpha)\alpha = (\alpha^2\beta)\alpha = \alpha^2(\beta\alpha) = \alpha^2(\alpha^2\beta) = \alpha^4\beta = \alpha\beta$.

Example 1 can be generalized as follows.

▮ EXAMPLE 2 Symmetric Group S_n Let $A = \{1, 2, \ldots, n\}$. The set of all permutations of A is called the *symmetric group of degree n* and is denoted by S_n. Elements of S_n have the form

$$\alpha = \begin{bmatrix} 1 & 2 & \ldots & n \\ \alpha(1) & \alpha(2) & \ldots & \alpha(n) \end{bmatrix}.$$

It is easy to compute the order of S_n. There are n choices of $\alpha(1)$. Once $\alpha(1)$ has been determined, there are $n - 1$ possibilities for $\alpha(2)$ [since α is one-to-one, we must have $\alpha(1) \neq \alpha(2)$]. After choosing $\alpha(2)$, there are exactly $n - 2$ possibilities for $\alpha(3)$. Continuing along in this fashion, we see that S_n has $n(n-1) \cdots 3 \cdot 2 \cdot 1 = n!$ elements. We leave it to the reader to prove that S_n is non-Abelian when $n \geq 3$ (Exercise 45). ∎

The symmetric groups are rich in subgroups. The group S_4 has 30 subgroups, and S_5 has well over 100 subgroups.

▮ EXAMPLE 3 Symmetries of a Square As a third example, we associate each motion in D_4 with the permutation of the locations of each of the four corners of a square. For example, if we label the four corner positions as in the figure below and keep these labels fixed for reference, we may describe a 90° counterclockwise rotation by the permutation

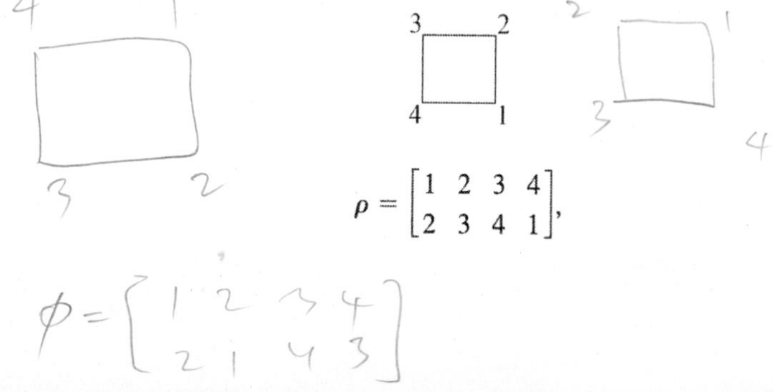

$$\rho = \begin{bmatrix} 1 & 2 & 3 & 4 \\ 2 & 3 & 4 & 1 \end{bmatrix},$$

whereas a reflection across a horizontal axis yields

$$\phi = \begin{bmatrix} 1 & 2 & 3 & 4 \\ 2 & 1 & 4 & 3 \end{bmatrix}.$$

These two elements generate the entire group (that is, every element is some combination of the ρ's and ϕ's).

When D_4 is represented in this way, we see that it is a subgroup of S_4. ∎

Cycle Notation

There is another notation commonly used to specify permutations. It is called *cycle notation* and was first introduced by the great French mathematician Cauchy in 1815. Cycle notation has theoretical advantages in that certain important properties of the permutation can be readily determined when cycle notation is used.

As an illustration of cycle notation, let us consider the permutation

$$\alpha = \begin{bmatrix} 1 & 2 & 3 & 4 & 5 & 6 \\ 2 & 1 & 4 & 6 & 5 & 3 \end{bmatrix}.$$

This assignment of values could be presented schematically as follows.

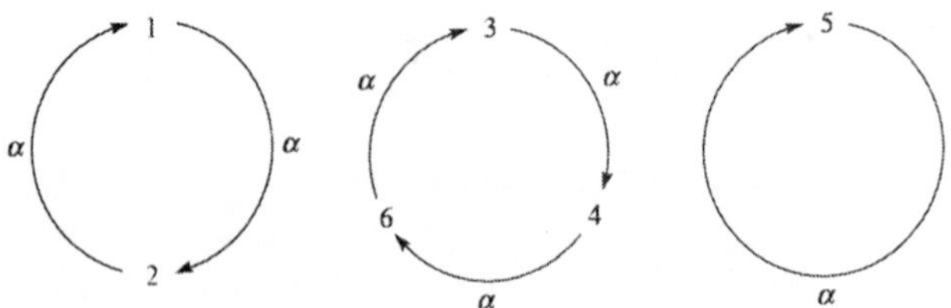

Although mathematically satisfactory, such diagrams are cumbersome. Instead, we leave out the arrows and simply write $\alpha = (1, 2)(3, 4, 6)(5)$. As a second example, consider

$$\beta = \begin{bmatrix} 1 & 2 & 3 & 4 & 5 & 6 \\ 5 & 3 & 1 & 6 & 2 & 4 \end{bmatrix}.$$

In cycle notation, β can be written $(2, 3, 1, 5)(6, 4)$ or $(4, 6)(3, 1, 5, 2)$, since both of these unambiguously specify the function β. An expression of the form $(a_1, a_2, \ldots, a_m)$ is called a *cycle of length m* or an *m-cycle.*

A multiplication of cycles can be introduced by thinking of a cycle as a permutation that fixes any symbol not appearing in the cycle. Thus, the cycle (4, 6) can be thought of as representing the permutation $\begin{bmatrix} 1 & 2 & 3 & 4 & 5 & 6 \\ 1 & 2 & 3 & 6 & 5 & 4 \end{bmatrix}$. In this way, we can multiply cycles by thinking of them as permutations given in array form. Consider the following example from S_8. Let $\alpha = (13)(27)(456)(8)$ and $\beta = (1237)(648)(5)$. (When the domain consists of single-digit integers, it is common practice to omit the commas between the digits.) What is the cycle form of $\alpha\beta$? Of course, one could say that $\alpha\beta = (13)(27)(456)(8)(1237)(648)(5)$, but it is usually more desirable to express a permutation in a *disjoint* cycle form (that is, the various cycles have no number in common). Well, keeping in mind that function composition is done from right to left and that each cycle that does not contain a symbol fixes the symbol, we observe that (5) fixes 1; (648) fixes 1; (1237) sends 1 to 2; (8) fixes 2; (456) fixes 2; (27) sends 2 to 7; and (13) fixes 7. So the net effect of $\alpha\beta$ is to send 1 to 7. Thus, we begin $\alpha\beta = (17 \cdots) \cdots$. Now, repeating the entire process beginning with 7, we have, cycle by cycle, right to left,

$$7 \to 7 \to 7 \to 1 \to 1 \to 1 \to 1 \to 3,$$

so that $\alpha\beta = (173 \cdots) \cdots$. Ultimately, we have $\alpha\beta = (1732)(48)(56)$. The important thing to bear in mind when multiplying cycles is to "keep moving" from one cycle to the next from right to left. (*Warning:* Some authors compose cycles from left to right. When reading another text, be sure to determine which convention is being used.)

To be sure you understand how to switch from one notation to the other and how to multiply permutations, we will do one more example of each.

If array notations for α and β, respectively, are

$$\begin{bmatrix} 1 & 2 & 3 & 4 & 5 \\ 2 & 1 & 3 & 5 & 4 \end{bmatrix} \quad \text{and} \quad \begin{bmatrix} 1 & 2 & 3 & 4 & 5 \\ 5 & 4 & 1 & 2 & 3 \end{bmatrix},$$

then, in cycle notation, $\alpha = (12)(3)(45)$, $\beta = (153)(24)$, and $\alpha\beta = (12)(3)(45)(153)(24)$.

To put $\alpha\beta$ in disjoint cycle form, observe that (24) fixes 1; (153) sends 1 to 5; (45) sends 5 to 4; and (3) and (12) both fix 4. So, $\alpha\beta$ sends 1 to 4. Continuing in this way we obtain $\alpha\beta = (14)(253)$.

One can convert $\alpha\beta$ back to array form without converting each cycle of $\alpha\beta$ into array form by simply observing that (14) means 1 goes to 4 and 4 goes to 1; (253) means $2 \to 5$, $5 \to 3$, $3 \to 2$.

One final remark about cycle notation: Mathematicians prefer not to write cycles that have only one entry. In this case, it is understood that any missing element is mapped to itself. With this convention, the permutation α above can be written as (12)(45). Similarly,

$$\alpha = \begin{bmatrix} 1 & 2 & 3 & 4 & 5 \\ 3 & 2 & 4 & 1 & 5 \end{bmatrix}$$

can be written $\alpha = (134)$. Of course, the identity permutation consists only of cycles with one entry, so we cannot omit all of these! In this case, one usually writes just one cycle. For example,

$$\varepsilon = \begin{bmatrix} 1 & 2 & 3 & 4 & 5 \\ 1 & 2 & 3 & 4 & 5 \end{bmatrix}$$

can be written as $\varepsilon = (5)$ or $\varepsilon = (1)$. Just remember that missing elements are mapped to themselves.

Properties of Permutations

We are now ready to state several theorems about permutations and cycles. The proof of the first theorem is implicit in our discussion of writing permutations in cycle form.

Theorem 5.1 Products of Disjoint Cycles

Every permutation of a finite set can be written as a cycle or as a product of disjoint cycles.

PROOF Let α be a permutation on $A = \{1, 2, \ldots, n\}$. To write α in disjoint cycle form, we start by choosing any member of A, say a_1, and let

$$a_2 = \alpha(a_1), \qquad a_3 = \alpha(\alpha(a_1)) = \alpha^2(a_1),$$

and so on, until we arrive at $a_1 = \alpha^m(a_1)$ for some m. We know that such an m exists because the sequence $a_1, \alpha(a_1), \alpha^2(a_1), \ldots$ must be finite; so there must eventually be a repetition, say $\alpha^i(a_1) = \alpha^j(a_1)$ for some i and j with $i < j$. Then $a_1 = \alpha^m(a_1)$, where $m = j - i$. We express this relationship among $a_1, a_2, \ldots, a_m$ as

$$\alpha = (a_1, a_2, \ldots, a_m) \cdots.$$

The three dots at the end indicate the possibility that we may not have exhausted the set A in this process. In such a case, we merely choose any element b_1 of A not appearing in the first cycle and proceed to create a new cycle as before. That is, we let $b_2 = \alpha(b_1)$, $b_3 = \alpha^2(b_1)$, and so on, until we reach $b_1 = \alpha^k(b_1)$ for some k. This new cycle will have no elements in common with the previously constructed cycle. For, if so, then $\alpha^i(a_1) = \alpha^j(b_1)$ for some i and j. But then $\alpha^{i-j}(a_1) = b_1$, and therefore $b_1 = a_t$ for some t. This contradicts the way b_1 was chosen. Continuing this process until we run out of elements of A, our permutation will appear as

$$\alpha = (a_1, a_2, \ldots, a_m)(b_1, b_2, \ldots, b_k) \cdots (c_1, c_2, \ldots, c_s).$$

In this way, we see that every permutation can be written as a product of disjoint cycles. ∎

Theorem 5.2 Disjoint Cycles Commute

If the pair of cycles $\alpha = (a_1, a_2, \ldots, a_m)$ and $\beta = (b_1, b_2, \ldots, b_n)$ have no entries in common, then $\alpha\beta = \beta\alpha$.

PROOF For definiteness, let us say that α and β are permutations of the set

$$S = \{a_1, a_2, \ldots, a_m, b_1, b_2, \ldots, b_n, c_1, c_2, \ldots, c_k\},$$

where the c's are the members of S left fixed by both α and β (there may not be any c's). To prove that $\alpha\beta = \beta\alpha$, we must show that $(\alpha\beta)(x) = (\beta\alpha)(x)$ for all x in S. If x is one of the a elements, say a_i, then

$$(\alpha\beta)(a_i) = \alpha(\beta(a_i)) = \alpha(a_i) = a_{i+1},$$

since β fixes all a elements. (We interpret a_{i+1} as a_1 if $i = m$.) For the same reason,

$$(\beta\alpha)(a_i) = \beta(\alpha(a_i)) = \beta(a_{i+1}) = a_{i+1}.$$

Hence, the functions of $\alpha\beta$ and $\beta\alpha$ agree on the a elements. A similar argument shows that $\alpha\beta$ and $\beta\alpha$ agree on the b elements as well. Finally, suppose that x is a c element, say c_i. Then, since both α and β fix c elements, we have

$$(\alpha\beta)(c_i) = \alpha(\beta(c_i)) = \alpha(c_i) = c_i$$

and

$$(\beta\alpha)(c_i) = \beta(\alpha(c_i)) = \beta(c_i) = c_i.$$

This completes the proof. ∎

In demonstrating how to multiply cycles, we showed that the product (13)(27)(456)(8)(1237)(648)(5) can be written in disjoint cycle form as (1732)(48)(56). Is economy in expression the only advantage to writing a permutation in disjoint cycle form? No. The next theorem shows that the disjoint cycle form has the enormous advantage of allowing us to "eyeball" the order of the permutation.

Theorem 5.3 Order of a Permutation (Ruffini, 1799)

The order of a permutation of a finite set written in disjoint cycle form is the least common multiple of the lengths of the cycles.

PROOF First, observe that a cycle of length n has order n. (Verify this yourself.) Next, suppose that α and β are disjoint cycles of lengths m and n, and let k be the least common multiple of m and n. It follows from Theorem 4.1 that both α^k and β^k are the identity permutation ε and, since α and β commute, $(\alpha\beta)^k = \alpha^k\beta^k$ is also the identity. Thus, we know by Corollary 2 to Theorem 4.1 ($a^k = e$ implies that $|a|$ divides k) that the order of $\alpha\beta$—let us call it t—must divide k. But then $(\alpha\beta)^t = \alpha^t\beta^t = \varepsilon$, so that $\alpha^t = \beta^{-t}$. However, it is clear that if α and β have no common symbol, the same is true for α^t and β^{-t}, since raising a cycle to a power does not introduce new symbols. But, if α^t and β^{-t} are equal and have no common symbol, they must both be the identity, because every symbol in α^t is fixed by β^{-t} and vice versa (remember that a symbol not appearing in a permutation is fixed by the permutation). It follows, then, that both m and n must divide t. This means that k, the least common multiple of m and n, divides t also. This shows that $k = t$.

Thus far, we have proved that the theorem is true in the cases where the permutation is a single cycle or a product of two disjoint cycles. The general case involving more than two cycles can be handled in an analogous way. ∎

Theorem 5.3 is an enomously powerful tool for calculating the orders of permutations and the number of permutations of a particular order. We demonstrate this in the next two examples.

EXAMPLE 4 To determine the orders of the $7! = 5040$ elements of S_7, we need only consider the possible disjoint cycle structures of the elements of S_7. For convenience, we denote an n-cycle by (n). Then, arranging all possible disjoint cycle structures of elements of S_7 according to longest cycle lengths left to right, we have

(7)
(6) (1)
(5) (2)
(5) (1) (1)
(4) (3)
(4) (2) (1)
(4) (1) (1) (1)
(3) (3) (1)
(3) (2) (2)
(3) (2) (1) (1)
(3) (1) (1) (1) (1)
(2) (2) (2) (1)
(2) (2) (1) (1) (1)
(2) (1) (1) (1) (1) (1)
(1) (1) (1) (1) (1) (1) (1).

Now, from Theorem 5.3 we see that the orders of the elements of S_7 are 7, 6, 10, 5, 12, 4, 3, 2, and 1. To do the same for the $10! = 3628800$ elements of S_{10} would be nearly as simple. ∎

∎ **EXAMPLE 5** We determine the number of elements of S_7 of order 3. By Theorem 5.3, we need only count the number of permutations of the forms $(a_1a_2a_3)$ and $(a_1a_2a_3)\,(a_4a_5a_6)$. In the first case consider the triple $a_1a_2a_3$. Clearly there are $7 \cdot 6 \cdot 5$ such triples. But this product counts the permutation $(a_1a_2a_3)$ three times (for example, it counts 134, 341, 413 as distinct triples whereas the cycles (134), (341), and (413) are the same group element). Thus, the number of permutations in S_7 for the form $(a_1a_2a_3)$ is $(7 \cdot 6 \cdot 5)/3 = 70$. For elements of S_7 of the form $(a_1a_2a_3)\,(a_4a_5a_6)$ there are $(7 \cdot 6 \cdot 5)/3$ ways to create the first cycle and $(4 \cdot 3 \cdot 2)/3$ to create the second cycle but the product of $(7 \cdot 6 \cdot 5)/3$ and $(4 \cdot 3 \cdot 2)/3)$ counts $(a_1a_2a_3)\,(a_4a_5a_6)$ and $(a_4a_5a_6)(a_3a_2a_1)$ as distinct when they are equal group elements. Thus, the number of elements in S_7 for the form $(a_1a_2a_3)\,(a_4a_5a_6)$ is $(7 \cdot 6 \cdot 5)(4 \cdot 3 \cdot 2)/(3 \cdot 3 \cdot 2) = 280$. This gives us 350 elements of order 3 in S_7. ∎

As we will soon see, it is often greatly advantageous to write a permutation as a product of cycles of length 2—that is, as permutations of the form (ab) where $a \neq b$. Many authors call these permutations *transpositions*, since the effect of (ab) is to interchange or transpose a and b. Example 6 and Theorem 5.4 show how this can always be done.

EXAMPLE 6

$$(12345) = (15)(14)(13)(12)$$
$$(1632)(457) = (12)(13)(16)(47)(45)$$

Theorem 5.4 Product of 2-Cycles

Every permutation in S_n, $n > 1$, is a product of 2-cycles.

PROOF First, note that the identity can be expressed as (12)(12), and so it is a product of 2-cycles. By Theorem 5.1, we know that every permutation can be written in the form

$$(a_1a_2 \cdots a_k)(b_1b_2 \cdots b_t) \cdots (c_1c_2 \cdots c_s).$$

A direct computation shows that this is the same as

$$(a_1a_k)(a_1a_{k-1}) \cdots (a_1a_2)(b_1b_t)(b_1b_{t-1}) \cdots (b_1b_2)$$
$$\cdots (c_1c_s)(c_1c_{s-1}) \cdots (c_1c_2).$$

This completes the proof.

The decomposition of a permutation into a product of 2-cycles given in Example 6 and in the proof of Theorem 5.4 is not the only way a permutation can be written as a product of 2-cycles. Although the next example shows that even the *number* of 2-cycles may vary from one decomposition to another, we will prove in Theorem 5.5 (first proved by Cauchy) that there is one aspect of a decomposition that never varies.

EXAMPLE 7

$$(12345) = (54)(53)(52)(51)$$
$$(12345) = (54)(52)(21)(25)(23)(13)$$

We isolate a special case of Theorem 5.5 as a lemma.

Lemma

If $\varepsilon = \beta_1\beta_2 \cdots \beta_r$, where the β's are 2-cycles, then r is even.

PROOF Clearly, $r \neq 1$, since a 2-cycle is not the identity. If $r = 2$, we are done. So, we suppose that $r > 2$, and we proceed by induction.

Suppose that the rightmost 2-cycle is (ab). Then, since $(ij) = (ji)$, the product $\beta_{r-1}\beta_r$ can be expressed in one of the following forms shown on the right:

$$\begin{aligned} \varepsilon &= (ab)(ab), \\ (ab)(bc) &= (ac)(ab), \\ (ac)(cb) &= (bc)(ab), \\ (ab)(cd) &= (cd)(ab). \end{aligned}$$

If the first case occurs, we may delete $\beta_{r-1}\beta_r$ from the original product to obtain $\varepsilon = \beta_1\beta_2 \cdots \beta_{r-2}$, and therefore, by the Second Principle of Mathematical Induction, $r - 2$ is even. In the other three cases, we replace the form of $\beta_{r-1}\beta_r$ on the right by its counterpart on the left to obtain a new product of r 2-cycles that is still the identity, but where the rightmost occurrence of the integer a is in the second-from-the-rightmost 2-cycle of the product instead of the rightmost 2-cycle. We now repeat the procedure just described with $\beta_{r-2}\beta_{r-1}$, and, as before, we obtain a product of $(r - 2)$ 2-cycles equal to the identity or a new product of r 2-cycles, where the rightmost occurrence of a is in the third 2-cycle from the right. Continuing this process, we must obtain a product of $(r - 2)$ 2-cycles equal to the identity, because otherwise we have a product equal to the identity in which the only occurrence of the integer a is in the leftmost 2-cycle, and such a product does not fix a, whereas the identity does. Hence, by the Second Principle of Mathematical Induction, $r - 2$ is even, and r is even as well. ∎

Theorem 5.5 Always Even or Always Odd

If a permutation α can be expressed as a product of an even (odd) number of 2-cycles, then every decomposition of α into a product of 2-cycles must have an even (odd) number of 2-cycles. In symbols, if

$$\alpha = \beta_1\beta_2 \cdots \beta_r \quad \text{and} \quad \alpha = \gamma_1\gamma_2 \cdots \gamma_s,$$

where the β's and the γ's are 2-cycles, then r and s are both even or both odd.

PROOF Observe that $\beta_1\beta_2 \cdots \beta_r = \gamma_1\gamma_2 \cdots \gamma_s$ implies

$$\begin{aligned} \varepsilon &= \gamma_1\gamma_2 \cdots \gamma_s\beta_r^{-1} \cdots \beta_2^{-1}\beta_1^{-1} \\ &= \gamma_1\gamma_2 \cdots \gamma_s\beta_r \cdots \beta_2\beta_1, \end{aligned}$$

since a 2-cycle is its own inverse. Thus, the lemma on page 108 guarantees that $s + r$ is even. It follows that r and s are both even or both odd. ∎

Definition Even and Odd Permutations

A permutation that can be expressed as a product of an even number of 2-cycles is called an *even* permutation. A permutation that can be expressed as a product of an odd number of 2-cycles is called an *odd* permutation.

Theorems 5.4 and 5.5 together show that every permutation can be unambiguously classified as either even or odd. The significance of this observation is given in Theorem 5.6.

Theorem 5.6 Even Permutations Form a Group

The set of even permutations in S_n forms a subgroup of S_n.

PROOF This proof is left to the reader (Exercise 17). ∎

The subgroup of even permutations in S_n arises so often that we give it a special name and notation.

Definition Alternating Group of Degree *n*

The group of even permutations of n symbols is denoted by A_n and is called the *alternating group of degree n*.

The next result shows that exactly half of the elements of $S_n (n > 1)$ are even permutations.

Theorem 5.7

For $n > 1$, A_n has order $n!/2$.

PROOF For each odd permutation α, the permutation $(12)\alpha$ is even and, by the cancellation property in groups, $(12)\alpha \neq (12)\beta$ when $\alpha \neq \beta$. Thus, there are at least as many even permutations as there are odd ones. On the other hand, for each even permutation α, the permutation $(12)\alpha$ is odd and $(12)\alpha \neq (12)\beta$ when $\alpha \neq \beta$. Thus, there are at least as many odd permutations as there are even ones. It follows that there are equal numbers of even and odd permutations. Since $|S_n| = n!$, we have $|A_n| = n!/2$. ∎

The names for the symmetric group and the alternating group of degree n come from the study of polynomials over n variables. A *symmetric* polynomial in the variables $x_1, x_2, \ldots, x_n$ is one that is unchanged under any transposition of two of the variables. An *alternating* polynomial is one that changes signs under any transposition of two of the variables. For example, the polynomial $x_1x_2x_3$ is unchanged by any transposition of two of the three variables, whereas the polynomial $(x_1-x_2)(x_1-x_3)(x_2-x_3)$ changes signs when any two of the variables are transposed. Since every member of the symmetric group is the product of transpositions, the symmetric polynomials are those that are unchanged by members of the symmetric group. Likewise, since any member of the alternating group is the product of an even number of transpositions, the alternating polynomials are those that are unchanged by members of the alternating group.

The alternating groups are among the most important examples of groups. The groups A_4 and A_5 will arise on several occasions in later chapters. In particular, A_5 has great historical significance.

A geometric interpretation of A_4 is given in Example 8, and a multiplication table for A_4 is given as Table 5.1.

EXAMPLE 8 Rotations of a Tetrahedron

The 12 rotations of a regular tetrahedron can be conveniently described with the elements of A_4. The top row of Figure 5.1 illustrates the identity and three 180° "edge" rotations about axes joining midpoints of two edges. The second row consists of 120° "face" rotations about axes joining a vertex to the center of the opposite face. The third row consists of −120° (or 240°) "face" rotations. Notice that the four rotations in the second row can be obtained from those in the first row by left-multiplying the four in the first row by the rotation (123), whereas those in the third row can be obtained from those in the first row by left-multiplying the ones in the first row by (132). ∎

Table 5.1 The Alternating Group A_4 of Even Permutations of {1, 2, 3, 4}

(In this table, the permutations of A_4 are designated as $\alpha_1, \alpha_2, \ldots, \alpha_{12}$ and an entry k inside the table represents α_k. For example, $\alpha_3\,\alpha_8 = \alpha_6$.)

	α_1	α_2	α_3	α_4	α_5	α_6	α_7	α_8	α_9	α_{10}	α_{11}	α_{12}
$(1) = \alpha_1$	1	2	3	4	5	6	7	8	9	10	11	12
$(12)(34) = \alpha_2$	2	1	4	3	6	5	8	7	10	9	12	11
$(13)(24) = \alpha_3$	3	4	1	2	7	8	5	6	11	12	9	10
$(14)(23) = \alpha_4$	4	3	2	1	8	7	6	5	12	11	10	9
$(123) = \alpha_5$	5	8	6	7	9	12	10	11	1	4	2	3
$(243) = \alpha_6$	6	7	5	8	10	11	9	12	2	3	1	4
$(142) = \alpha_7$	7	6	8	5	11	10	12	9	3	2	4	1
$(134) = \alpha_8$	8	5	7	6	12	9	11	10	4	1	3	2
$(132) = \alpha_9$	9	11	12	10	1	3	4	2	5	7	8	6
$(143) = \alpha_{10}$	10	12	11	9	2	4	3	1	6	8	7	5
$(234) = \alpha_{11}$	11	9	10	12	3	1	2	4	7	5	6	8
$(124) = \alpha_{12}$	12	10	9	11	4	2	1	3	8	6	5	7

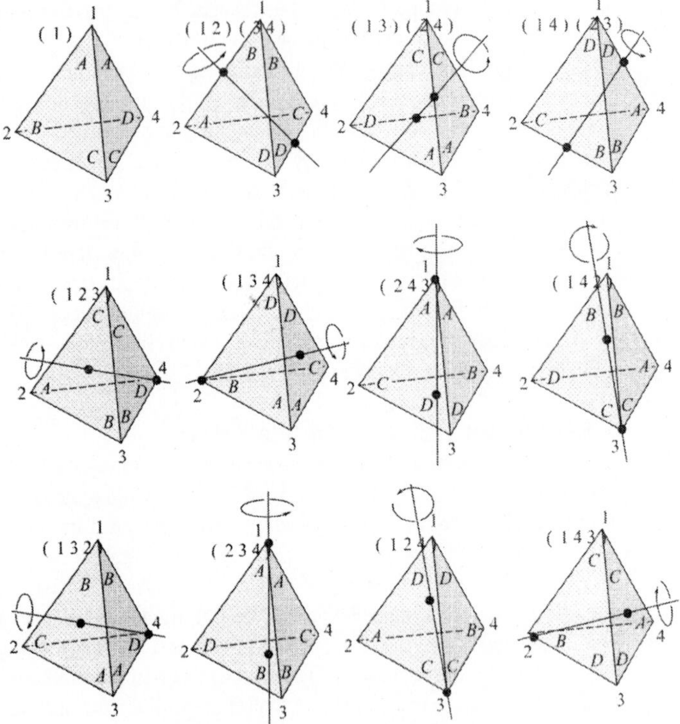

Figure 5.1 Rotations of a regular tetrahedron.

Many molecules with chemical formulas of the form AB_4, such as methane (CH_4) and carbon tetrachloride (CCl_4), have A_4 as their symmetry group. Figure 5.2 shows the form of one such molecule.

Many games and puzzles can be analyzed using permutations.

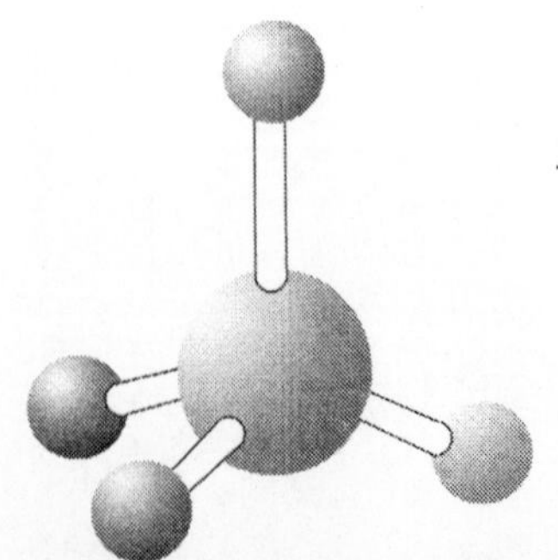

Figure 5.2 A tetrahedral AB_4 molecule.

EXAMPLE 9 (Loren Larson) A Sliding Disk Puzzle

Consider the puzzle shown below (the space in the middle is empty).

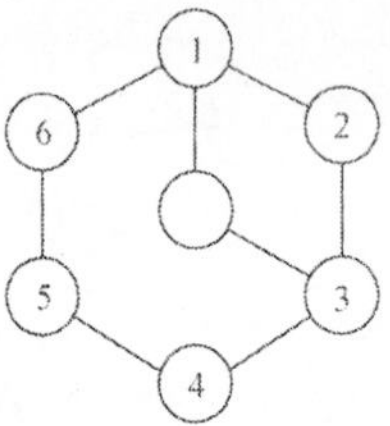

By sliding disks from one position to another along the lines indicated without lifting or jumping them, can we obtain the following arrangement?

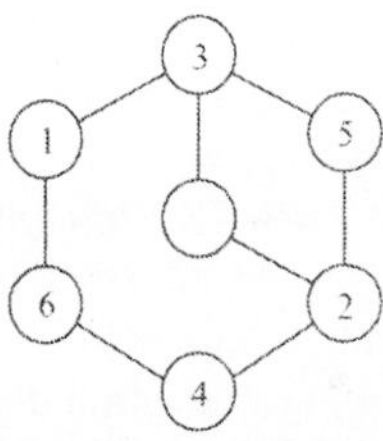

To answer this question, we view the positions as numbered in the first figure above and consider two basic operations. Let r denote the following operation: Move the disk in position 1 to the center position, then move the disk in position 6 to position 1, the disk in position 5 to position 6, the disk in position 4 to position 5, the disk in position 3 to position 4, then the disk in the middle position to position 3. Let s denote the operation: Move the disk in position 1 to the center position, then move the disk in position 2 to position 1, then move the disk in position 3 to position 2, and finally move the disk in the center to position 3. In permutation notation, we have $r = (13456)$ and $s = (132)$. The permutation for the arrangement we seek is (16523). Clearly, if we can express (16523) as a string of r's and s's, we can achieve the desired arangement. Rather than attempt to find an appropriate combination of r's and s's by hand, it is easier to employ computer software that is designed for this kind of problem. One such software program is GAP (see Suggested Software at the end of this chapter). With GAP, all we need to do is use the following commands:

```
gap> G := SymmetricGroup(6);
gap> r := (1,3,4,5,6); s := (1, 3, 2);
gap> K := Subgroup(G,[r,s]);
gap> Factorization(K,(1,6,5,2,3));
```

The first three lines inform the computer that our group is the subgroup of S_6 generated by $r = (13456)$ and $s = (132)$. The fourth line requests that (16523) be expressed in terms of r and s. If we include the command

gap> Size (K)

we would find that the order of the subgroup generated by r and s is 360. Then, observing that r and s are even permutations and that $|A_6| = 360$, we deduce that r and s can achieve any arrangement that corresponds to an even permutation. ∎

Rubik's Cube

The Rubik's Cube made from 48 cubes called "facets" is the quintessential example of a group theory puzzle. It was invented in 1974 by the Hungarian Errő Rubik. By 2009 more than 350 million Rubik's Cubes had been sold. The current record time for solving it is under 7 seconds; under 31 seconds blindfolded. Although it was proved in 1995 that there was a starting configuration that required at least 20 moves to solve, it was not until 2010 that it was determined that every cube could be solved in at most 20 moves. This computer calculation utilized about 35 CPU-years donated by Google to complete. In early discussions about the minimum number of moves to solve the cube in the worst possible case, someone called it "God's number," and the name stuck. A history of the quest to find God's number is given at the website at **http://www.cube20.org/**.

The set of all configuration of the Rubik's Cube form a group of permutations of order 43,252,003,274,489,856,00. This order can be computed using GAP by labeling the faces of the cube as shown here.

			1	2	3						
			4	top	5						
			6	7	8						
9	10	11	17	18	19	25	26	27	33	34	35
12	left	13	20	front	21	28	right	29	36	rear	37
14	15	16	22	23	24	30	31	32	38	39	40
			41	42	43						
			44	bottom	45						
			46	47	48						

The group of permutations of the cube is generated by the following rotations of the six layers.

top = (1,3,8,6)(2,5,7,4)(9,33,25,17)(10,34,26,18)(11,35,27,19)
left = (9,11,16,14)(10,13,15,12)(1,17,41,40)(4,20,44,37)(6,22,46,35)
front = (17,19,24,22)(18,21,23,20)(6,25,43,16)(7,28,42,13)(8,30,41,11)
right = (25,27,32,30)(26,29,31,28)(3,38,43,19)(5,36,45,21)(8,33,48,24)
rear = (33,35,40,38)(34,37,39,36)(3,9,46,32)(2,12,47,29)(1,14,48,27)
bottom = (41,43,48,46)(42,45,47,44)(14,22,30,38)(15,23,31,39)
(16,24,32,40)

A Check-Digit Scheme Based on D_5

In Chapter 0, we presented several schemes for appending a check digit to an identification number. Among these schemes, only the International Standard Book Number method was capable of detecting all single-digit errors and all transposition errors involving adjacent digits. However, recall that this success was achieved by introducing the alphabetical character X to handle the case where 10 was required to make the dot product 0 modulo 11.

In contrast, in 1969, J. Verhoeff [2] devised a method utilizing the dihedral group of order 10 that detects all single-digit errors and all transposition errors involving adjacent digits without the necessity of avoiding certain numbers or introducing a new character. To describe this method, consider the permutation $\sigma = (01589427)(36)$ and the dihedral group of order 10 as represented in Table 5.2. (Here we use 0 through 4 for the rotations, 5 through 9 for the reflections, and $*$ for the operation of D_5.)

Table 5.2 Multiplication for D_5

*	0	1	2	3	4	5	6	7	8	9
0	0	1	2	3	4	5	6	7	8	9
1	1	2	3	4	0	6	7	8	9	5
2	2	3	4	0	1	7	8	9	5	6
3	3	4	0	1	2	8	9	5	6	7
4	4	0	1	2	3	9	5	6	7	8
5	5	9	8	7	6	0	4	3	2	1
6	6	5	9	8	7	1	0	4	3	2
7	7	6	5	9	8	2	1	0	4	3
8	8	7	6	5	9	3	2	1	0	4
9	9	8	7	6	5	4	3	2	1	0

Verhoeff's idea was to view the digits 0 through 9 as the elements of the group D_5 and to replace ordinary addition with calculations done in D_5. In particular, to any string of digits $a_1a_2 \ldots a_{n-1}$, we append the check digit a_n so that $\sigma(a_1) * \sigma^2(a_2) * \cdots * \sigma^{n-2}(a_{n-2}) * \sigma^{n-1}(a_{n-1}) * \sigma^n(a_n) = 0$. [Here $\sigma^2(x) = \sigma(\sigma(x))$, $\sigma^3(x) = \sigma(\sigma^2(x))$, and so on.] Since σ has the property that $\sigma^i(a) \neq \sigma^i(b)$ if $a \neq b$, all single-digit errors are detected. Also, because

$$a * \sigma(b) \neq b * \sigma(a) \qquad \text{if } a \neq b, \tag{1}$$

as can be checked on a case-by-case basis (see Exercise 67), it follows that all transposition errors involving adjacent digits are detected [since Equation (1) implies that $\sigma^i(a) * \sigma^{i+1}(b) \neq \sigma^i(b) * \sigma^{i+1}(a)$ if $a \neq b$].

From 1990 until 2002, the German government used a minor modification of Verhoeff's check-digit scheme to append a check digit to the serial numbers on German banknotes. Table 5.3 gives the values of the functions $\sigma, \sigma^2, \ldots, \sigma^{10}$ needed for the computations. [The functional value $\sigma^i(j)$ appears in the row labeled with σ^i and the column labeled j.] Since the serial numbers on the banknotes use 10 letters of the alphabet in addition to the 10 decimal digits, it is necessary to assign numerical values to the letters to compute the check digit. This assignment is shown in Table 5.4.

Table 5.3 Powers of σ

	0	1	2	3	4	5	6	7	8	9
σ	1	5	7	6	2	8	3	0	9	4
σ^2	5	8	0	3	7	9	6	1	4	2
σ^3	8	9	1	6	0	4	3	5	2	7
σ^4	9	4	5	3	1	2	6	8	7	0
σ^5	4	2	8	6	5	7	3	9	0	1
σ^6	2	7	9	3	8	0	6	4	1	5
σ^7	7	0	4	6	9	1	3	2	5	8
σ^8	0	1	2	3	4	5	6	7	8	9
σ^9	1	5	7	6	2	8	3	0	9	4
σ^{10}	5	8	0	3	7	9	6	1	4	2

Table 5.4 Letter Values

A	D	G	K	L	N	S	U	Y	Z
0	1	2	3	4	5	6	7	8	9

To any string of digits $a_1a_2 \ldots a_{10}$ corresponding to a banknote serial number, the check digit a_{11} is chosen such that $\sigma(a_1) * \sigma^2(a_2) * \cdots * \sigma^9(a_9) * \sigma^{10}(a_{10}) * a_{11} = 0$ [instead of $\sigma(a_1) * \sigma^2(a_2) * \cdots * \sigma^{10}(a_{10}) * \sigma^{11}(a_{11}) = 0$ as in the Verhoeff scheme].

To trace through a specific example, consider the banknote (featuring the mathematician Gauss) shown in Figure 5.3 with the number AG8536827U7. To verify that 7 is the appropriate check digit, we observe that $\sigma(0) * \sigma^2(2) * \sigma^3(8) * \sigma^4(5) * \sigma^5(3) * \sigma^6(6) * \sigma^7(8) * \sigma^8(2) * \sigma^9(7) * \sigma^{10}(7) * 7 = 1 * 0 * 2 * 2 * 6 * 6 * 5 * 2 * 0 * 1 * 7 = 0$, as it should be. [To illustrate how to use the multiplication table for D_5, we compute $1 * 0 * 2 * 2 = (1 * 0) * 2 * 2 = 1 * 2 * 2 = (1 * 2) * 2 = 3 * 2 = 0$.]

Figure 5.3 German banknote with serial number AG8536827U and check digit 7.

One shortcoming of the German banknote scheme is that it does not distinguish between a letter and its assigned numerical value. Thus, a substitution of 7 for U (or vice versa) and the transposition of 7 and U are not detected by the check digit. Moreover, the banknote scheme does not detect all transpositions of adjacent characters involving the check digit itself. For example, the transposition of D and 8 in positions 10 and 11 is not detected. Both of these defects can be avoided by using the Verhoeff method with D_{18}, the dihedral group of order 36, to assign every letter and digit a distinct value together with an appropriate function σ [1]. Using this method to append a check character, all single-position errors and all transposition errors involving adjacent digits will be detected.

Exercises

When you feel how depressingly
slowly you climb,
it's well to remember that
Things Take Time.

PIET HEIN, "T. T. T.," *Grooks (1966)*[†*]

1. Let

$$\alpha = \begin{bmatrix} 1 & 2 & 3 & 4 & 5 & 6 \\ 2 & 1 & 3 & 5 & 4 & 6 \end{bmatrix} \text{ and } \beta = \begin{bmatrix} 1 & 2 & 3 & 4 & 5 & 6 \\ 6 & 1 & 2 & 4 & 3 & 5 \end{bmatrix}.$$

Compute each of the following.
a. α^{-1}
b. $\beta\alpha$
c. $\alpha\beta$

2. Let

$$\alpha = \begin{bmatrix} 1 & 2 & 3 & 4 & 5 & 6 & 7 & 8 \\ 2 & 3 & 4 & 5 & 1 & 7 & 8 & 6 \end{bmatrix} \text{ and } \beta = \begin{bmatrix} 1 & 2 & 3 & 4 & 5 & 6 & 7 & 8 \\ 1 & 3 & 8 & 7 & 6 & 5 & 2 & 4 \end{bmatrix}.$$

Write α, β, and $\alpha\beta$ as
a. products of disjoint cycles;
b. products of 2-cycles.

3. Write each of the following permutations as a product of disjoint cycles.
a. (1235)(413)
b. (13256)(23)(46512)
c. (12)(13)(23)(142)

4. Find the order of each of the following permutations.
a. (14)
b. (147)
c. (14762)
d. $(a_1a_2 \cdots a_k)$

5. What is the order of each of the following permutations?
a. (124)(357)
b. (124)(3567)
c. (124)(35)
d. (124)(357869)
e. (1235)(24567)
f. (345)(245)

[†]Hein is a Danish engineer and poet and is the inventor of the game *Hex*.

6. What is the order of each of the following permutations?

a. $\begin{bmatrix} 1 & 2 & 3 & 4 & 5 & 6 \\ 2 & 1 & 5 & 4 & 6 & 3 \end{bmatrix}$

b. $\begin{bmatrix} 1 & 2 & 3 & 4 & 5 & 6 & 7 \\ 7 & 6 & 1 & 2 & 3 & 4 & 5 \end{bmatrix}$

7. What is the order of the product of a pair of disjoint cycles of lengths 4 and 6?

8. Show that A_8 contains an element of order 15.

9. What are the possible orders for the elements of S_6 and A_6? What about A_7? (This exercise is referred to in Chapter 25.)

10. What is the maximum order of any element in A_{10}?

11. Determine whether the following permutations are even or odd.
    a. (135)
    b. (1356)
    c. (13567)
    d. (12)(134)(152)
    e. (1243)(3521)

12. Show that a function from a finite set S to itself is one-to-one if and only if it is onto. Is this true when S is infinite? (This exercise is referred to in Chapter 6.)

13. Suppose that α is a mapping from a set S to itself and $\alpha(\alpha(x)) = x$ for all x in S. Prove that α is one-to-one and onto.

14. Find eight elements in S_6 that commute with (12)(34)(56). Do they form a subgroup of S_6?

15. Let n be a positive integer. If n is odd, is an n-cycle an odd or an even permutation? If n is even, is an n-cycle an odd or an even permutation?

16. If α is even, prove that α^{-1} is even. If α is odd, prove that α^{-1} is odd.

17. Prove Theorem 5.6.

18. In S_n, let α be an r-cycle, β an s-cycle, and γ a t-cycle. Complete the following statements: $\alpha\beta$ is even if and only if $r + s$ is . . . ; $\alpha\beta\gamma$ is even if and only if $r + s + t$ is . . . .

19. Let α and β belong to S_n. Prove that $\alpha\beta$ is even if and only if α and β are both even or both odd.

20. Associate an even permutation with the number $+1$ and an odd permutation with the number -1. Draw an analogy between the result of multiplying two permutations and the result of multiplying their corresponding numbers $+1$ or -1.

21. Let σ be the permutation of the letters A through Z that takes each letter to the one directly below it in the display following. Write σ in cycle form.

ABCDEFGHIJ KLMNOPQRST UVWXYZ
HDBGJ ECMILONP FKRUS AWQTVZXY

22. If α and β are distinct 2-cycles, what are the possibilities for $|\alpha\beta|$?
23. Show that if H is a subgroup of S_n, then either every member of H is an even permutation or exactly half of the members are even. (This exercise is referred to in Chapter 25.)
24. Suppose that H is a subgroup of S_n of odd order. Prove that H is a subgroup of A_n.
25. Give two reasons why the set of odd permutations in S_n is not a subgroup.
26. Let α and β belong to S_n. Prove that $\alpha^{-1}\beta^{-1}\alpha\beta$ is an even permutation.
27. Use Table 5.1 to compute the following.
    a. The centralizer of $\alpha_3 = (13)(24)$
    b. The centralizer of $\alpha_{12} = (124)$
28. How many elements of order 5 are in S_7?
29. How many elements of order 4 does S_6 have? How many elements of order 2 does S_6 have?
30. Prove that (1234) is not the product of 3-cycles.
31. Let $\beta \in S_7$ and suppose $\beta^4 = (2143567)$. Find β. What are the possibilities for β if $\beta \in S_9$?
32. Let $\beta = (123)(145)$. Write β^{99} in disjoint cycle form.
33. Find three elements σ in S_9 with the property that $\sigma^3 = (157)(283)(469)$.
34. What cycle is $(a_1a_2 \cdots a_n)^{-1}$?
35. Let G be a group of permutations on a set X. Let $a \in X$ and define $\text{stab}(a) = \{\alpha \in G \mid \alpha(a) = a\}$. We call stab($a$) the *stabilizer of a in G* (since it consists of all members of G that leave a fixed). Prove that stab(a) is a subgroup of G. (This subgroup was introduced by Galois in 1832.) This exercise is referred to in Chapter 7.
36. Let $\beta = (1,3,5,7,9,8,6)(2,4,10)$. What is the smallest positive integer n for which $\beta^n = \beta^{-5}$?
37. Let $\alpha = (1,3,5,7,9)(2,4,6)(8,10)$. If α^m is a 5-cycle, what can you say about m?
38. Let $H = \{\beta \in S_5 \mid \beta(1) = 1 \text{ and } \beta(3) = 3\}$. Prove that H is a subgroup of S_5. How many elements are in H? Is your argument valid when S_5 is replaced by S_n for $n \geq 3$? How many elements are in H when S_5 is replaced by A_n for $n \geq 4$?

39. How many elements of order 5 are there in A_6?

40. In S_4, find a cyclic subgroup of order 4 and a noncyclic subgroup of order 4.

41. Suppose that β is a 10-cycle. For which integers i between 2 and 10 is β^i also a 10-cycle?

42. In S_3, find elements α and β such that $|\alpha| = 2$, $|\beta| = 2$, and $|\alpha\beta| = 3$.

43. Find group elements α and β in S_5 such that $|\alpha| = 3$, $|\beta| = 3$, and $|\alpha\beta| = 5$.

44. Represent the symmetry group of an equilateral triangle as a group of permutations of its vertices (see Example 3).

45. Prove that S_n is non-Abelian for all $n \geq 3$.

46. Prove that A_n is non-Abelian for all $n \geq 4$.

47. For $n \geq 3$, let $H = \{\beta \in S_n \mid \beta(1) = 1 \text{ or } 2 \text{ and } \beta(2) = 1 \text{ or } 2\}$. Prove that H is a subgroup of S_n. Determine $|H|$.

48. Show that in S_7, the equation $x^2 = (1234)$ has no solutions but the equation $x^3 = (1234)$ has at least two.

49. If (ab) and (cd) are distinct 2-cycles in S_n, prove that (ab) and (cd) commute if and only if they are disjoint.

50. Let α be a 2-cycle and β be a t-cycle in S_n. Prove that $\alpha\beta\alpha$ is a t-cycle.

51. Use the previous exercise to prove that, if α and β belong to S_n and β is the product of k-cycles of lengths $n_1, n_2, \ldots, n_k$, then $\alpha\beta\alpha^{-1}$ is the product of k-cycles of lengths $n_1, n_2, \ldots n_k$.

52. Let α and β belong to S_n. Prove that $\beta\alpha\beta^{-1}$ and α are both even or both odd.

53. What is the smallest positive integer n such that S_n has an element of order greater than $2n$?

54. Let n be an even positive integer. Prove that A_n has an element of order greater than n if and only if $n \geq 8$.

55. Let n be an odd positive integer. Prove that A_n has an element of order greater than $2n$ if and only if $n \geq 13$.

56. Let n be an even positive integer. Prove that A_n has an element of order greater than $2n$ if and only if $n \geq 14$.

57. Viewing the members of D_4 as a group of permutations of a square labeled 1, 2, 3, 4 as described in Example 3, which geometric symmetries correspond to even permutations?

58. Viewing the members of D_5 as a group of permutations of a regular pentagon with consecutive vertices labeled 1, 2, 3, 4, 5, what geometric symmetry corresponds to the permutation (14253)? Which symmetry corresponds to the permutation (25)(34)?

59. Let n be an odd integer greater than 1. Viewing D_n as a group of permutations of a regular n-gon with consecutive vertices labeled $1, 2, \ldots, n$, explain why the rotation subgroup of D_n is a subgroup of A_n.

60. Let n be an integer greater than 1. Viewing D_n as a group of permutations of a regular n-gon with consecutive vertices labeled $1, 2, \ldots, n$, determine for which n all the permutations corresponding to reflections in D_n are even permutations. Hint: Consider the fours cases for n mod 4.

61. Show that A_5 has 24 elements of order 5, 20 elements of order 3, and 15 elements of order 2. (This exercise is referred to in Chapter 25.)

62. Find a cyclic subgroup of A_8 that has order 4.

63. Find a noncyclic subgroup of A_8 that has order 4.

64. Compute the order of each member of A_4. What arithmetic relationship do these orders have with the order of A_4?

65. Show that every element in A_n for $n \geq 3$ can be expressed as a 3-cycle or a product of 3-cycle.

66. Show that for $n \geq 3$, $Z(S_n) = \{\varepsilon\}$.

67. Verify the statement made in the discussion of the Verhoeff check digit scheme based on D_5 that $a * \sigma(b) \neq b * \sigma(a)$ for distinct a and b. Use this to prove that $\sigma^i(a) * \sigma^{i+1}(b) \neq \sigma^i(b) * \sigma^{i+1}(a)$ for all i. Prove that this implies that all transposition errors involving adjacent digits are detected.

68. Use the Verhoeff check-digit scheme based on D_5 to append a check digit to 45723.

69. Prove that every element of S_n $(n > 1)$ can be written as a product of elements of the form $(1k)$.

70. (Indiana College Mathematics Competition) A card-shuffling machine always rearranges cards in the same way relative to the order in which they were given to it. All of the hearts arranged in order from ace to king were put into the machine, and then the shuffled cards were put into the machine again to be shuffled. If the cards emerged in the order 10, 9, Q, 8, K, 3, 4, A, 5, J, 6, 2, 7, in what order were the cards after the first shuffle?

71. Show that a permutation with odd order must be an even permutation.

72. Let G be a group. Prove or disprove that $H = \{g^2 \mid g \in G\}$ is a subgroup of G. (Compare with Example 5 in Chapter 3.)

73. Let $H = \{\alpha^2 \mid \alpha \in S_4\}$ and $K = \{\alpha^2 \mid \alpha \in S_5\}$. Prove $H = A_4$ and $K = A_5$.

74. Let $H = \{\alpha^2 \mid \alpha \in S_6\}$. Prove $H \neq A_6$.

75. Determine integers n for which $H = \{\alpha \in A_n \mid \alpha^2 = \varepsilon\}$ is a subgroup of A_n.

76. Given that β and γ are in S_4 with $\beta\gamma = (1432)$, $\gamma\beta = (1243)$, and $\beta(1) = 4$, determine β and γ.

77. Why does the fact that the orders of the elements of A_4 are 1, 2, and 3 imply that $|Z(A_4)| = 1$?

78. Find five subgroups of S_5 of order 24.

79. Find six subgroups of order 60 in S_6.

80. For $n > 1$, let H be the set of all permutations in S_n that can be expressed as a product of a multiple of four transpositions. Show that $H = A_n$.

81. Shown below are four tire rotation patterns recommended by the Dunlop Tire Company. Explain how these patterns can be represented as permutations in S_4 and find the smallest subgroup of S_4 that contains these four patterns. Is the subgroup Abelian?

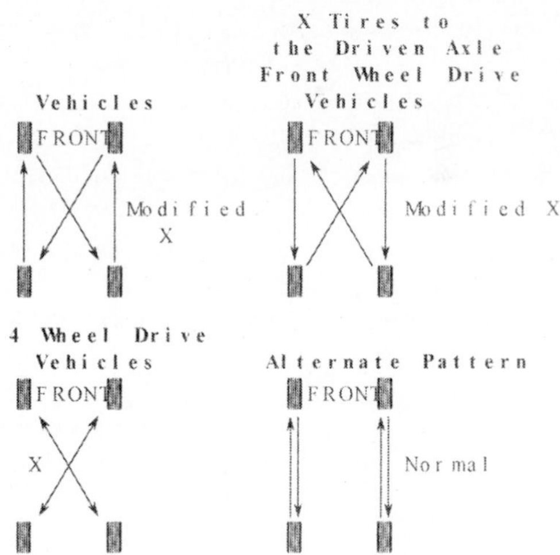

82. Label the four locations of tires on an automobile with the labels 1, 2, 3, and 4, clockwise. Let a represent the operation of switching the tires in positions 1 and 3 and switching the tires in positions 2 and 4. Let b represent the operation of rotating the tires in positions 2, 3, and 4 clockwise and leaving the tire in position 1 as is. Let G be the group of all possible combinations of a and b. How many elements are in G?

83. What would be wrong with using the 2-cycle notation (11) instead of the 1-cycle (1) to indicate that a cycle sends 1 to 1?

Computer Exercises

Computer exercises for this chapter are available at the website:

http://www.d.umn.edu/~jgallian

References

1. J. A. Gallian, "The Mathematics of Identification Numbers," *The College Mathematics Journal* 22 (1991): 194–202.
2. J. Verhoeff, *Error Detecting Decimal Codes,* Amsterdam: Mathematisch Centrum, 1969.

Suggested Readings

Douglas E. Ensley, "Invariants Under Actions to Amaze Your Friends," *Mathematics Magazine,* Dec. 1999: 383–387.

This article explains some card tricks that are based on permutation groups.

Dmitry Fomin, "Getting It Together with 'Polyominoes,'" *Quantum,* Nov./Dec. 1991: 20–23.

In this article, permutation groups are used to analyze various sorts of checkerboard tiling problems.

J. A. Gallian, "Error Detection Methods," *ACM Computing Surveys* 28 (1996): 504–517.

This article gives a comprehensive survey of error-detection methods that use check digits. This article can be downloaded at **http://www.d.umn.edu/~jgallian/detection.pdf**

I. N. Herstein and I. Kaplansky, *Matters Mathematical,* New York: Chelsea, 1978.

Chapter 3 of this book discusses several interesting applications of permutations to games.

Douglas Hofstadter, "The Magic Cube's Cubies Are Twiddled by Cubists and Solved by Cubemeisters," *Scientific American* 244 (1981): 20–39.

This article, written by a Pulitzer Prize recipient, discusses the group theory involved in the solution of the Magic (Rubik's) Cube. In particular, permutation groups, subgroups, conjugates (elements of the form xyx^{-1}), commutators (elements of the form $xyx^{-1}y^{-1}$), and the "always even or always odd" theorem (Theorem 5.5) are prominently mentioned. At one point, Hofstadter says, "It is this kind of marvelously concrete illustration of an abstract notion of group theory that makes the Magic Cube one of the most amazing things ever invented for teaching mathematical ideas."

John O. Kiltinen, *Oval Track & Other Permutation Puzzles & Just Enough Group Theory to Solve Them*, Washington, D.C.: Mathematical Association of America, 2003.

This book and the software that comes with it present the user with an array of computerized puzzles, plus tools to vary them in thousands of ways. The book provides the background needed to use the puzzle software to its fullest potential, and also gives the reader a gentle, not-too-technical introduction to the theory of permutation groups that is a prerequisite to a full understanding of how to solve puzzles of this type. The website **http://www-instruct.nmu.edu/math_cs/kiltinen/web/mathpuzzles/** provides resources that expand upon the book. It also has news about puzzle software—modules that add functionality and fun to puzzles.

Vladimir Dubrovsky, "Portrait of Three Puzzle Graces," *Quantum*, Nov./Dec. 1991: 63–66.

The author uses permutation groups to analyze solutions to the 15 puzzle, Rubik's Cube, and Rubik's Clock.

A. White and R. Wilson, "The Hunting Group," *Mathematical Gazette* 79 (1995): 5–16.

This article explains how permutation groups are used in bell ringing.

S. Winters, "Error-Detecting Schemes Using Dihedral Groups," *UMAP Journal* 11, no. 4 (1990): 299–308.

This article discusses error-detection schemes based on D_n for odd n. Schemes for both one and two check digits are analyzed.

Suggested Software

GAP is free for downloading. Versions are available for Unix, Windows, and Macintosh at:

http://www.gap-system.org

Augustin Cauchy

You see that little young man? Well! He will supplant all of us in so far as we are mathematicians.

Spoken by Lagrange to Laplace about the 11-year-old Cauchy

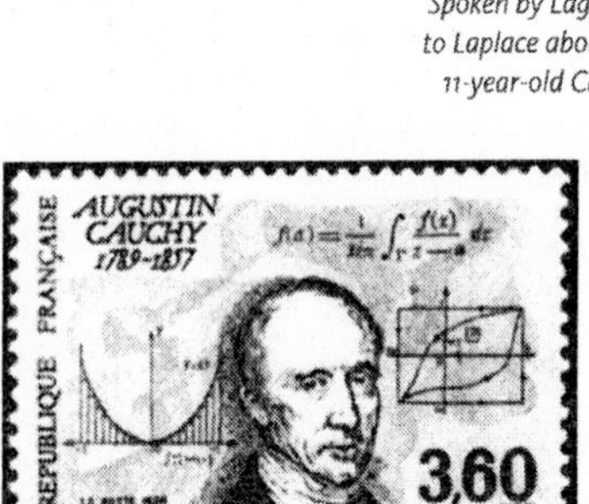

This stamp was issued by France in Cauchy's honor.

AUGUSTIN LOUIS CAUCHY was born on August 21, 1789, in Paris. By the time he was 11, both Laplace and Lagrange had recognized Cauchy's extraordinary talent for mathematics. In school he won prizes for Greek, Latin, and the humanities. At the age of 21, he was given a commission in Napoleon's army as a civil engineer. For the next few years, Cauchy attended to his engineering duties while carrying out brilliant mathematical research on the side.

In 1815, at the age of 26, Cauchy was made Professor of Mathematics at the École Polytechnique and was recognized as the leading mathematician in France. Cauchy and his contemporary Gauss were among the last mathematicians to know the whole of mathematics as known at their time, and both made important contributions to nearly every branch, both pure and applied, as well as to physics and astronomy.

Cauchy introduced a new level of rigor into mathematical analysis. We owe our contemporary notions of limit and continuity to him. He gave the first proof of the Fundamental Theorem of Calculus. Cauchy was the founder of complex function theory and a pioneer in the theory of permutation groups and determinants. His total written output of mathematics fills 24 large volumes. He wrote more than 500 research papers after the age of 50. Cauchy died at the age of 67 on May 23, 1857.

For more information about Cauchy, visit:

http://www-groups.dcs.st-and.ac.uk/~history/

6 Isomorphisms

> The basis for poetry and scientific discovery is the ability to comprehend the unlike in the like and the like in the unlike.
>
> JACOB BRONOWSKI

Motivation

Suppose an American and a German are asked to count a handful of objects. The American says, "One, two, three, four, five, . . . ," whereas the German says, "Eins, zwei, drei, vier, fünf, ..." Are the two doing different things? No. They are both counting the objects, but they are using different terminology to do so. Similarly, when one person says, "Two plus three is five" and another says, "Zwei und drei ist fünf," the two are in agreement on the *concept* they are describing, but they are using different terminology to describe the concept. An analogous situation often occurs with groups; the same group is described with different terminology. We have seen two examples of this so far. In Chapter 1, we described the symmetries of a square in geometric terms (e.g., R_{90}), whereas in Chapter 5 we described the *same* group by way of permutations of the corners. In both cases, the underlying group was the symmetries of a square. In Chapter 4, we observed that when we have a cyclic group of order n generated by a, the operation turns out to be essentially that of addition modulo n, since $a^r a^s = a^k$, where $k = (r + s) \bmod n$. For example, each of $U(43)$ and $U(49)$ is cyclic of order 42. So, each has the form $\langle a \rangle$, where $a^r a^s = a^{(r+s) \bmod 42}$.

Definition and Examples

In this chapter, we give a formal method for determining whether two groups defined in different terms are really the same. When this is the case, we say that there is an isomorphism between the two groups. This notion was first introduced by Galois about 180 years ago. The term *isomorphism* is derived from the Greek words *isos*, meaning "same" or "equal," and *morphe*, meaning "form." R. Allenby has colorfully defined an algebraist as "a person who can't tell the difference between isomorphic systems."

Definition Group Isomorphism

An *isomorphism* ϕ from a group G to a group $\overline{G}$ is a one-to-one mapping (or function) from G onto $\overline{G}$ that preserves the group operation. That is,

$$\phi(ab) = \phi(a)\phi(b) \qquad \text{for all } a, b \text{ in } G.$$

If there is an isomorphism from G onto $\overline{G}$, we say that G and $\overline{G}$ are *isomorphic* and write $G \approx \overline{G}$.

This definition can be visualized as shown in Figure 6.1. The pairs of dashed arrows represent the group operations.

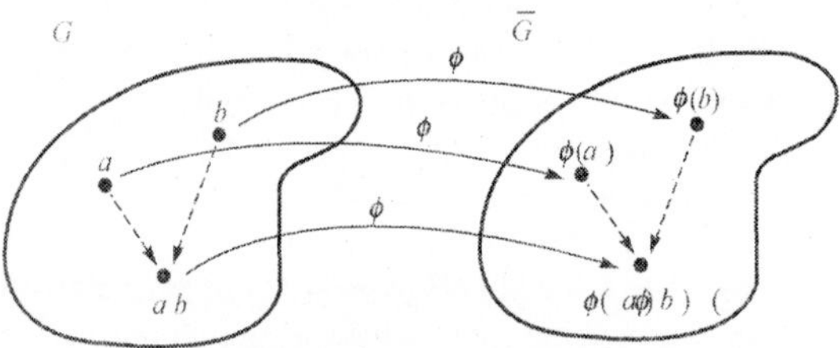

Figure 6.1

It is implicit in the definition of isomorphism that isomorphic groups have the same order. It is also implicit in the definition of isomorphism that the operation on the left side of the equal sign is that of G, whereas the operation on the right side is that of $\overline{G}$. The four cases involving $\cdot$ and $+$ are shown in Table 6.1.

Table 6.1

G Operation	$\overline{G}$ Operation	Operation Preservation
$\cdot$	$\cdot$	$\phi(a \cdot b) = \phi(a) \cdot \phi(b)$
$\cdot$	$+$	$\phi(a \cdot b) = \phi(a) + \phi(b)$
$+$	$\cdot$	$\phi(a + b) = \phi(a) \cdot \phi(b)$
$+$	$+$	$\phi(a + b) = \phi(a) + \phi(b)$

There are four separate steps involved in proving that a group G is isomorphic to a group $\overline{G}$.

Step 1 "Mapping." Define a candidate for the isomorphism; that is, define a function ϕ from G to $\overline{G}$.

Step 2 "1–1." Prove that ϕ is one-to-one; that is, assume that $\phi(a) = \phi(b)$ and prove that $a = b$.

Step 3 "Onto." Prove that ϕ is onto; that is, for any element $\overline{g}$ in $\overline{G}$, find an element g in G such that $\phi(g) = \overline{g}$.

Step 4 "O.P." Prove that ϕ is operation-preserving; that is, show that $\phi(ab) = \phi(a)\phi(b)$ for all a and b in G.

None of these steps is unfamiliar to you. The only one that may appear novel is the fourth one. It requires that one be able to obtain the same result by combining two elements and then mapping, or by mapping two elements and then combining them. Roughly speaking, this says that the two processes—operating and mapping—can be done in either order without affecting the result. This same concept arises in calculus when we say

$$\lim_{x\to a}(f(x) \,.\, g(x)) = \lim_{x\to a} f(x) \lim_{x\to a} g(x)$$

or

$$\int_a^b (f + g)\, dx = \int_a^b f\, dx + \int_a^b g\, dx.$$

Before going any further, let's consider some examples.

EXAMPLE 1 Let G be the real numbers under addition and let $\overline{G}$ be the positive real numbers under multiplication. Then G and $\overline{G}$ are isomorphic under the mapping $\phi(x) = 2^x$. Certainly, ϕ is a function from G to $\overline{G}$. To prove that it is one-to-one, suppose that $2^x = 2^y$. Then $\log_2 2^x = \log_2 2^y$, and therefore $x = y$. For "onto," we must find for any positive real number y some real number x such that $\phi(x) = y$; that is, $2^x = y$. Well, solving for x gives $\log_2 y$. Finally,

$$\phi(x + y) = 2^{x+y} = 2^x \cdot 2^y = \phi(x)\phi(y)$$

for all x and y in G, so that ϕ is operation-preserving as well. ▮

EXAMPLE 2 Any infinite cyclic group is isomorphic to Z. Indeed, if a is a generator of the cyclic group, the mapping $a^k \to k$ is an isomorphism. Any finite cyclic group $\langle a \rangle$ of order n is isomorphic to Z_n under the mapping $a^k \to k \bmod n$. That these correspondences are functions and are one-to-one is the essence of Theorem 4.1. Obviously, the mappings are onto. That the mappings are operation-preserving follows from Exercise 9 in Chapter 0 in the finite case and from the definitions in the infinite case. ▮

EXAMPLE 3 The mapping from **R** under addition to itself given by $\phi(x) = x^3$ is *not* an isomorphism. Although ϕ is one-to-one and onto, it is not operation-preserving, since it is not true that $(x + y)^3 = x^3 + y^3$ for all x and y. ∎

EXAMPLE 4 $U(10) \approx Z_4$ and $U(5) \approx Z_4$. To verify this, one need only observe that both $U(10)$ and $U(5)$ are cyclic of order 4. Then appeal to Example 2. ∎

EXAMPLE 5 $U(10) \not\approx U(12)$. This is a bit trickier to prove. First, note that $x^2 = 1$ for all x in $U(12)$. Now, suppose that ϕ is an isomorphism from $U(10)$ onto $U(12)$. Then,

$$\phi(9) = \phi(3 \cdot 3) = \phi(3)\phi(3) = 1$$

and

$$\phi(1) = \phi(1 \cdot 1) = \phi(1)\phi(1) = 1.$$

Thus, $\phi(9) = \phi(1)$, but $9 \neq 1$, which contradicts the assumption that ϕ is one-to-one. ∎

EXAMPLE 6 There is no isomorphism from Q, the group of rational numbers under addition, to Q^*, the group of nonzero rational numbers under multiplication. If ϕ were such a mapping, there would be a rational number a such that $\phi(a) = -1$. But then

$$-1 = \phi(a) = \phi(\tfrac{1}{2}a + \tfrac{1}{2}a) = \phi(\tfrac{1}{2}a)\phi(\tfrac{1}{2}a) = [\phi(\tfrac{1}{2}a)]^2.$$

However, no rational number squared is -1. ∎

EXAMPLE 7 Let $G = SL(2, \mathbf{R})$, the group of 2×2 real matrices with determinant 1. Let M be any 2×2 real matrix with determinant 1. Then we can define a mapping from G to G itself by $\phi_M(A) = MAM^{-1}$ for all A in G. To verify that ϕ_M is an isomorphism, we carry out the four steps.

Step 1 ϕ_M is a function from G to G. Here, we must show that $\phi_M(A)$ is indeed an element of G whenever A is. This follows from properties of determinants:

$$\det(MAM^{-1}) = (\det M)(\det A)(\det M)^{-1} = 1 \cdot 1 \cdot 1^{-1} = 1.$$

Thus, MAM^{-1} is in G.

Step 2 ϕ_M is one-to-one. Suppose that $\phi_M(A) = \phi_M(B)$. Then $MAM^{-1} = MBM^{-1}$ and, by left and right cancellation, $A = B$.

Step 3 ϕ_M is onto. Let B belong to G. We must find a matrix A in G such that $\phi_M(A) = B$. How shall we do this? If such a matrix A is to exist, it must have the property that $MAM^{-1} = B$. But this tells us exactly what A must be! For we can solve for A to obtain $A = M^{-1}BM$ and verify that $\phi_M(A) = MAM^{-1} = M(M^{-1}BM)M^{-1} = B$.

Step 4 ϕ_M is operation-preserving. Let A and B belong to G. Then,

$$\begin{aligned}\phi_M(AB) &= M(AB)M^{-1} = MA(M^{-1}M)BM^{-1} \\ &= (MAM^{-1})(MBM^{-1}) = \phi_M(A)\phi_M(B).\end{aligned}$$

The mapping ϕ_M is called *conjugation* by M. ∎

Cayley's Theorem

Our first theorem is a classic result of Cayley. An important generalization of it will be given in Chapter 25.

Theorem 6.1 Cayley's Theorem (1854)

Every group is isomorphic to a group of permutations.

PROOF To prove this, let G be any group. We must find a group $\overline{G}$ of permutations that we believe is isomorphic to G. Since G is all we have to work with, we will have to use it to construct $\overline{G}$. For any g in G, define a function T_g from G to G by

$$T_g(x) = gx \qquad \text{for all } x \text{ in } G.$$

(In words, T_g is just multiplication by g on the left.) We leave it as an exercise (Exercise 33) to prove that T_g is a permutation on the set of elements of G. Now, let $\overline{G} = \{T_g \mid g \in G\}$. Then, $\overline{G}$ is a group under the operation of function composition. To verify this, we first observe that for any g and h in G we have $T_gT_h(x) = T_g(T_h(x)) = T_g(hx) = g(hx) = (gh)x = T_{gh}(x)$, so that $T_gT_h = T_{gh}$. From this it follows that T_e is the identity and $(T_g)^{-1} = T_{g^{-1}}$ (see Exercise 9). Since function composition is associative, we have verified all the conditions for $\overline{G}$ to be a group.

The isomorphism ϕ between G and $\overline{G}$ is now ready-made. For every g in G, define $\phi(g) = T_g$. If $T_g = T_h$, then $T_g(e) = T_h(e)$ or $ge = he$. Thus, $g = h$ and ϕ is one-to-one. By the way $\overline{G}$ was constructed, we see that ϕ is onto. The only condition that remains to be checked is that ϕ is operation-preserving. To this end, let a and b belong to G. Then

$$\phi(ab) = T_{ab} = T_aT_b = \phi(a)\phi(b).$$ ∎

The group $\overline{G}$ constructed previously is called the *left regular representation of G.*

EXAMPLE 8 For concreteness, let us calculate the left regular representation $\overline{U(12)}$ for $U(12) = \{1, 5, 7, 11\}$. Writing the permutations of $U(12)$ in array form, we have (remember, T_x is just multiplication by x)

$$T_1 = \begin{bmatrix} 1 & 5 & 7 & 11 \\ 1 & 5 & 7 & 11 \end{bmatrix}, \quad T_5 = \begin{bmatrix} 1 & 5 & 7 & 11 \\ 5 & 1 & 11 & 7 \end{bmatrix},$$

$$T_7 = \begin{bmatrix} 1 & 5 & 7 & 11 \\ 7 & 11 & 1 & 5 \end{bmatrix}, \quad T_{11} = \begin{bmatrix} 1 & 5 & 7 & 11 \\ 11 & 7 & 5 & 1 \end{bmatrix}.$$

It is instructive to compare the Cayley tables for $U(12)$ and its left regular representation $\overline{U(12)}$.

$U(12)$	1	5	7	11
1	1	5	7	11
5	5	1	11	7
7	7	11	1	5
11	11	7	5	1

$\overline{U(12)}$	T_1	T_5	T_7	T_{11}
T_1	T_1	T_5	T_7	T_{11}
T_5	T_5	T_1	T_{11}	T_7
T_7	T_7	T_{11}	T_1	T_5
T_{11}	T_{11}	T_7	T_5	T_1

It should be abundantly clear from these tables that $U(12)$ and $\overline{U(12)}$ are only notationally different. ∎

Cayley's Theorem is important for two contrasting reasons. One is that it allows us to represent an abstract group in a concrete way. A second is that it shows that the present-day set of axioms we have adopted for a group is the correct abstraction of its much earlier predecessor—a group of permutations. Indeed, Cayley's Theorem tells us that abstract groups are not different from permutation groups. Rather, it is the viewpoint that is different. It is this difference of viewpoint that has stimulated the tremendous progress in group theory and many other branches of mathematics in the 20th century.

It is sometimes very difficult to prove or disprove, whichever the case may be, that two particular groups are isomorphic. For example, it requires somewhat sophisticated techniques to prove the surprising fact that the group of real numbers under addition is isomorphic to the group of complex numbers under addition. Likewise, it is not easy to prove the fact that the group of nonzero complex numbers under multiplication is isomorphic to the group of complex numbers with absolute value of 1 under multiplication. In geometric terms, this says that, as groups, the punctured plane and the unit circle are isomorphic [1].

Properties of Isomorphisms

Our next two theorems give a catalog of properties of isomorphisms and isomorphic groups.

Theorem 6.2 Properties of Isomorphisms Acting on Elements

Suppose that ϕ is an isomorphism from a group G onto a group $\overline{G}$. Then

1. *ϕ carries the identity of G to the identity of $\overline{G}$.*
2. *For every integer n and for every group element a in G, $\phi(a^n) = [\phi(a)]^n$.*
3. *For any elements a and b in G, a and b commute if and only if $\phi(a)$ and $\phi(b)$ commute.*
4. *$G = \langle a \rangle$ if and only if $\overline{G} = \langle \phi(a) \rangle$.*
5. *$|a| = |\phi(a)|$ for all a in G (isomorphisms preserve orders).*
6. *For a fixed integer k and a fixed group element b in G, the equation $x^k = b$ has the same number of solutions in G as does the equation $x^k = \phi(b)$ in $\overline{G}$.*
7. *If G is finite, then G and $\overline{G}$ have exactly the same number of elements of every order.*

PROOF We will restrict ourselves to proving only properties 1, 2, and 4, but observe that property 5 follows from properties 1 and 2, property 6 follows from property 2, and property 7 follows from property 5. For convenience, let us denote the identity in G by e and the identity in $\overline{G}$ by $\overline{e}$. Then, since $e = ee$, we have

$$\phi(e) = \phi(ee) = \phi(e)\phi(e).$$

Also, because $\phi(e) \in \overline{G}$, we have $\phi(e) = \overline{e}\phi(e)$, as well. Thus, by cancellation, $\overline{e} = \phi(e)$. This proves property 1.

For positive integers, property 2 follows from the definition of an isomorphism and mathematical induction. If n is negative, then $-n$ is positive, and we have from property 1 and the observation about the positive integer case that $e = \phi(e) = \phi(g^n g^{-n}) = \phi(g^n)\phi(g^{-n}) = \phi(g^n)(\phi(g))^{-n}$. Thus, multiplying both sides on the right by $(\phi(g))^n$, we have $(\phi(g))^n = \phi(g^n)$. Property 1 takes care of the case $n = 0$.

To prove property 4, let $G = \langle a \rangle$ and note that, by closure, $\langle \phi(a) \rangle \subseteq \overline{G}$. Because ϕ is onto, for any element b in $\overline{G}$, there is an element a^k in G such that $\phi(a^k) = b$. Thus, $b = (\phi(a))^k$ and so $b \in \langle \phi(a) \rangle$. This proves that $\overline{G} = \langle \phi(a) \rangle$.

Now suppose that $\overline{G} = \langle \phi(a) \rangle$. Clearly, $\langle a \rangle \subseteq G$. For any element b in G, we have $\phi(b) \in \langle \phi(a) \rangle$. So, for some integer k we have

$\phi(b) = (\phi(a))^k = \phi(a^k)$. Because ϕ is one-to-one, $b = a^k$. This proves that $\langle a \rangle = G$. ∎

When the group operation is addition, property 2 of Theorem 6.2 is $\phi(na) = n\phi(a)$; property 4 says that an isomorphism between two cyclic groups takes a generator to a generator.

Property 6 is quite useful for showing that two groups are *not* isomorphic. Often b is picked to be the identity. For example, consider $\mathbf{C}^*$ and $\mathbf{R}^*$. Because the equation $x^4 = 1$ has four solutions in $\mathbf{C}^*$ but only two in $\mathbf{R}^*$, no matter how one attempts to define an isomorphism from $\mathbf{C}^*$ to $\mathbf{R}^*$, property 6 cannot hold.

Theorem 6.3 Properties of Isomorphisms Acting on Groups

Suppose that ϕ is an isomorphism from a group G onto a group $\overline{G}$. Then

1. ***ϕ^{-1} is an isomorphism from $\overline{G}$ onto G.***
2. ***G is Abelian if and only if $\overline{G}$ is Abelian.***
3. ***G is cyclic if and only if $\overline{G}$ is cyclic.***
4. ***If K is a subgroup of G, then $\phi(K) = \{\phi(k) \mid k \in K\}$ is a subgroup of $\overline{G}$.***
5. ***If $\overline{K}$ is a subgroup of $\overline{G}$, then $\phi^{-1}(\overline{K}) = \{g \in G \mid \phi(g) \in \overline{K}\}$ is a subgroup of G.***
6. ***$\phi(Z(G)) = Z(\overline{G})$.***

PROOF Properties 1 and 4 are left as exercises (Exercises 31 and 32). Properties 2 and 6 are a direct consequence of property 3 of Theorem 6.2. Property 3 follows from property 4 of Theorem 6.2 and property 1 of Theorem 6.3. Property 5 follows from properties 1 and 4. ∎

Theorems 6.2 and 6.3 show that isomorphic groups have many properties in common. Actually, the definition is precisely formulated so that isomorphic groups have *all* group theoretic properties in common. By this we mean that if two groups are isomorphic, then any property that can be expressed in the language of group theory is true for one if and only if it is true for the other. This is why algebraists speak of isomorphic groups as "equal" or "the same." Admittedly, calling such groups equivalent, rather than the same, might be more appropriate, but we bow to long-standing tradition.

Automorphisms

Certain kinds of isomorphisms are referred to so often that they have been given special names.

Definition Automorphism
An isomorphism from a group G onto itself is called an *automorphism* of G.

The isomorphism in Example 7 is an automorphism of $SL(2, \mathbf{R})$. Two more examples follow.

EXAMPLE 9 The function ϕ from $\mathbf{C}$ to $\mathbf{C}$ given by $\phi(a + bi) = a - bi$ is an automorphism of the group of complex numbers under addition. The restriction of ϕ to $\mathbf{C}^*$ is also an automorphism of the group of nonzero complex numbers under multiplication. (See Exercise 35.) ■

EXAMPLE 10 Let $\mathbf{R}^2 = \{(a, b) \mid a, b \in \mathbf{R}\}$. Then $\phi(a, b) = (b, a)$ is an automorphism of the group $\mathbf{R}^2$ under componentwise addition. Geometrically, ϕ reflects each point in the plane across the line $y = x$. More generally, any reflection across a line passing through the origin or any rotation of the plane about the origin is an automorphism of $\mathbf{R}^2$. ■

The isomorphism in Example 7 is a particular instance of an automorphism that arises often enough to warrant a name and notation of its own.

Definition Inner Automorphism Induced by *a*
Let G be a group, and let $a \in G$. The function ϕ_a defined by $\phi_a(x) = axa^{-1}$ for all x in G is called the *inner automorphism of G induced by a.*

We leave it for the reader to show that ϕ_a is actually an automorphism of G. (Use Example 7 as a model.)

EXAMPLE 11 The action of the inner automorphism of D_4 induced by R_{90} is given in the following table.

x	$\xrightarrow{\phi_{R_{90}}}$	$R_{90} x R_{90}^{-1}$
R_0	$\rightarrow$	$R_{90}R_0R_{90}^{-1} = R_0$
R_{90}	$\rightarrow$	$R_{90}R_{90}R_{90}^{-1} = R_{90}$
R_{180}	$\rightarrow$	$R_{90}R_{180}R_{90}^{-1} = R_{180}$
R_{270}	$\rightarrow$	$R_{90}R_{270}R_{90}^{-1} = R_{270}$
H	$\rightarrow$	$R_{90}HR_{90}^{-1} = V$
V	$\rightarrow$	$R_{90}VR_{90}^{-1} = H$
D	$\rightarrow$	$R_{90}DR_{90}^{-1} = D'$
D'	$\rightarrow$	$R_{90}D'R_{90}^{-1} = D$

■

When G is a group, we use Aut(G) to denote the set of all automorphisms of G and Inn(G) to denote the set of all inner automorphisms of G. The reason these sets are noteworthy is demonstrated by the next theorem.

Theorem 6.4 Aut(G) and Inn(G) Are Groups[†]

The set of automorphisms of a group and the set of inner automorphisms of a group are both groups under the operation of function composition.

PROOF The proof of Theorem 6.4 is left as an exercise (Exercise 15). ∎

The determination of Inn(G) is routine. If $G = \{e, a, b, c. \ldots\}$, then Inn($G$) $= \{\phi_e, \phi_a, \phi_b, \phi_c, \ldots\}$. This latter list may have duplications, however, since ϕ_a may be equal to ϕ_b even though $a \neq b$ (see Exercise 43). Thus, the only work involved in determining Inn(G) is deciding which distinct elements give the distinct automorphisms. On the other hand, the determination of Aut(G) is, in general, quite involved.

EXAMPLE 12 Inn(D_4)

To determine Inn(D_4), we first observe that the complete list of inner automorphisms is ϕ_{R_0}, $\phi_{R_{90}}$, $\phi_{R_{180}}$, $\phi_{R_{270}}$, ϕ_H, ϕ_V, ϕ_D, and $\phi_{D'}$. Our job is to determine the repetitions in this list. Since $R_{180} \in Z(D_4)$, we have $\phi_{R_{180}}(x) = R_{180}xR_{180}^{-1} = x$, so that $\phi_{R_{180}} = \phi_{R_0}$. Also, $\phi_{R_{270}}(x) = R_{270}xR_{270}^{-1} = R_{90}R_{180}xR_{180}^{-1}R_{90}^{-1} = R_{90}xR_{90}^{-1} = \phi_{R_{90}}(x)$. Similarly, since $H = R_{180}V$ and $D' = R_{180}D$, we have $\phi_H = \phi_V$ and $\phi_D = \phi_{D'}$. This proves that the previous list can be pared down to ϕ_{R_0}, $\phi_{R_{90}}$, ϕ_H, and ϕ_D. We leave it to the reader to show that these are distinct (Exercise 13). ∎

EXAMPLE 13 Aut(Z_{10})

To compute Aut(Z_{10}), we try to discover enough information about an element α of Aut(Z_{10}) to determine how α must be defined. Because Z_{10} is so simple, this is not difficult to do. To begin with, observe that once we know $\alpha(1)$, we know $\alpha(k)$ for any k, because

[†]The group Aut(G) was first studied by O. Hölder in 1893 and, independently, by E. H. Moore in 1894.

$$\alpha(k) = \alpha(\underbrace{1 + 1 + \cdots + 1}_{k \text{ terms}})$$
$$= \underbrace{\alpha(1) + \alpha(1) + \cdots + \alpha(1)}_{k \text{ terms}} = k\alpha(1).$$

So, we need only determine the choices for $\alpha(1)$ that make α an automorphism of Z_{10}. Since property 5 of Theorem 6.2 tells us that $|\alpha(1)| = 10$, there are four candidates for $\alpha(1)$:

$$\alpha(1) = 1, \quad \alpha(1) = 3, \quad \alpha(1) = 7, \quad \alpha(1) = 9.$$

To distinguish among the four possibilities, we refine our notation by denoting the mapping that sends 1 to 1 by α_1, 1 to 3 by α_3, 1 to 7 by α_7, and 1 to 9 by α_9. So the only possibilities for Aut(Z_{10}) are α_1, α_3, α_7, and α_9. But are all these automorphisms? Clearly, α_1 is the identity. Let us check α_3. Since x mod 10 = y mod 10 implies $3x$ mod 10 = $3y$ mod 10, α_3 is well defined. Moreover, because $\alpha_3(1) = 3$ is a generator of Z_{10}, it follows that α_3 is onto (and, by Exercise 12 in Chapter 5, it is also one-to-one). Finally, since $\alpha_3(a + b) = 3(a + b) = 3a + 3b = \alpha_3(a) + \alpha_3(b)$, we see that α_3 is operation-preserving as well. Thus, $\alpha_3 \in$ Aut(Z_{10}). The same argument shows that α_7 and α_9 are also automorphisms.

This gives us the elements of Aut(Z_{10}) but not the structure. For instance, what is $\alpha_3\alpha_3$? Well, $(\alpha_3\alpha_3)(1) = \alpha_3(3) = 3 \cdot 3 = 9 = \alpha_9(1)$, so $\alpha_3\alpha_3 = \alpha_9$. Similar calculations show that $\alpha_3^3 = \alpha_7$ and $\alpha_3^4 = \alpha_1$, so that $|\alpha_3| = 4$. Thus, Aut(Z_{10}) is cyclic. Actually, the following Cayley tables reveal that Aut(Z_{10}) is isomorphic to $U(10)$.

$U(10)$	1	3	7	9
1	1	3	7	9
3	3	9	1	7
7	7	1	9	3
9	9	7	3	1

Aut(Z_{10})	α_1	α_3	α_7	α_9
α_1	α_1	α_3	α_7	α_9
α_3	α_3	α_9	α_1	α_7
α_7	α_7	α_1	α_9	α_3
α_9	α_9	α_7	α_3	α_1

■

With Example 13 as a guide, we are now ready to tackle the group Aut(Z_n). The result is particularly nice, since it relates the two kinds of groups we have most frequently encountered thus far—the cyclic groups Z_n and the U-groups $U(n)$.

Theorem 6.5 Aut(Z_n) ≈ $U(n)$

For every positive integer n, Aut(Z_n) is isomorphic to $U(n)$.

PROOF As in Example 13, any automorphism α is determined by the value of $\alpha(1)$, and $\alpha(1) \in U(n)$. Now consider the correspondence from Aut(Z_n) to $U(n)$ given by $T: \alpha \to \alpha(1)$. The fact that $\alpha(k) = k\alpha(1)$ (see Example 13) implies that T is a one-to-one mapping. For if α and β belong to Aut(Z_n) and $\alpha(1) = \beta(1)$, then $\alpha(k) = k\alpha(1) = k\beta(1) = \beta(k)$ for all k in Z_n, and therefore $\alpha = \beta$.

To prove that T is onto, let $r \in U(n)$ and consider the mapping α from Z_n to Z_n defined by $\alpha(s) = sr \pmod{n}$ for all s in Z_n. We leave it as an exercise to verify that α is an automorphism of Z_n (see Exercise 27). Then, since $T(\alpha) = \alpha(1) = r$, T is onto $U(n)$.

Finally, we establish the fact that T is operation-preserving. Let α, $\beta \in$ Aut(Z_n). We then have

$$\begin{aligned} T(\alpha\beta) &= (\alpha\beta)(1) = \alpha(\beta(1)) = \alpha(\underbrace{1 + 1 + \cdots + 1}_{\beta(1)}) \\ &= \underbrace{\alpha(1) + \alpha(1) + \cdots + \alpha(1)}_{\beta(1)} = \alpha(1)\beta(1) \\ &= T(\alpha)T(\beta). \end{aligned}$$

This completes the proof. ■

Exercises

Being a mathematician is a bit like being a manic depressive: you spend your life alternating between giddy elation and black despair.

STEVEN G. KRANTZ, *A Primer of Mathematical Writing*

1. Find an isomorphism from the group of integers under addition to the group of even integers under addition.
2. Find Aut(Z).
3. Let $\mathbf{R}^+$ be the group of positive real numbers under multiplication. Show that the mapping $\phi(x) = \sqrt{x}$ is an automorphism of $\mathbf{R}^+$.
4. Show that $U(8)$ is not isomorphic to $U(10)$.
5. Show that $U(8)$ is isomorphic to $U(12)$.
6. Prove that isomorphism is an equivalence relation. That is, for any groups G, H, and K, $G \approx G$, $G \approx H$ implies $H \approx G$, and $G \approx H$ and $H \approx K$ implies $G \approx K$.
7. Prove that S_4 is not isomorphic to D_{12}.
8. Show that the mapping $a \to \log_{10} a$ is an isomorphism from $\mathbf{R}^+$ under multiplication to $\mathbf{R}$ under addition.
9. In the notation of Theorem 6.1, prove that T_e is the identity and that $(T_g)^{-1} = T_{g^{-1}}$.

10. Let G be a group. Prove that the mapping $\alpha(g) = g^{-1}$ for all g in G is an automorphism if and only if G is Abelian.
11. If g and h are elements from a group, prove that $\phi_g\phi_h = \phi_{gh}$.
12. Find two groups G and H such that $G \not\approx H$, but Aut(G) $\approx$ Aut(H).
13. Prove the assertion in Example 12 that the inner automorphisms ϕ_{R_0}, $\phi_{R_{90}}$, ϕ_H, and ϕ_D of D_4 are distinct.
14. Find Aut(Z_6).
15. If G is a group, prove that Aut(G) and Inn(G) are groups.
16. If a group G is isomorphic to H, prove that Aut(G) is isomorphic to Aut(H).
17. Suppose ϕ belongs to Aut(Z_n) and a is relatively prime to n. If $\phi(a) = b$, determine a formula for $\phi(x)$.
18. Let H be the subgroup of all rotations in D_n and let ϕ be an automorphism of D_n. Prove that $\phi(H) = H$. (In words, an automorphism of D_n carries rotations to rotations.)
19. Let $H = \{\beta \in S_5 \mid \beta(1) = 1\}$ and $K = \{\beta \in S_5 \mid \beta(2) = 2\}$. Prove that H is isomorphic to K. Is the same true if S_5 is replaced by S_n, where $n \geq 3$?
20. Show that Z has infinitely many subgroups isomorphic to Z.
21. Let n be an even integer greater than 2 and let ϕ be an automorphism of D_n. Determine $\phi(R_{180})$.
22. Let ϕ be an automorphism of a group G. Prove that $H = \{x \in G \mid \phi(x) = x\}$ is a subgroup of G.
23. Give an example of a cyclic group of smallest order that contains a subgroup isomorphic to Z_{12} and a subgroup isomorphic to Z_{20}. No need to prove anything, but explain your reasoning.
24. Suppose that ϕ: $Z_{20} \to Z_{20}$ is an automorphism and $\phi(5) = 5$. What are the possibilities for $\phi(x)$?
25. Identify a group G that has subgroups isomorphic to Z_n for all positive integers n.
26. Prove that the mapping from $U(16)$ to itself given by $x \to x^3$ is an automorphism. What about $x \to x^5$ and $x \to x^7$? Generalize.
27. Let $r \in U(n)$. Prove that the mapping α: $Z_n \to Z_n$ defined by $\alpha(s) = sr \bmod n$ for all s in Z_n is an automorphism of Z_n. (This exercise is referred to in this chapter.)
28. The group $\left\{\begin{bmatrix} 1 & a \\ 0 & 1 \end{bmatrix} \middle| \, a \in Z\right\}$ is isomorphic to what familiar group? What if Z is replaced by **R**?

29. If ϕ and γ are isomorphisms from the cyclic group $\langle a \rangle$ to some group and $\phi(a) = \gamma(a)$, prove that $\phi = \gamma$.

30. Suppose that ϕ: $Z_{50} \to Z_{50}$ is an automorphism with $\phi(11) = 13$. Determine a formula for $\phi(x)$.

31. Prove property 1 of Theorem 6.3.

32. Prove property 4 of Theorem 6.3.

33. Referring to Theorem 6.1, prove that T_g is indeed a permutation on the set G.

34. Prove or disprove that $U(20)$ and $U(24)$ are isomorphic.

35. Show that the mapping $\phi(a + bi) = a - bi$ is an automorphism of the group of complex numbers under addition. Show that ϕ preserves complex multiplication as well—that is, $\phi(xy) = \phi(x)\phi(y)$ for all x and y in **C**. (This exercise is referred to in Chapter 15.)

36. Let

$$G = \{a + b\sqrt{2} \mid a, b \text{ are rational}\}$$

and

$$H = \left\{ \begin{bmatrix} a & 2b \\ b & a \end{bmatrix} \middle| \, a, b \text{ are rational} \right\}.$$

Show that G and H are isomorphic under addition. Prove that G and H are closed under multiplication. Does your isomorphism preserve multiplication as well as addition? (G and H are examples of rings—a topic we will take up in Part 3.)

37. Prove that Z under addition is not isomorphic to Q under addition.

38. Prove that the quaternion group (see Exercise 4, Supplementary Exercises for Chapters 1–4) is not isomorphic to the dihedral group D_4.

39. Let **C** be the complex numbers and

$$M = \left\{ \begin{bmatrix} a & -b \\ b & a \end{bmatrix} \middle| \, a, b \in \mathbf{R} \right\}.$$

Prove that **C** and M are isomorphic under addition and that $\mathbf{C}^*$ and M^*, the nonzero elements of M, are isomorphic under multiplication.

40. Let $\mathbf{R}^n = \{(a_1, a_2, \ldots, a_n) \mid a_i \in \mathbf{R}\}$. Show that the mapping ϕ: $(a_1, a_2, \ldots, a_n) \to (-a_1, -a_2, \ldots, -a_n)$ is an automorphism of the group $\mathbf{R}^n$ under componentwise addition. This automorphism is called *inversion*. Describe the action of ϕ geometrically.

41. Consider the following statement: The order of a subgroup divides the order of the group. Suppose you could prove this for finite permutation groups. Would the statement then be true for all finite groups? Explain.

42. Suppose that G is a finite Abelian group and G has no element of order 2. Show that the mapping $g \to g^2$ is an automorphism of G. Show, by example, that there is an infinite Abelian group for which the mapping $g \to g^2$ is one-to-one and operation-preserving but not an automorphism.
43. Let G be a group and let $g \in G$. If $z \in Z(G)$, show that the inner automorphism induced by g is the same as the inner automorphism induced by zg (that is, that the mappings ϕ_g and ϕ_{zg} are equal).
44. Show that the mapping $a \to \log_{10} a$ is an isomorphism from $\mathbf{R}^+$ under multiplication to $\mathbf{R}$ under addition.
45. Suppose that g and h induce the same inner automorphism of a group G. Prove that $h^{-1}g \in Z(G)$.
46. Combine the results of Exercises 43 and 45 into a single "if and only if" theorem.
47. If x and y are elements in S_n ($n \geq 3$), prove that $\phi_x = \phi_y$ implies $x = y$. (Here, ϕ_x is the inner automorphism of S_n induced by x.)
48. Let ϕ be an isomorphism from a group G to a group $\overline{G}$ and let a belong to G. Prove that $\phi(C(a)) = C(\phi(a))$.
49. Suppose the ϕ and γ are isomorphisms of some group G to the same group. Prove that $H = \{g \in G \mid \phi(g) = \gamma(g)\}$ is a subgroup of G.
50. Suppose that β is an automorphism of a group G. Prove that $H = \{g \in G \mid \beta^2(g) = g\}$ is a subgroup of G. Generalize.
51. Suppose that G is an Abelian group and ϕ is an automorphism of G. Prove that $H = \{x \in G \mid \phi(x) = x^{-1}\}$ is a subgroup of G.
52. Given a group G, define a new group G^* that has the same elements as G with the operation $*$ defines by $a * b = ba$ for all a and b in G^*. Prove that the mapping from G to G^* defined by $\phi(x) = x^{-1}$ for all x in G is an isomorphism from G onto G^*.
53. Let a belong to a group G and let $|a|$ be finite. Let ϕ_a be the automorphism of G given by $\phi_a(x) = axa^{-1}$. Show that $|\phi_a|$ divides $|a|$. Exhibit an element a from a group for which $1 < |\phi_a| < |a|$.
54. Let $G = \{0, \pm 2, \pm 4, \pm 6, \ldots\}$ and $H = \{0, \pm 3, \pm 6, \pm 9, \ldots\}$. Show that G and H are isomorphic groups under addition. Does your isomorphism preserve multiplication? Generalize to the case when $G = \langle m \rangle$ and $H = \langle n \rangle$, where m and n are integers.
55. Suppose that ϕ is an automorphism of D_4 such that $\phi(R_{90}) = R_{270}$ and $\phi(V) = V$. Determine $\phi(D)$ and $\phi(H)$.
56. In Aut(Z_9), let α_i denote the automorphism that sends 1 to i where $\gcd(i, 9) = 1$. Write α_5 and α_8 as permutations of $\{0, 1, \ldots, 8\}$ in disjoint cycle form. [For example, $\alpha_2 = (0)(124875)(36)$.]

57. Write the permutation corresponding to R_{90} in the left regular representation of D_4 in cycle form.
58. Show that every automorphism ϕ of the rational numbers Q under addition to itself has the form $\phi(x) = x\phi(1)$.
59. Prove that Q^+, the group of positive rational numbers under multiplication, is isomorphic to a proper subgroup.
60. Prove that Q, the group of rational numbers under addition, is not isomorphic to a proper subgroup of itself.
61. Prove that every automorphism of $\mathbf{R}^*$, the group of nonzero real numbers under multiplication, maps positive numbers to positive numbers and negative numbers to negative numbers.
62. Let G be a finite group. Show that in the disjoint cycle form of the right regular representation $T_g(x) = xg$ of G, each cycle has length $|g|$.
63. Give a group theoretic proof that Q under addition is not isomorphic to $\mathbf{R}^+$ under multiplication.

Reference

1. J. R. Clay, "The Punctured Plane Is Isomorphic to the Unit Circle," *Journal of Number Theory* 1 (1969): 500–501.

Computer Exercises

Software for the computer exercise in this chapter is available at the website:

http://www.d.umn.edu/~jgallian

Arthur Cayley

Cayley is forging the weapons for future generations of physicists.

PETER TAIT

The Granger Collection, New York

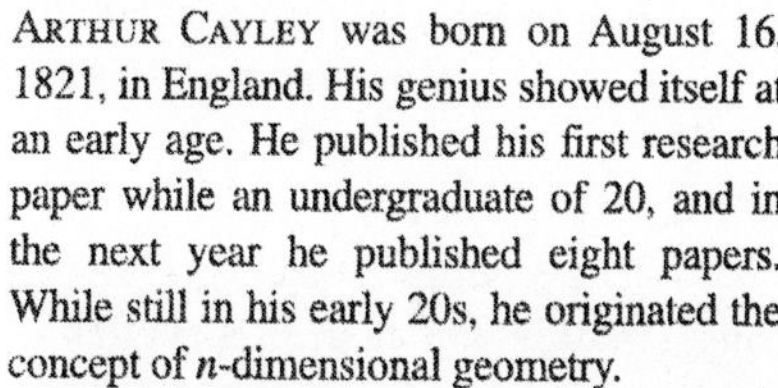

ARTHUR CAYLEY was born on August 16, 1821, in England. His genius showed itself at an early age. He published his first research paper while an undergraduate of 20, and in the next year he published eight papers. While still in his early 20s, he originated the concept of n-dimensional geometry.

After graduating from Trinity College, Cambridge, Cayley stayed on for three years as a tutor. At the age of 25, he began a 14-year career as a lawyer. During this period, he published approximately 200 mathematical papers, many of which are now classics.

In 1863, Cayley accepted the newly established Sadlerian professorship of mathematics at Cambridge University. He spent the rest of his life in that position. One of his notable accomplishments was his role in the successful effort to have women admitted to Cambridge.

Among Cayley's many innovations in mathematics were the notions of an abstract group and a group algebra, and the matrix concept. He made major contributions to geometry and linear algebra. Cayley and his lifelong friend and collaborator J. J. Sylvester were the founders of the theory of invariants, which was later to play an important role in the theory of relativity.

Cayley's collected works comprise 13 volumes, each about 600 pages in length. He died on January 26, 1895.

To find more information about Cayley, visit:

http://www-groups.dcs.st-and.ac.uk/~history/

7 Cosets and Lagrange's Theorem

It might be difficult, at this point, for students to see the extreme importance of this result [Lagrange's Theorem]. As we penetrate the subject more deeply they will become more and more aware of its basic character.

I. N. HERSTEIN, *Topics in Algebra*

Properties of Cosets

In this chapter, we will prove the single most important theorem in finite group theory—Lagrange's Theorem. In his book on abstract algebra, I. N. Herstein likened it to the ABC's for finite groups. But first we introduce a new and powerful tool for analyzing a group—the notion of a coset. This notion was invented by Galois in 1830, although the term was coined by G. A. Miller in 1910.

Definition Coset of *H* in *G*

Let G be a group and let H be a nonempty subset of G. For any $a \in G$, the set $\{ah \mid h \in H\}$ is denoted by aH. Analogously, $Ha = \{ha \mid h \in H\}$ and $aHa^{-1} = \{aha^{-1} \mid h \in H\}$. When H is a subgroup of G, the set aH is called the *left coset of H in G containing a*, whereas Ha is called the *right coset of H in G containing a*. In this case, the element a is called the *coset representative of aH* (or Ha). We use $|aH|$ to denote the number of elements in the set aH, and $|Ha|$ to denote the number of elements in Ha.

EXAMPLE 1 Let $G = S_3$ and $H = \{(1), (13)\}$. Then the left cosets of H in G are

$$\begin{aligned}
(1)H &= H, \\
(12)H &= \{(12), (12)(13)\} = \{(12), (132)\} = (132)H, \\
(13)H &= \{(13), (1)\} = H, \\
(23)H &= \{(23), (23)(13)\} = \{(23), (123)\} = (123)H.
\end{aligned}$$ ∎

EXAMPLE 2 Let $\mathcal{H} = \{R_0, R_{180}\}$ in D_4, the dihedral group of order 8. Then,

$$\begin{aligned} R_0\mathcal{H} &= \mathcal{H}, \\ R_{90}\mathcal{H} &= \{R_{90}, R_{270}\} = R_{270}\mathcal{H}, \\ R_{180}\mathcal{H} &= \{R_{180}, R_0\} = \mathcal{H}, \\ V\mathcal{H} &= \{V, H\} = H\mathcal{H}, \\ D\mathcal{H} &= \{D, D'\} = D'\mathcal{H}. \end{aligned}$$

EXAMPLE 3 Let $H = \{0, 3, 6\}$ in Z_9 under addition. In the case that the group operation is addition, we use the notation $a + H$ instead of aH. Then the cosets of H in Z_9 are

$$\begin{aligned} 0 + H &= \{0, 3, 6\} = 3 + H = 6 + H, \\ 1 + H &= \{1, 4, 7\} = 4 + H = 7 + H, \\ 2 + H &= \{2, 5, 8\} = 5 + H = 8 + H. \end{aligned}$$

The three preceding examples illustrate a few facts about cosets that are worthy of our attention. First, cosets are usually not subgroups. Second, aH may be the same as bH, even though a is not the same as b. Third, since in Example 1 $(12)H = \{(12), (132)\}$ whereas $H(12) = \{(12), (123)\}$, aH need not be the same as Ha.

These examples and observations raise many questions. When does $aH = bH$? Do aH and bH have any elements in common? When does $aH = Ha$? Which cosets are subgroups? Why are cosets important? The next lemma and theorem answer these questions. (Analogous results hold for right cosets.)

Lemma Properties of Cosets

Let H be a subgroup of G, and let a and b belong to G. Then,

1. $a \in aH$.
2. $aH = H$ *if and only if* $a \in H$.
3. $(ab)H = a(bH)$ *and* $H(ab) = (Ha)b$.
4. $aH = bH$ *if and only if* $a \in bH$.
5. $aH = bH$ *or* $aH \cap bH = \emptyset$.
6. $aH = bH$ *if and only if* $a^{-1}b \in H$.
7. $|aH| = |bH|$.
8. $aH = Ha$ *if and only if* $H = aHa^{-1}$.
9. aH *is a subgroup of G if and only if* $a \in H$.

PROOF

1. $a = ae \in aH$.
2. To verify property 2, we first suppose that $aH = H$. Then $a = ae \in aH = H$. Next, we assume that $a \in H$ and show that $aH \subseteq H$

and $H \subseteq aH$. The first inclusion follows directly from the closure of H. To show that $H \subseteq aH$, let $h \in H$. Then, since $a \in H$ and $h \in H$, we know that $a^{-1}h \in H$. Thus, $h = eh = (aa^{-1})h = a(a^{-1}h) \in aH$.

3. This follows directly from $(ab)h = a(bh)$ and $h(ab) = (ha)b$.
4. If $aH = bH$, then $a = ae \in aH = bH$. Conversely, if $a \in bH$ we have $a = bh$ where $h \in H$, and therefore $aH = (bh)H = b(hH) = bH$.
5. Property 5 follows directly from property 4, for if there is an element c in $aH \cap bH$, then $cH = aH$ and $cH = bH$.
6. Observe that $aH = bH$ if and only if $H = a^{-1}bH$. The result now follows from property 2.
7. To prove that $|aH| = |bH|$, it suffices to define a one-to-one mapping from aH onto bH. Obviously, the correspondence $ah \to bh$ maps aH onto bH. That it is one-to-one follows directly from the cancellation property.
8. Note that $aH = Ha$ if and only if $(aH)a^{-1} = (Ha)a^{-1} = H(aa^{-1}) = H$—that is, if and only if $aHa^{-1} = H$.
9. If aH is a subgroup, then it contains the identity e. Thus, $aH \cap eH \neq \varnothing$; and, by property 5, we have $aH = eH = H$. Thus, from property 2, we have $a \in H$. Conversely, if $a \in H$, then, again by property 2, $aH = H$. ∎

Although most mathematical theorems are written in symbolic form, one should also know what they say *in words*. In the preceding lemma, property 1 says simply that the left coset of H containing a does contain a. Property 2 says that the H "absorbs" an element if and only if the element belongs to H. Property 3 says that the left coset of H created by multiplying H on the left by ab is the same as the one created by multiplying H on the left by b then multiplying the resulting coset bH on the left by a (and analogously for multiplication on the right by ab). Property 4 shows that a left coset of H is uniquely determined by any one of its elements. In particular, any element of a left coset can be used to represent the coset. Property 5 says—and this is very important—that two left cosets of H are either identical or disjoint. Thus, a left coset of H is uniquely determined by any one of its elements. In particular, any element of a left coset can be used to represent the coset. Property 6 shows how we may transfer a question about equality of left cosets of H to a question about H itself and vice versa. Property 7 says that all left cosets of H have the same size. Property 8 is analogous to property 6 in that it shows how a question about the equality of the left and right cosets of H containing a is equivalent to a question about the equality of two subgroups of G. The last property of the lemma says that H itself is the only coset of H that is a subgroup of G.

Note that properties 1, 5, and 7 of the lemma guarantee that the left cosets of a subgroup H of G partition G into blocks of equal size. Indeed, we may view the cosets of H as a partitioning of G into equivalence classes under the equivalence relation defined by $a \sim b$ if $aH = bH$ (see Theorem 0.7).

In practice, the subgroup H is often chosen so that the cosets partition the group in some highly desirable fashion. For example, if G is 3-space $\mathbf{R}^3$ and H is a plane through the origin, then the coset $(a, b, c) + H$ (addition is done componentwise) is the plane passing through the point (a, b, c) and parallel to H. Thus, the cosets of H constitute a partition of 3-space into planes parallel to H. If $G = GL(2, \mathbf{R})$ and $H = SL(2, \mathbf{R})$, then for any matrix A in G, the coset AH is the set of *all* 2×2 matrices with the same determinant as A. Thus,

$$\begin{bmatrix} 2 & 0 \\ 0 & 1 \end{bmatrix} H \quad \text{is the set of all } 2 \times 2 \text{ matrices of determinant } 2$$

and

$$\begin{bmatrix} 1 & 2 \\ 2 & 1 \end{bmatrix} H \quad \text{is the set of all } 2 \times 2 \text{ matrices of determinant } -3.$$

Property 5 of the lemma is useful for actually finding the distinct cosets of a subgroup. We illustrate this in the next example.

EXAMPLE 4 To find the cosets of $H = \{1, 15\}$ in $G = U(32) = \{1, 3, 5, 7, 9, 11, 13, 15, 17, 19, 21, 23, 25, 27, 29, 31\}$, we begin with $H = \{1, 15\}$. We can find a second coset by choosing any element not in H, say 3, as a coset representative. This gives the coset $3H = \{3, 13\}$. We find our next coset by choosing a representative not already appearing in the two previously chosen cosets, say 5. This gives us the coset $5H = \{5, 11\}$. We continue to form cosets by picking elements from $U(32)$ that have not yet appeared in the previous cosets as representatives of the cosets until we have accounted for every element of $U(32)$. We then have the complete list of all distinct cosets of H. ■

Lagrange's Theorem and Consequences

We are now ready to prove a theorem that has been around for more than 200 years—longer than group theory itself! (This theorem was not originally stated in group theoretic terms.) At this stage, it should come as no surprise.

Theorem 7.1 Lagrange's Theorem[†]: $|H|$ Divides $|G|$

If G is a finite group and H is a subgroup of G, then $|H|$ divides $|G|$. Moreover, the number of distinct left (right) cosets of H in G is $|G|/|H|$.

[†]Lagrange stated his version of this theorem in 1770, but the first complete proof was given by Pietro Abbati some 30 years later.

PROOF Let $a_1H, a_2H, \ldots, a_rH$ denote the distinct left cosets of H in G. Then, for each a in G, we have $aH = a_iH$ for some i. Also, by property 1 of the lemma, $a \in aH$. Thus, each member of G belongs to one of the cosets a_iH. In symbols,

$$G = a_1H \cup \cdots \cup a_rH.$$

Now, property 5 of the lemma shows that this union is disjoint, so that

$$|G| = |a_1H| + |a_2H| + \cdots + |a_rH|.$$

Finally, since $|a_iH| = |H|$ for each i, we have $|G| = r|H|$. ∎

We pause to emphasize that Lagrange's Theorem is a subgroup candidate criterion; that is, it provides a list of candidates for the orders of the subgroups of a group. Thus, a group of order 12 may have subgroups of order 12, 6, 4, 3, 2, 1, but no others. *Warning!* The converse of Lagrange's Theorem is false. For example, a group of order 12 need not have a subgroup of order 6. We prove this in Example 5.

A special name and notation have been adopted for the number of left (or right) cosets of a subgroup in a group. The *index* of a subgroup H in G is the number of distinct left cosets of H in G. This number is denoted by $|G:H|$. As an immediate consequence of the proof of Lagrange's Theorem, we have the following useful formula for the number of distinct left (or right) cosets of H in G.

Corollary 1 $|G:H| = |G|/|H|$

If G is a finite group and H is a subgroup of G, then $|G:H| = |G|/|H|$.

Corollary 2 $|a|$ Divides $|G|$

In a finite group, the order of each element of the group divides the order of the group.

PROOF Recall that the order of an element is the order of the subgroup generated by that element. ∎

Corollary 3 Groups of Prime Order Are Cyclic

A group of prime order is cyclic.

PROOF Suppose that G has prime order. Let $a \in G$ and $a \neq e$. Then, $|\langle a \rangle|$ divides $|G|$ and $|\langle a \rangle| \neq 1$. Thus, $|\langle a \rangle| = |G|$ and the corollary follows. ∎

Corollary 4 $a^{|G|} = e$

Let G be a finite group, and let $a \in G$. Then, $a^{|G|} = e$.

PROOF By Corollary 2, $|G| = |a|k$ for some positive integer k. Thus, $a^{|G|} = a^{|a|k} = e^k = e$. ∎

Corollary 5 Fermat's Little Theorem

For every integer a and every prime p, $a^p \bmod p = a \bmod p$.

PROOF By the division algorithm, $a = pm + r$, where $0 \leq r < p$. Thus, $a \bmod p = r$, and it suffices to prove that $r^p \bmod p = r$. If $r = 0$, the result is trivial, so we may assume that $r \in U(p)$. [Recall that $U(p) = \{1, 2, \ldots, p - 1\}$ under multiplication modulo p.] Then, by the preceding corollary, $r^{p-1} \bmod p = 1$ and, therefore, $r^p \bmod p = r$. ∎

Fermat's Little Theorem has been used in conjunction with computers to test for primality of certain numbers. One case concerned the number $p = 2^{257} - 1$. If p is prime, then we know from Fermat's Little Theorem that $10^p \bmod p = 10 \bmod p$ and, therefore, $10^{p+1} \bmod p = 100 \bmod p$. Using multiple precision and a simple loop, a computer was able to calculate $10^{p+1} \bmod p = 10^{2^{257}} \bmod p$ in a few seconds. The result was not 100, and so p is not prime.

EXAMPLE 5 The Converse of Lagrange's Theorem Is False.†
The group A_4 of order 12 has no subgroups of order 6. To verify this, recall that A_4 has eight elements of order 3 (α_5 through α_{12}, in the notation of Table 5.1) and suppose that H is a subgroup of order 6. Let a be any element of order 3 in A_4. If a is not in H, then $A_4 = H \cup aH$. But then a^2 is in H or a^2 is in aH. If a^2 is in H then so is $(a^2)^2 = a^4 = a$, so this case is ruled out. If a^2 is in aH, then $a^2 = ah$ for some h in H, but this also implies that a is in H. This argument shows that any subgroup of A_4 of order 6 must contain all eight elements of A_4 of order 3, which is absurd. ∎

†The first counterexample to the converse of Lagrange's Theorem was given by Paolo Ruffini in 1799.

Lagrange's Theorem demonstrates that the finiteness of a group imposes severe restrictions on the possible orders of subgroups. The next theorem also places powerful limits on the existence of certain subgroups in finite groups.

Theorem 7.2 $|HK| = |H||K|/|H \cap K|$

For two finite subgroups H and K of a group, define the set $HK = \{hk \mid h \in H, k \in K\}$. Then $|HK| = |H||K|/|H \cap K|$.

PROOF Although the set HK has $|H||K|$ products, not all of these products need represent distinct group elements. That is, we may have $hk = h'k'$ where $h \neq h'$ and $k \neq k'$. To determine $|HK|$, we must find the extent to which this happens. For every t in $H \cap K$, the product $hk = (ht)(t^{-1}k)$, so each group element in HK is represented by at least $|H \cap K|$ products in HK. But $hk = h'k'$ implies $t = h^{-1}h' = kk'^{-1} \in H \cap K$, so that $h' = ht$ and $k' = t^{-1}k$. Thus, each element in HK is represented by exactly $|H \cap K|$ products. So, $|HK| = |H||K|/|H \cap K|$. ■

EXAMPLE 6 A group of order 75 can have at most one subgroup of order 25. (It is shown in Chapter 24 that every group of order 75 has a subgroup of order 25). To see that a group of order 75 cannot have two subgroups of order 25, suppose H and K are two such subgroups. Since $|H \cap K|$ divides $|H| = 25$ and $|H \cap K| = 1$ or 5 results in $|HK| = |H||K|/|H \cap K| = 25 \cdot 25/|H \cap K| = 625$ or 125 elements, we have that $|H \cap K| = 25$ and therefore $H = K$. ■

For any prime $p > 2$, we know that Z_{2p} and D_p are nonisomorphic groups of order $2p$. This naturally raises the question of whether there could be other possible groups of these orders. Remarkably, with just the simple machinery available to us at this point, we can answer this question.

Theorem 7.3 Classification of Groups of Order $2p$

Let G be a group of order 2p, where p is a prime greater than 2. Then G is isomorphic to Z_{2p} or D_p.

PROOF We assume that G does not have an element of order $2p$ and show that $G \approx D_p$. We begin by first showing that G must have an element of order p. By our assumption and Lagrange's Theorem, any nonidentity element of G must have order 2 or p. Thus, to verify our assertion, we may assume that every nonidentity element of G has order 2.

In this case, we have for all a and b in the group $ab = (ab)^{-1} = b^{-1}a^{-1} = ba$, so that G is Abelian. Then, for any nonidentity elements $a, b \in G$ with $a \neq b$, the set $\{e, a, b, ab\}$ is closed and therefore is a subgroup of G of order 4. Since this contradicts Lagrange's Theorem, we have proved that G must have an element of order p; call it a.

Now let b be any element not in $\langle a \rangle$. Then by Lagrange's Theorem and our assumption that G does not have an element of order $2p$, we have that $|b| = 2$ or p. Because $|\langle a \rangle \cap \langle b \rangle|$ divides $|\langle a \rangle| = p$ and $\langle a \rangle \neq \langle b \rangle$ we have that $|\langle a \rangle \cap \langle b \rangle| = 1$. But then $|b| = 2$, for otherwise, by Theorem 7.2 $|\langle a \rangle\langle b \rangle| = |\langle a \rangle||\langle b \rangle| = p^2 > 2p = |G|$, which is impossible. So, any element of G not in $\langle a \rangle$ has order 2.

Next consider ab. Since $ab \notin \langle a \rangle$, our argument above shows that $|ab| = 2$. Then $ab = (ab)^{-1} = b^{-1}a^{-1} = ba^{-1}$. Moreover, this relation completely determines the multiplication table for G. [For example, $a^3(ba^4) = a^2(ab)a^4 = a^2(ba^{-1})a^4 = a(ab)a^3 = a(ba^{-1})a^3 = (ab)a^2 = (ba^{-1})a^2 = ba$.] Since the multiplication table for all noncyclic groups of order $2p$ is uniquely determined by the relation $ab = ba^{-1}$, all noncyclic groups of order $2p$ must be isomorphic to each other. But of course, D_p, the dihedral group of order $2p$, is one such group. ∎

As an immediate corollary, we have that the non-Abelian groups S_3, the symmetric group of degree 3, and GL(2, Z_2), the group of 2×2 matrices with nonzero determinants with entries from Z_2 (see Example 19 and Exercise 51 in Chapter 2) are isomorphic to D_3.

An Application of Cosets to Permutation Groups

Lagrange's Theorem and its corollaries dramatically demonstrate the fruitfulness of the coset concept. We next consider an application of cosets to permutation groups.

Definition Stabilizer of a Point

Let G be a group of permutations of a set S. For each i in S, let $\text{stab}_G(i) = \{\phi \in G \mid \phi(i) = i\}$. We call $\text{stab}_G(i)$ the *stabilizer of i in G*.

The student should verify that $\text{stab}_G(i)$ is a subgroup of G. (See Exercise 35 in Chapter 5.)

Definition Orbit of a Point

Let G be a group of permutations of a set S. For each s in S, let $\text{orb}_G(s) = \{\phi(s) \mid \phi \in G\}$. The set $\text{orb}_G(s)$ is a subset of S called the *orbit of s under G*. We use $|\text{orb}_G(s)|$ to denote the number of elements in $\text{orb}_G(s)$.

Example 7 should clarify these two definitions.

EXAMPLE 7 Let

$$G = \{(1), (132)(465)(78), (132)(465), (123)(456), (123)(456)(78), (78)\}.$$

Then,

$$\begin{aligned} &\text{orb}_G(1) = \{1, 3, 2\}, && \text{stab}_G(1) = \{(1), (78)\}, \\ &\text{orb}_G(2) = \{2, 1, 3\}, && \text{stab}_G(2) = \{(1), (78)\}, \\ &\text{orb}_G(4) = \{4, 6, 5\}, && \text{stab}_G(4) = \{(1), (78)\}, \\ &\text{orb}_G(7) = \{7, 8\}, && \text{stab}_G(7) = \{(1), (132)(465), (123)(456)\}. \end{aligned}$$

EXAMPLE 8 We may view D_4 as a group of permutations of a square region. Figure 7.1(a) illustrates the orbit of the point p under D_4, and Figure 7.1(b) illustrates the orbit of the point q under D_4. Observe that $\text{stab}_{D_4}(p) = \{R_0, D\}$, whereas $\text{stab}_{D_4}(q) = \{R_0\}$.

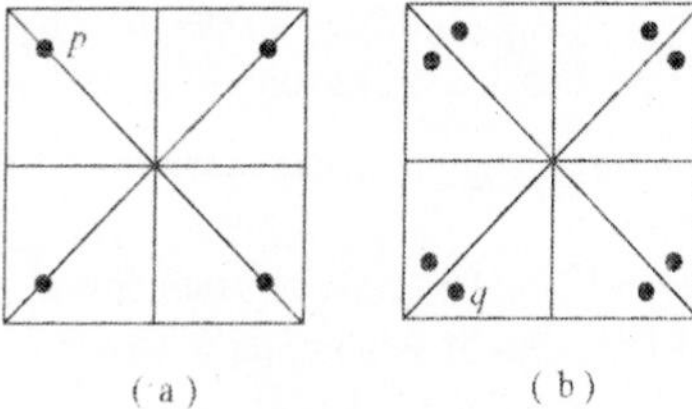

Figure 7.1

The preceding two examples also illustrate the following theorem.

Theorem 7.4 Orbit-Stabilizer Theorem

Let G be a finite group of permutations of a set S. Then, for any i from S, $|G| = |\text{orb}_G(i)|\,|\text{stab}_G(i)|$.

PROOF By Lagrange's Theorem, $|G|/|\text{stab}_G(i)|$ is the number of distinct left cosets of $\text{stab}_G(i)$ in G. Thus, it suffices to establish a one-to-one correspondence between the left cosets of $\text{stab}_G(i)$ and the elements in the orbit of i. To do this, we define a correspondence T by mapping the coset $\phi\text{stab}_G(i)$ to $\phi(i)$ under T. To show that T is a well-defined function, we must show that $\alpha\text{stab}_G(i) = \beta\text{stab}_G(i)$ implies $\alpha(i) = \beta(i)$. But $\alpha\text{stab}_G(i) = \beta\text{stab}_G(i)$ implies $\alpha^{-1}\beta \in \text{stab}_G(i)$, so that $(\alpha^{-1}\beta)(i) = i$ and, therefore, $\beta(i) = \alpha(i)$. Reversing the argument from the last step to the first step shows that T is also one-to-one. We conclude the proof by showing that T is onto $\text{orb}_G(i)$. Let $j \in \text{orb}_G(i)$. Then $\alpha(i) = j$ for some $\alpha \in G$ and clearly $T(\alpha\text{stab}_G(i)) = \alpha(i) = j$, so that T is onto. ∎

We leave as an exercise the proof of the important fact that the orbits of the elements of a set S under a group partition S (Exercise 43).

The Rotation Group of a Cube and a Soccer Ball

It cannot be overemphasized that Theorem 7.4 and Lagrange's Theorem (Theorem 7.1) are *counting* theorems.[†] They enable us to determine the numbers of elements in various sets. To see how Theorem 7.4 works, we will determine the order of the rotation group of a cube and a soccer ball. That is, we wish to find the number of essentially different ways in which we can take a cube or a soccer ball in a certain location in space, physically rotate it, and then have it still occupy its original location.

EXAMPLE 9 Let G be the rotation group of a cube. Label the six faces of the cube 1 through 6. Since any rotation of the cube must carry each face of the cube to exactly one other face of the cube and different rotations induce different permutations of the faces, G can be viewed as a group of permutations on the set $\{1, 2, 3, 4, 5, 6\}$. Clearly, there is some rotation about a central horizontal or vertical axis that carries face number 1 to any other face, so that $|\text{orb}_G(1)| = 6$. Next, we consider $\text{stab}_G(1)$. Here, we are asking for all rotations of a cube that leave face number 1 where it is. Surely, there are only four such motions—rotations of 0°, 90°, 180°, and 270°—about the line perpendicular to the face and passing through its center (see Figure 7.2). Thus, by Theorem 7.4, $|G| = |\text{orb}_G(1)|\,|\text{stab}_G(1)| = 6 \cdot 4 = 24$. ∎

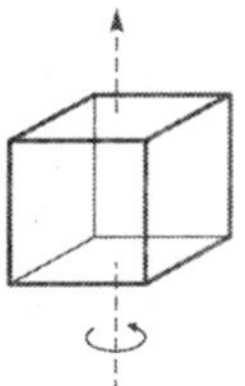

Figure 7.2 Axis of rotation of a cube.

Now that we know how many rotations a cube has, it is simple to determine the actual structure of the rotation group of a cube. Recall that S_4 is the symmetric group of degree 4.

[†]"People who don't count won't count" (Anatole France).

Theorem 7.5 The Rotation Group of a Cube

The group of rotations of a cube is isomorphic to S_4.

PROOF Since the group of rotations of a cube has the same order as S_4, we need only prove that the group of rotations is isomorphic to a subgroup of S_4. To this end, observe that a cube has four diagonals and that the rotation group induces a group of permutations on the four diagonals. But we must be careful not to assume that different rotations correspond to different permutations. To see that this is so, all we need do is show that all 24 permutations of the diagonals arise from rotations. Labeling the consecutive diagonals 1, 2, 3, and 4, it is obvious that there is a 90° rotation that yields the permutation $\alpha = (1234)$; another 90° rotation about an axis perpendicular to our first axis yields the permutation $\beta = (1423)$. See Figure 7.3. So, the group of permutations induced by the rotations contains the eight-element subgroup $\{\varepsilon, \alpha, \alpha^2, \alpha^3, \beta^2, \beta^2\alpha, \beta^2\alpha^2, \beta^2\alpha^3\}$ (see Exercise 63) and $\alpha\beta$, which has order 3. Clearly, then, the rotations yield all 24 permutations, since the order of the rotation group must be divisible by both 8 and 3. ∎

EXAMPLE 10 A traditional soccer ball has 20 faces that are regular hexagons and 12 faces that are regular pentagons. (The technical term for this solid is *truncated icosahedron*.) To determine the number of rotational symmetries of a soccer ball using Theorem 7.4, we may choose our set S to be the 20 hexagons or the 12 pentagons. Let us say that S is the set of 12 pentagons. Since any pentagon can be carried to any other

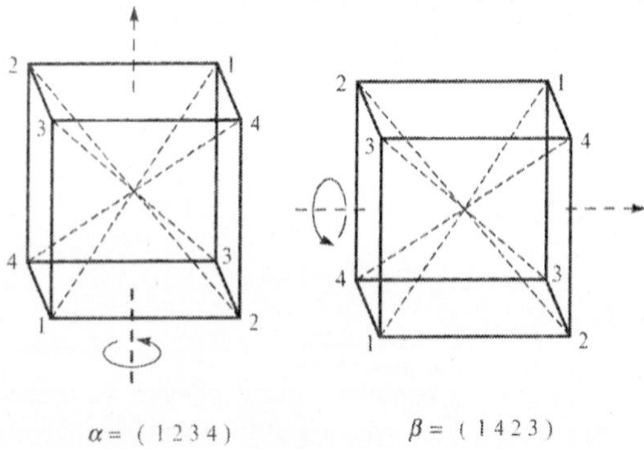

Figure 7.3

pentagon by some rotation, the orbit of any pentagon is S. Also, there are five rotations that fix (stabilize) any particular pentagon. Thus, by the Orbit-Stabilizer Theorem, there are $12 \cdot 5 = 60$ rotational symmetries. (In case you are interested, the rotation group of a soccer ball is isomorphic to A_5.) ∎

In 1985, chemists Robert Curl, Richard Smalley, and Harold Kroto caused tremendous excitement in the scientific community when they created a new form of carbon by using a laser beam to vaporize graphite. The structure of the new molecule was composed of 60 carbon atoms arranged in the shape of a soccer ball! Because the shape of the new molecule reminded them of the dome structures built by the architect R. Buckminster Fuller, Curl, Smalley, and Kroto named their discovery "buckyballs." Buckyballs are the roundest, most symmetric large molecules known. Group theory has been particularly useful in illuminating the properties of buckyballs, since the absorption spectrum of a molecule depends on its symmetries and chemists classify various molecular states according to their symmetry properties. The buckyball discovery spurred a revolution in carbon chemistry. In 1996, Curl, Smalley, and Kroto received the Nobel Prize in chemistry for their discovery.

An Application of Cosets to the Rubik's Cube

Recall from Chapter 5 that in 2010 it was proved via a computer computation, which took 35 CPU-years to complete, that every Rubik's cube could be solved in at most 20 moves. To carry out this effort, the research team of Morley Davidson, John Dethridge, Herbert Kociemba, and Tomas Rokicki applied a program of Rokicki, which built on early work of Kociemba, that checked the elements of the cosets of a subgroup H of order $(8! \cdot 8! \cdot 4!)/2 = 19{,}508{,}428{,}800$ to see if each cube in a position corresponding to the elements in a coset could be solved within 20 moves. In the rare cases where Rokicki's program did not work, an alternate method was employed. Using symmetry considerations, they were able to reduce the approximately 2 billion cosets of H

to about 56 million cosets for testing. Cosets played a role in this effort because Rokicki's program could handle the 19.5+ billion elements in the same coset in about 20 seconds.

Exercises

I don't know, Marge. Trying is the first step towards failure.

HOMER SIMPSON

1. Let $H = \{(1), (12)(34), (13)(24), (14)(23)\}$. Find the left cosets of H in A_4 (see Table 5.1 on page 111).
2. Let H be as in Exercise 1. How many left cosets of H in S_4 are there? (Determine this without listing them.)
3. Let $H = \{0, \pm 3, \pm 6, \pm 9, \ldots\}$. Find all the left cosets of H in Z.
4. Rewrite the condition $a^{-1}b \in H$ given in property 5 of the lemma on page 145 in additive notation. Assume that the group is Abelian.
5. Let H be as in Exercise 3. Use Exercise 4 to decide whether or not the following cosets of H are the same.
   a. $11 + H$ and $17 + H$
   b. $-1 + H$ and $5 + H$
   c. $7 + H$ and $23 + H$
6. Let n be a positive integer. Let $H = \{0, \pm n, \pm 2n, \pm 3n, \ldots\}$. Find all left cosets of H in Z. How many are there?
7. Find all of the left cosets of $\{1, 11\}$ in $U(30)$.
8. Suppose that a has order 15. Find all of the left cosets of $\langle a^5 \rangle$ in $\langle a \rangle$.
9. Let $|a| = 30$. How many left cosets of $\langle a^4 \rangle$ in $\langle a \rangle$ are there? List them.
10. Give an example of a group G and subgroups H and K such that $HK = \{h \in H, k \in K\}$ is not a subgroup of G.
11. If H and K are subgroups of G and g belongs to G, show that $g(H \cap K) = gH \cap gK$.
12. Let a and b be nonidentity elements of different orders in a group G of order 155. Prove that the only subgroup of G that contains a and b is G itself.
13. Let H be a subgroup of $\mathbf{R}^*$, the group of nonzero real numbers under multiplication. If $\mathbf{R}^+ \subseteq H \subseteq \mathbf{R}^*$, prove that $H = \mathbf{R}^+$ or $H = \mathbf{R}^*$.
14. Let $\mathbf{C}^*$ be the group of nonzero complex numbers under multiplication and let $H = \{a + bi \in \mathbf{C}^* \mid a^2 + b^2 = 1\}$. Give a geometric description of the coset $(3 + 4i)H$. Give a geometric description of the coset $(c + di)H$.

15. Let G be a group of order 60. What are the possible orders for the subgroups of G?

16. Suppose that K is a proper subgroup of H and H is a proper subgroup of G. If $|K| = 42$ and $|G| = 420$, what are the possible orders of H?

17. Let G be a group with $|G| = pq$, where p and q are prime. Prove that every proper subgroup of G is cyclic.

18. Recall that, for any integer n greater than 1, $\phi(n)$ denotes the number of positive integers less than n and relatively prime to n. Prove that if a is any integer relatively prime to n, then $a^{\phi(n)} \bmod n = 1$.

19. Compute $5^{15} \bmod 7$ and $7^{13} \bmod 11$.

20. Use Corollary 2 of Lagrange's Theorem (Theorem 7.1) to prove that the order of $U(n)$ is even when $n > 2$.

21. Suppose G is a finite group of order n and m is relatively prime to n. If $g \in G$ and $g^m = e$, prove that $g = e$.

22. Suppose H and K are subgroups of a group G. If $|H| = 12$ and $|K| = 35$, find $|H \cap K|$. Generalize.

23. Suppose that H is a subgroup of S_4 and that H contains (12) and (234). Prove that $H = S_4$.

24. Suppose that H and K are subgroups of G and there are elements a and b in G such that $aH \subseteq bK$. Prove that $H \subseteq K$.

25. Suppose that G is an Abelian group with an odd number of elements. Show that the product of all of the elements of G is the identity.

26. Suppose that G is a group with more than one element and G has no proper, nontrivial subgroups. Prove that $|G|$ is prime. (Do not assume at the outset that G is finite.)

27. Let $|G| = 15$. If G has only one subgroup of order 3 and only one of order 5, prove that G is cyclic. Generalize to $|G| = pq$, where p and q are prime.

28. Let G be a group of order 25. Prove that G is cyclic or $g^5 = e$ for all g in G. Generalize to any group of order p^2 where p is prime. Does your proof work for this generalization?

29. Let $|G| = 33$. What are the possible orders for the elements of G? Show that G must have an element of order 3.

30. Let $|G| = 8$. Show that G must have an element of order 2.

31. Can a group of order 55 have exactly 20 elements of order 11? Give a reason for your answer.

32. Determine all finite subgroups of $\mathbf{C}^*$, the group of nonzero complex numbers under multiplication.

33. Let H and K be subgroups of a finite group G with $H \subseteq K \subseteq G$. Prove that $|G:H| = |G:K|\,|K:H|$.

34. Suppose that a group contains elements of orders 1 through 10. What is the minimum possible order of the group?

35. Give an example of the dihedral group of smallest order that contains a subgroup isomorphic to Z_{12} and a subgroup isomorphic to Z_{20}. No need to prove anything, but explain your reasoning.

36. Show that in any group of order 100, either every element has order that is a power of a prime or there is an element of order 10.

37. Suppose that a finite Abelian group G has at least three elements of order 3. Prove that 9 divides $|G|$.

38. Prove that if G is a finite group, the index of $Z(G)$ cannot be prime.

39. Find an example of a subgroup H of a group G and elements a and b in G such that $aH \neq Hb$ and $aH \cap Hb \neq \phi$. (Compare with property 5 of cosets.)

40. Prove that a group of order 63 must have an element of order 3.

41. Let G be a group of order 100 that has a subgroup H of order 25. Prove that every element of G of order 5 is in H.

42. Let G be a group of order n and k be any integer relatively prime to n. Show that the mapping from G to G given by $g \to g^k$ is one-to-one. If G is also Abelian, show that the mapping given by $g \to g^k$ is an automorphism of G.

43. Let G be a group of permutations of a set S. Prove that the orbits of the members of S constitute a partition of S. (This exercise is referred to in this chapter and in Chapter 29.)

44. Prove that every subgroup of D_n of odd order is cyclic.

45. Let $G = \{(1), (12)(34), (1234)(56), (13)(24), (1432)(56), (56)(13), (14)(23), (24)(56)\}$.
    a. Find the stabilizer of 1 and the orbit of 1.
    b. Find the stabilizer of 3 and the orbit of 3.
    c. Find the stabilizer of 5 and the orbit of 5.

46. Prove that a group of order 12 must have an element of order 2.

47. Show that in a group G of odd order, the equation $x^2 = a$ has a unique solution for all a in G.

48. Let G be a group of order pqr, where p, q, and r are distinct primes. If H and K are subgroups of G with $|H| = pq$ and $|K| = qr$, prove that $|H \cap K| = q$.

49. Prove that a group that has more than one subgroup of order 5 must have order at least 25.

50. Prove that A_5 has a subgroup of order 12.

51. Prove that A_5 has no subgroup of order 30.

52. Prove that A_5 has no subgroup of order 15 to 20.

53. Suppose that α is an element from a permutation group G and one of its cycles in disjoint cycle form is $(a_1a_2\cdots a_k)$. Show that $\{a_1, a_2, \ldots, a_k\} \subseteq \text{orb}_G(a_i)$ for $1 = 1, 2, \ldots, k$.

54. Let G be a group and suppose that H is a subgroup of G with the property that for any a in G we have $aH = Ha$. (That is, every element of the form ah where h is some element of H can be written in the form h_1a for some $h_1 \in H$.) If a has order 2, prove that the set $K = H \cup aH$ is a subgroup of G. Generalize to the case that $|a| = k$.

55. Prove that A_5 is the only subgroup of S_5 of order 60.

56. Why does the fact that A_4 has no subgroup of order 6 imply that $|Z(A_4)| = 1$?

57. Let $G = GL(2, \mathbf{R})$ and $H = SL(2, \mathbf{R})$. Let $A \in G$ and suppose that $\det A = 2$. Prove that AH is the set of all 2×2 matrices in G that have determinant 2.

58. Let G be the group of rotations of a plane about a point P in the plane. Thinking of G as a group of permutations of the plane, describe the orbit of a point Q in the plane. (This is the motivation for the name "orbit.")

59. Let G be the rotation group of a cube. Label the faces of the cube 1 through 6, and let H be the subgroup of elements of G that carry face 1 to itself. If σ is a rotation that carries face 2 to face 1, give a physical description of the coset $H\sigma$.

60. The group D_4 acts as a group of permutations of the square regions shown below. (The axes of symmetry are drawn for reference purposes.) For each square region, locate the points in the orbit of the indicated point under D_4. In each case, determine the stabilizer of the indicated point.

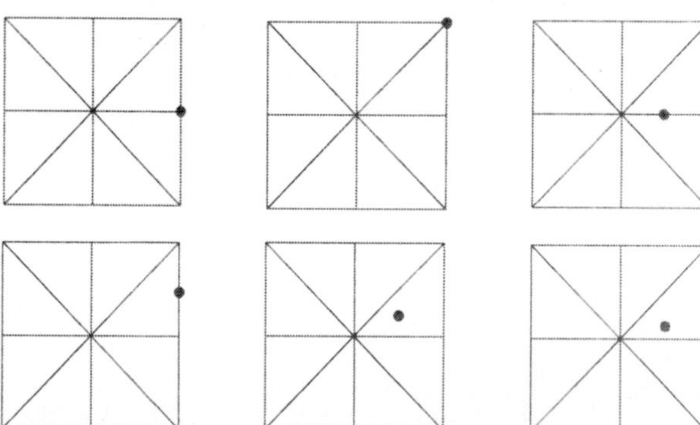

61. Let $G = GL(2, \mathbf{R})$, the group of 2×2 matrices over $\mathbf{R}$ with nonzero determinant. Let H be the subgroup of matrices of determinant ± 1. If $a, b \in G$ and $aH = bH$, what can be said about det (a) and det (b)? Prove or disprove the converse. [Determinants have the property that det (xy) = det (x)det (y).]

62. Calculate the orders of the following (refer to Figure 27.5 for illustrations).

a. The group of rotations of a regular tetrahedron (a solid with four congruent equilateral triangles as faces)

b. The group of rotations of a regular octahedron (a solid with eight congruent equilateral triangles as faces)

c. The group of rotations of a regular dodecahedron (a solid with 12 congruent regular pentagons as faces)

d. The group of rotations of a regular icosahedron (a solid with 20 congruent equilateral triangles as faces)

63. Prove that the eight-element set in the proof of Theorem 7.5 is a group.

64. A soccer ball has 20 faces that are regular hexagons and 12 faces that are regular pentagons. Use Theorem 7.4 to explain why a soccer ball cannot have a 60° rotational symmetry about a line through the centers of two opposite hexagonal faces.

65. If G is a finite group with fewer than 100 elements and G has subgroups of orders 10 and 25, what is the order of G?

Computer Exercises

A computer exercise for this chapter is available at the website:

http://www.d.umn.edu/~jgallian

Joseph Lagrange

Lagrange is the Lofty Pyramid of the Mathematical Sciences.

NAPOLEON BONAPARTE

The Granger Collection, New York

JOSEPH LOUIS LAGRANGE was born in Italy of French ancestry on January 25, 1736. He became captivated by mathematics at an early age when he read an essay by Halley on Newton's calculus. At the age of 19, he became a professor of mathematics at the Royal Artillery School in Turin. Lagrange made significant contributions to many branches of mathematics and physics, among them the theory of numbers, the theory of equations, ordinary and partial differential equations, the calculus of variations, analytic geometry, fluid dynamics, and celestial mechanics. His methods for solving third- and fourth-degree polynomial equations by radicals laid the groundwork for the group theoretic approach to solving polynomials taken by Galois. Lagrange was a very careful writer with a clear and elegant style.

This stamp was issued by France in Lagrange's honor in 1958.

At the age of 40, Lagrange was appointed head of the Berlin Academy, succeeding Euler. In offering this appointment, Frederick the Great proclaimed that the "greatest king in Europe" ought to have the "greatest mathematician in Europe" at his court. In 1787, Lagrange was invited to Paris by Louis XVI and became a good friend of the king and his wife, Marie Antoinette. In 1793, Lagrange headed a commission, which included Laplace and Lavoisier, to devise a new system of weights and measures. Out of this came the metric system. Late in his life he was made a count by Napoleon. Lagrange died on April 10, 1813.

To find more information about Lagrange, visit:

http://www-groups.dcs.st-and.ac.uk/~history/

8 External Direct Products

The universe is an enormous direct product of representations of symmetry groups.

STEVEN WEINBERG†

Definition and Examples

In this chapter, we show how to piece together groups to make larger groups. In Chapter 9, we will show that we can often start with one large group and decompose it into a product of smaller groups in much the same way as a composite positive integer can be broken down into a product of primes. These methods will later be used to give us a simple way to construct all finite Abelian groups.

Definition External Direct Product

Let $G_1, G_2, \ldots, G_n$ be a finite collection of groups. The *external direct product* of $G_1, G_2, \ldots, G_n$, written as $G_1 \oplus G_2 \oplus \cdots \oplus G_n$, is the set of all n-tuples for which the ith component is an element of G_i and the operation is componentwise.

In symbols,

$$G_1 \oplus G_2 \oplus \cdots \oplus G_n = \{(g_1, g_2, \ldots, g_n) \mid g_i \in G_i\},$$

where $(g_1, g_2, \ldots, g_n)(g_1', g_2', \ldots, g_n')$ is defined to be $(g_1g_1', g_2g_2', \ldots, g_ng_n')$. It is understood that each product g_ig_i' is performed with the operation of G_i. Note that in the case that each G_i is finite, we have by properties of sets that $|G_1 \oplus G_2 \oplus \cdots \oplus G_n| = |G_1||G_2| \cdots |G_n|$. We leave it to the reader to show that the external direct product of groups is itself a group (Exercise 1).

This construction is not new to students who have had linear algebra or physics. Indeed, $\mathbf{R}^2 = \mathbf{R} \oplus \mathbf{R}$ and $\mathbf{R}^3 = \mathbf{R} \oplus \mathbf{R} \oplus \mathbf{R}$—the operation being componentwise addition. Of course, there is also scalar multiplication, but

†Weinberg received the 1979 Nobel Prize in physics with Sheldon Glashow and Abdus Salam for their construction of a single theory incorporating weak and electromagnetic interactions.

we ignore this for the time being, since we are interested only in the group structure at this point.

EXAMPLE 1

$$U(8) \oplus U(10) = \{(1, 1), (1, 3), (1, 7), (1, 9), (3, 1), (3, 3), (3, 7), (3, 9), (5, 1), (5, 3), (5, 7), (5, 9), (7, 1), (7, 3), (7, 7), (7, 9)\}.$$

The product $(3, 7)(7, 9) = (5, 3)$, since the first components are combined by multiplication modulo 8, whereas the second components are combined by multiplication modulo 10. ∎

EXAMPLE 2

$$Z_2 \oplus Z_3 = \{(0, 0), (0, 1), (0, 2), (1, 0), (1, 1), (1, 2)\}.$$

Clearly, this is an Abelian group of order 6. Is this group related to another Abelian group of order 6 that we know, namely, Z_6? Consider the subgroup of $Z_2 \oplus Z_3$ generated by $(1, 1)$. Since the operation in each component is addition, we have $(1, 1) = (1, 1)$, $2(1, 1) = (0, 2)$, $3(1, 1) = (1, 0)$, $4(1, 1) = (0, 1)$, $5(1, 1) = (1, 2)$, and $6(1, 1) = (0, 0)$. Hence $Z_2 \oplus Z_3$ is cyclic. It follows that $Z_2 \oplus Z_3$ is isomorphic to Z_6. ∎

In Theorem 7.3 we classified the groups of order $2p$ where p is an odd prime. Now that we have defined $Z_2 \oplus Z_2$, it is easy to classify the groups of order 4.

EXAMPLE 3 Classification of Groups of Order 4

A group of order 4 is isomorphic to Z_4 or $Z_2 \oplus Z_2$. To verify this, let $G = \{e, a, b, ab\}$. If G is not cyclic, then it follows from Lagrange's Theorem that $|a| = |b| = |ab| = 2$. Then the mapping $e \rightarrow (0, 0)$, $a \rightarrow (1, 0)$, $b \rightarrow (0, 1)$, and $ab \rightarrow (1, 1)$ is an isomorphism from G onto $Z_2 \oplus Z_2$. ∎

We see from Examples 2 and 3 that in some cases $Z_m \oplus Z_n$ is isomorphic to Z_{mn} and in some cases it is not. Theorem 8.2 provides a simple characterization for when the isomorphism holds.

Properties of External Direct Products

Our first theorem gives a simple method for computing the order of an element in a direct product in terms of the orders of the component pieces.

Theorem 8.1 Order of an Element in a Direct Product

The order of an element in a direct product of a finite number of finite groups is the least common multiple of the orders of the components of the element. In symbols,

$$|(g_1, g_2, \ldots, g_n)| = \text{lcm}(|g_1|, |g_2|, \ldots, |g_n|).$$

PROOF Denote the identity of G_i by e_i. Let $s = \text{lcm}(|g_1|, |g_2|, \ldots, |g_n|)$ and $t = |(g_1, g_2, \ldots, g_n)|$. Because the fact that s is a multiple of each $|g_i|$ implies that $(g_1, g_2, \ldots, g_n)^s = (g_1^s, g_2^s, \ldots, g_n^s) = (e_1, e_2, \ldots, e_n)$, we know that $t \leq s$. On the other hand, from $(g_1^t, g_2^t, \ldots, g_n^t) = (g_1, g_2, \ldots, g_n)^t = (e_1, e_2, \ldots, e_n)$ we see that t is a common multiple of $|g_1|, |g_2|, \ldots, |g_n|$. Thus, $s \leq t$. ∎

The next two examples are applications of Theorem 8.1.

EXAMPLE 4 We determine the number of elements of order 5 in $Z_{25} \oplus Z_5$. By Theorem 8.1, we may count the number of elements (a, b) in $Z_{25} \oplus Z_5$ with the property that $5 = |(a, b)| = \text{lcm}(|a|, |b|)$. Clearly this requires that either $|a| = 5$ and $|b| = 1$ or 5, or $|b| = 5$ and $|a| = 1$ or 5. We consider two mutually exclusive cases.

Case 1 $|a| = 5$ and $|b| = 1$ or 5. Here there are four choices for a (namely, 5, 10, 15, and 20) and five choices for b. This gives 20 elements of order 5.

Case 2 $|a| = 1$ and $|b| = 5$. This time there is one choice for a and four choices for b, so we obtain four more elements of order 5.

Thus, $Z_{25} \oplus Z_5$ has 24 elements of order 5. ∎

EXAMPLE 5 We determine the number of cyclic subgroups of order 10 in $Z_{100} \oplus Z_{25}$. We begin by counting the number of elements (a, b) of order 10.

Case 1 $|a| = 10$ and $|b| = 1$ or 5. Since Z_{100} has a unique cyclic subgroup of order 10 and any cyclic group of order 10 has four generators (Theorem 4.4), there are four choices for a. Similarly, there are five choices for b. This gives 20 possibilities for (a, b).

Case 2 $|a| = 2$ and $|b| = 5$. Since any finite cyclic group of even order has a unique subgroup of order 2 (Theorem 4.4), there is only one choice for a. Obviously, there are four choices for b. So, this case yields four more possibilities for (a, b).

Thus, $Z_{100} \oplus Z_{25}$ has 24 elements of order 10. Because each cyclic subgroup of order 10 has four elements of order 10 and no two of the cyclic subgroups can have an element of order 10 in common, there must be $24/4 = 6$ cyclic subgroups of order 10. (This method is analogous to determining the number of sheep in a flock by counting legs and dividing by 4.) ■

The direct product notation is convenient for specifying certain subgroups of a direct product.

■ **EXAMPLE 6** For each divisor r of m and s of n, the group $Z_m \oplus Z_n$ has a subgroup isomorphic to $Z_r \oplus Z_s$ (see Exercise 19). To find a subgroup of, say, $Z_{30} \oplus Z_{12}$ isomorphic to $Z_6 \oplus Z_4$, we observe that $\langle 5 \rangle$ is a subgroup of Z_{30} of order 6 and $\langle 3 \rangle$ is a subgroup of Z_{12} of order 4, so $\langle 5 \rangle \oplus \langle 3 \rangle$ is the desired subgroup. ■

The next theorem and its first corollary characterize those direct products of cyclic groups that are themselves cyclic.

Theorem 8.2 Criterion for $G \oplus H$ to be Cyclic

Let G and H be finite cyclic groups. Then $G \oplus H$ is cyclic if and only if $|G|$ and $|H|$ are relatively prime.

PROOF Let $|G| = m$ and $|H| = n$, so that $|G \oplus H| = mn$. To prove the first half of the theorem, we assume $G \oplus H$ is cyclic and show that m and n are relatively prime. Suppose that $\gcd(m, n) = d$ and (g, h) is a generator of $G \oplus H$. Since $(g, h)^{mn/d} = ((g^m)^{n/d}, (h^n)^{m/d}) = (e, e)$, we have $mn = |(g, h)| \leq mn/d$. Thus, $d = 1$.

To prove the other half of the theorem, let $G = \langle g \rangle$ and $H = \langle h \rangle$ and suppose $\gcd(m, n) = 1$. Then, $|(g, h)| = \operatorname{lcm}(m, n) = mn = |G \oplus H|$, so that (g, h) is a generator of $G \oplus H$. ■

As a consequence of Theorem 8.2 and an induction argument, we obtain the following extension of Theorem 8.2.

Corollary 1 Criterion for $G_1 \oplus G_2 \oplus \cdots \oplus G_n$ to Be Cyclic

An external direct product $G_1 \oplus G_2 \oplus \cdots \oplus G_n$ of a finite number of finite cyclic groups is cyclic if and only if $|G_i|$ and $|G_j|$ are relatively prime when $i \neq j$.

Corollary 2 Criterion for $Z_{n_1n_2\cdots n_k} \approx Z_{n_1} \oplus Z_{n_2} \oplus \cdots \oplus Z_{n_k}$

Let $m = n_1n_2 \cdots n_k$. Then Z_m is isomorphic to $Z_{n_1} \oplus Z_{n_2} \oplus \cdots \oplus Z_{n_k}$ if and only if n_i and n_j are relatively prime when $i \neq j$.

By using the results above in an iterative fashion, one can express the same group (up to isomorphism) in many different forms. For example, we have

$$Z_2 \oplus Z_2 \oplus Z_3 \oplus Z_5 \approx Z_2 \oplus Z_6 \oplus Z_5 \approx Z_2 \oplus Z_{30}.$$

Similarly,

$$\begin{aligned} Z_2 \oplus Z_2 \oplus Z_3 \oplus Z_5 &\approx Z_2 \oplus Z_6 \oplus Z_5 \\ &\approx Z_2 \oplus Z_3 \oplus Z_2 \oplus Z_5 \approx Z_6 \oplus Z_{10}. \end{aligned}$$

Thus, $Z_2 \oplus Z_{30} \approx Z_6 \oplus Z_{10}$. Note, however, that $Z_2 \oplus Z_{30} \not\approx Z_{60}$.

The Group of Units Modulo n as an External Direct Product

The U-groups provide a convenient way to illustrate the preceding ideas. We first introduce some notation. If k is a divisor of n, let

$$U_k(n) = \{x \in U(n) \mid x \bmod k = 1\}.$$

For example, $U_7(105) = \{1, 8, 22, 29, 43, 64, 71, 92\}$. It can be readily shown that $U_k(n)$ is indeed a subgroup of $U(n)$. (See Exercise 31 in Chapter 3.)

Theorem 8.3 $U(n)$ as an External Direct Product

Suppose s and t are relatively prime. Then U(st) is isomorphic to the external direct product of U(s) and U(t). In short,

$$U(st) \approx U(s) \oplus U(t).$$

Moreover, $U_s(st)$ is isomorphic to $U(t)$ and $U_t(st)$ is isomorphic to $U(s)$.

PROOF An isomorphism from $U(st)$ to $U(s) \oplus U(t)$ is $x \to (x \bmod s, x \bmod t)$; an isomorphism from $U_s(st)$ to $U(t)$ is $x \to x \bmod t$; an isomorphism from $U_t(st)$ to $U(s)$ is $x \to x \bmod s$. We leave the verification that these mappings are operation-preserving, one-to-one, and onto to the reader. (See Exercises 9, 17, and 19 in Chapter 0; see also [1].) ∎

As a consequence of Theorem 8.3, we have the following result.

Corollary

Let $m = n_1n_2 \cdots n_k$, where $\gcd(n_i, n_j) = 1$ for $i \neq j$. Then,

$$U(m) \approx U(n_1) \oplus U(n_2) \oplus \cdots \oplus U(n_k).$$

To see how these results work, let's apply them to $U(105)$. We obtain

$$\begin{aligned} U(105) &\approx U(7) \oplus U(15), \\ U(105) &\approx U(21) \oplus U(5), \\ U(105) &\approx U(3) \oplus U(5) \oplus U(7). \end{aligned}$$

Moreover,

$$\begin{aligned} U(7) &\approx U_{15}(105) = \{1, 16, 31, 46, 61, 76\}, \\ U(15) &\approx U_7(105) = \{1, 8, 22, 29, 43, 64, 71, 92\}, \\ U(21) &\approx U_5(105) = \{1, 11, 16, 26, 31, 41, 46, 61, 71, 76, 86, 101\}, \\ U(5) &\approx U_{21}(105) = \{1, 22, 43, 64\}, \\ U(3) &\approx U_{35}(105) = \{1, 71\}. \end{aligned}$$

Among all groups, surely the cyclic groups Z_n have the simplest structures and, at the same time, are the easiest groups with which to compute. Direct products of groups of the form Z_n are only slightly more complicated in structure and computability. Because of this, algebraists endeavor to describe a finite Abelian group as such a direct product. Indeed, we shall soon see that every finite Abelian group can be so represented. With this goal in mind, let us reexamine the U-groups. Using the corollary to Theorem 8.3 and the facts (see [2, p. 93]), first proved by Carl Gauss in 1801, that

$$U(2) \approx \{0\}, \qquad U(4) \approx Z_2, \qquad U(2^n) \approx Z_2 \oplus Z_{2^{n-2}} \qquad \text{for } n \geq 3,$$

and

$$U(p^n) \approx Z_{p^n - p^{n-1}} \qquad \text{for } p \text{ an odd prime},$$

we now can write any U-group as an external direct product of cyclic groups. For example,

$$\begin{aligned} U(105) &= U(3 \cdot 5 \cdot 7) \approx U(3) \oplus U(5) \oplus U(7) \\ &\approx Z_2 \oplus Z_4 \oplus Z_6 \end{aligned}$$

and

$$\begin{aligned} U(720) &= U(16 \cdot 9 \cdot 5) \approx U(16) \oplus U(9) \oplus U(5) \\ &\approx Z_2 \oplus Z_4 \oplus Z_6 \oplus Z_4. \end{aligned}$$

What is the advantage of expressing a group in this form? Well, for one thing, we immediately see that the orders of the elements $U(720)$ can only be 1, 2, 3, 4, 6, and 12. This follows from the observations that an element from $Z_2 \oplus Z_4 \oplus Z_6 \oplus Z_4$ has the form (a, b, c, d), where $|a| = 1$ or 2, $|b| = 1, 2$, or 4, $|c| = 1, 2, 3$, or 6, and $|d| = 1, 2$, or 4, and that $|(a, b, c, d)| = \text{lcm}(|a|, |b|, |c|, |d|)$. For another thing, we can readily determine the number of elements of order 12, say, that $U(720)$ has. Because $U(720)$ is isomorphic to $Z_2 \oplus Z_4 \oplus Z_6 \oplus Z_4$, it suffices to calculate the number of elements of order 12 in $Z_2 \oplus Z_4 \oplus Z_6 \oplus Z_4$. But this is easy. By Theorem 8.1, an element (a, b, c, d) has order 12 if and only if $\text{lcm}(|a|, |b|, |c|, |d|) = 12$. Since $|a| = 1$ or 2, it does not matter how a is chosen. So, how can we have $\text{lcm}(|b|, |c|, |d|) = 12$? One way is to have $|b| = 4$, $|c| = 3$ or 6, and d arbitrary. By Theorem 4.4, there are two choices for b, four choices for c, and four choices for d. So, in this case, we have $2 \cdot 4 \cdot 4 = 32$ choices. The only other way to have $\text{lcm}(|b|, |c|, |d|) = 12$ is for $|d| = 4$, $|c| = 3$ or 6, and $|b| = 1$ or 2 (we exclude $|b| = 4$, since this was already accounted for). This gives $2 \cdot 4 \cdot 2 = 16$ new choices. Finally, since a can be either of the two elements in Z_2, we have a total of $2(32 + 16) = 96$ elements of order 12.

These calculations tell us more. Since $\text{Aut}(Z_{720})$ is isomorphic to $U(720)$, we also know that there are 96 automorphisms of Z_{720} of order 12. Imagine trying to deduce this information directly from $U(720)$ or, worse yet, from $\text{Aut}(Z_{720})$! These results beautifully illustrate the advantage of being able to represent a finite Abelian group as a direct product of cyclic groups. They also show the value of our theorems about $\text{Aut}(Z_n)$ and $U(n)$. After all, theorems are labor-saving devices. If you want to convince yourself of this, try to prove directly from the definitions that $\text{Aut}(Z_{720})$ has exactly 96 elements of order 12.

Applications

We conclude this chapter with five applications of the material presented here—three to cryptography, the science of sending and deciphering secret messages, one to genetics, and one to electric circuits.

Data Security

Because computers are built from two-state electronic components, it is natural to represent information as strings of 0s and 1s called *binary strings*. A binary string of length n can naturally be thought of as an element of $Z_2 \oplus Z_2 \oplus \cdots \oplus Z_2$ (n copies) where the parentheses and the commas have been deleted. Thus the binary string

11000110 corresponds to the element (1, 1, 0, 0, 0, 1, 1, 0) in $Z_2 \oplus Z_2 \oplus Z_2 \oplus Z_2 \oplus Z_2 \oplus Z_2 \oplus Z_2 \oplus Z_2$. Similarly, two binary strings $a_1a_2 \cdots a_n$ and $b_1b_2 \cdots b_n$ are added componentwise modulo 2 just as their corresponding elements in $Z_2 \oplus Z_2 \oplus \cdots \oplus Z_2$ are. For example,

$$11000111 + 01110110 = 10110001$$

and

$$10011100 + 10011100 = 00000000.$$

The fact that the sum of two binary sequences $a_1a_2 \cdots a_n + b_1b_2 \cdots b_n = 00 \cdots 0$ if and only if the sequences are identical is the basis for a data security system used to protect Internet transactions.

Suppose that you want to purchase a compact disc from **http://www.amazon.com**. Need you be concerned that a hacker will intercept your credit-card number during the transaction? As you might expect, your credit-card number is sent to Amazon in a way that protects the data. We explain one way to send credit-card numbers over the Web securely. When you place an order with Amazon, the company sends your computer a randomly generated string of 0's and 1's called a *key*. This key has the same length as the binary string corresponding to your credit-card number and the two strings are added (think of this process as "locking" the data). The resulting sum is then transmitted to Amazon. Amazon in turn adds the same key to the received string, which then produces the original string corresponding to your credit-card number (adding the key a second time "unlocks" the data).

To illustrate the idea, say you want to send an eight-digit binary string such as $s = 10101100$ to Amazon (actual credit-card numbers have very long strings) and Amazon sends your computer the key $k = 00111101$. Your computer returns the string $s + k = 10101100 + 00111101 = 10010001$ to Amazon, and Amazon adds k to this string to get $10010001 + 00111101 = 10101100$, which is the string representing your credit-card number. If someone intercepts the number $s + k = 10010001$ during transmission it is no value without knowing k.

The method is secure because the key sent by Amazon is randomly generated and used only one time. You can tell when you are using an encryption scheme on a Web transaction by looking to see if the Web address begins with "https" rather than the customary "http." You will also see a small padlock in the status bar at the bottom of the browser window.

Public Key Cryptography

Unlike auctions such as those on eBay, where each bid is known by everyone, a silent auction is one in which each bid is secret. Suppose that you wanted to use your Twitter account to run a silent auction.

How could a scheme be devised so that users could post their bids in such a way that the amounts are intelligible only to the account holder? In the mid-1970s, Ronald Rivest, Adi Shamir, and Leonard Adleman devised an ingenious method that permits each person who is to receive a secret message to tell publicly how to scramble messages sent to him or her. And even though the method used to scramble the message is known publicly, only the person for whom it is intended will be able to unscramble the message. The idea is based on the fact that there exist efficient methods for finding very large prime numbers (say about 100 digits long) and for multiplying large numbers, but no one knows an efficient algorithm for factoring large integers (say about 200 digits long). The person who is to receive the message chooses a pair of large primes p and q and chooses an integer e (called the *encryption exponent*) with $1 < e < m$, where $m = \text{lcm}(p - 1, q - 1)$, such that e is relatively prime to m (any such e will do). This person calculates $n = pq$ (n is called the *key*) and announces that a message M is to be sent to him or her publicly as $M^e \bmod n$. Although e, n, and M^e are available to everyone, only the person who knows how to factor n as pq will be able to decipher the message.

To present a simple example that nevertheless illustrates the principal features of the method, say we wish to send the messages "YES." We convert the message into a string of digits by replacing A by 01, B by 02, . . . , Z by 26, and a blank by 00. So, the message YES becomes 250519. To keep the numbers involved from becoming too unwieldy, we send the message in blocks of four digits and fill in with blanks when needed. Thus, the messages YES is represented by the two blocks 2505 and 1900. The person to whom the message is to be sent has picked two primes p and q, say $p = 37$ and $q = 73$, and a number e that has no prime divisors in common with $\text{lcm}(p - 1, q - 1) = 72$, say $e = 5$, and has published $n = 37 \cdot 73 = 2701$ and $e = 5$ in a public forum. We will send the "scrambled" numbers $(2505)^5 \bmod 2701$ and $(1900)^5 \bmod 2701$ rather than 2505 and 1900, and the receiver will unscramble them. We show the work involved for us and the receiver only for the block 2505. We determine $(2505)^5 \bmod 2701 = 2415$ by using a modular arithmetic calculator such as the one at **http://users.wpi.edu/~martin/mod.html**.[†]

[†]Provided that the numbers are not too large, the Google search engine at **http://www.google.com** will do modular arithmetic. For example, entering 2505^2 mod 2701 in the search box yields 602. Be careful, however: Entering 2505^5 mod 2701 does not return a value, because 2505^5 is too large. Instead, we can use Google to compute smaller powers such as 2505^2 mod 2701 and 2505^3 mod 2701 (which yields 852) and then enter $(852 \times 602) \bmod 2701$.

Thus, the number 2415 is sent to the receiver. Now the receiver must take this number and convert it back to 2505. To do so, the receiver takes the two factors of 2701, $p = 37$ and $q = 73$, and calculates the least common multiple of $p - 1 = 36$ and $q - 1 = 72$, which is 72. (This is where the knowledge of p and q is necessary.) Next, the receiver must find $e^{-1} = d$ (called the *decryption exponent*) in $U(72)$—that is, solve the equation $5 \cdot d = 1 \bmod 72$. This number is 29. See **http://www.d.umn.edu/~jgallian/msproject06/chap8.html #chap8ex5** or use a Google search box to compute 5^k for each divisor k of $|U(72)| = \phi(9) \cdot \phi(8) = 24$ starting with 2 until we reach $5^k \bmod 72 = 1$. Doing so, we obtain $5^6 \bmod 72 = 1$, which implies that $5^5 \bmod 72 = 29$ is 5^{-1} in $U(72)$.

Then the receiver takes the number received, 2415, and calculates $(2415)^{29} \bmod 2701 = 2505$, the encoded number. Thus, the receiver correctly determines the code for "YE." On the other hand, without knowing how pq factors, one cannot find the modulus (in our case, 72) that is needed to determine the decryption exponent d.

The procedure just described is called the *RSA public key encryption scheme* in honor of the three people (Rivest, Shamir, and Adleman) who discovered the method. It is widely used in conjunction with web servers and browsers, e-mail programs, remote login sessions, and electronic financial transactions. The algorithm is summarized below.

Receiver

1. Pick very large primes p and q and compute $n = pq$.
2. Compute the least common multiple of $p - 1$ and $q - 1$; let us call it m.
3. Pick e relatively prime to m.
4. Find d such that $ed \bmod m = 1$.
5. Publicly announce n and e.

Sender

1. Convert the message to a string of digits.
2. Break up the message into uniform blocks of digits; call them $M_1, M_2, \ldots, M_k$.
3. Check to see that the greatest common divisor of each M_i and n is 1. If not, n can be factored and our code is broken. (In practice, the primes p and q are so large that they exceed all M_i, so this step may be omitted.)
4. Calculate and send $R_i = M_i^e \bmod n$.

Receiver

1. For each received message R_i, calculate R_i^d mod n.
2. Convert the string of digits back to a string of characters.

Why does this method work? Well, we know that $U(n) \approx U(p) \oplus U(q) \approx Z_{p-1} \oplus Z_{q-1}$. Thus, an element of the form x^m in $U(n)$ corresponds under an isomorphism to one of the form (mx_1, mx_2) in $Z_{p-1} \oplus Z_{q-1}$. Since m is the least common multiple of $p - 1$ and $q - 1$, we may write $m = s(p - 1)$ and $m = t(q - 1)$ for some integers s and t. Then $(mx_1, mx_2) = (s(p - 1)x_1, t(q - 1)x_2) = (0, 0)$ in $Z_{p-1} \oplus Z_{q-1}$, and it follows that $x^m = 1$ for all x in $U(n)$. So, because each message M_i is an element of $U(n)$ and e was chosen so that $ed = 1 + km$ for some k, we have, modulo n,

$$R_i^d = (M_i^e)^d = M_i^{ed} = M_i^{1+km} = M_i(M_i^m)^k = M_i 1^k = M_i.$$

In 2002, Ronald Rivest, Adi Shamir, and Leonard Adleman received the Association for Computing Machinery A. M. Turing Award, which is considered the "Nobel Prize of computing," for their contribution to public key cryptography.

An RSA calculator that does all the calculations is provided at **http://www.d.umn.edu/~jgallian/msproject06/chap8.html#chap8ex5**. A list of primes can be found by searching the Web for "list of primes."

Digital Signatures

With so many financial transactions now taking place electronically, the problem of authenticity is paramount. How is a stockbroker to know that an electronic message she receives that tells her to sell one stock and buy another actually came from her client? The technique used in public key cryptography allows for digital signatures as well. Let us say that person A wants to send a secret message to person B in such a way that only B can decode the message and B will know that only A could have sent it. Abstractly, let E_A and D_A denote the algorithms that A uses for encryption and decryption, respectively, and let E_B and D_B denote the algorithms that B uses for encryption and decryption, respectively. Here we assume that E_A and E_B are available to the public, whereas D_A is known only to A and D_B is known only to B, and that D_BE_B and E_AD_A applied to any message leaves the message unchanged. Then A sends a message M to B as $E_B(D_A(M))$ and B decodes the received message by applying the function E_AD_B to it to obtain

$$(E_AD_B)(E_B(D_A(M)) = E_A(D_BE_B)(D_A(M)) = E_A(D_A(M)) = M.$$

Notice that only A can execute the first step (i.e., create $D_A(M)$) and only B can implement the last step (i.e., apply $E_A D_B$ to the received message).

Transactions using digital signatures became legally binding in the United States in October 2000.

Genetics[†]

The genetic code can be conveniently modeled using elements of $Z_4 \oplus Z_4 \oplus \cdots \oplus Z_4$, where we omit the parentheses and the commas and just use strings of 0's, 1's, 2's, and 3's and add componentwise modulo 4. A DNA molecule is composed of two long strands in the form of a double helix. Each strand is made up of strings of the four nitrogen bases adenine (A), thymine (T), guanine (G), and cytosine (C). Each base on one strand binds to a complementary base on the other strand. Adenine always is bound to thymine, and guanine always is bound to cytosine. To model this process, we identify A with 0, T with 2, G with 1, and C with 3. Thus, the DNA segment ACGTAACAGGA and its complement segment TGCATTGTCCT are denoted by 03120030110 and 21302212332. Noting that in Z_4, $0 + 2 = 2$, $2 + 2 = 0$, $1 + 2 = 3$, and $3 + 2 = 1$, we see that adding 2 to elements of Z_4 interchanges 0 and 2 and 1 and 3. So, for any DNA segment $a_1a_2 \cdots a_n$ represented by elements of $Z_4 \oplus Z_4 \oplus \cdots \oplus Z_4$, we see that its complementary segment is represented by $a_1a_2 \cdots a_n + 22 \cdots 2$.

Electric Circuits

Many homes have light fixtures that are operated by a pair of switches. They are wired so that when either switch is thrown, the light changes its status (from on to off or vice versa). Suppose the wiring is done so that the light is on when both switches are in the up position. We can conveniently think of the states of the two switches as being matched with the elements of $Z_2 \oplus Z_2$, with the two switches in the up position corresponding to $(0, 0)$ and the two switches in the down position corresponding to $(1, 1)$. Each time a switch is thrown, we add 1 to the corresponding component in the group $Z_2 \oplus Z_2$. We then see that the lights are on when the switches correspond to the elements of the subgroup $\langle(1, 1)\rangle$ and are off when the switches correspond to the elements in the coset $(1, 0) + \langle(1, 1)\rangle$. A similar analysis applies in the case of three switches, with the subgroup $\{(0, 0, 0), (1, 1, 0), (0, 1, 1), (1, 0, 1)\}$ corresponding to the lights-on situation.

[†]This discussion is adapted from [3].

Exercises

What's the most difficult aspect of your life as a mathematician, Diane Maclagan, an assistant professor at Rutgers, was asked. "Trying to prove theorems," she said. And the most fun? "Trying to prove theorems."

1. Prove that the external direct product of any finite number of groups is a group. (This exercise is referred to in this chapter.)
2. Show that $Z_2 \oplus Z_2 \oplus Z_2$ has seven subgroups of order 2.
3. Let G be a group with identity e_G and let H be a group with identity e_H. Prove that G is isomorphic to $G \oplus \{e_H\}$ and that H is isomorphic to $\{e_G\} \oplus H$.
4. Show that $G \oplus H$ is Abelian if and only if G and H are Abelian. State the general case.
5. Prove or disprove that $Z \oplus Z$ is a cyclic group.
6. Prove, by comparing orders of elements, that $Z_8 \oplus Z_2$ is not isomorphic to $Z_4 \oplus Z_4$.
7. Prove that $G_1 \oplus G_2$ is isomorphic to $G_2 \oplus G_1$. State the general case.
8. Is $Z_3 \oplus Z_9$ isomorphic to Z_{27}? Why?
9. Is $Z_3 \oplus Z_5$ isomorphic to Z_{15}? Why?
10. How many elements of order 9 does $Z_3 \oplus Z_9$ have? (Do not do this exercise by brute force.)
11. How many elements of order 4 does $Z_4 \oplus Z_4$ have? (Do not do this by examining each element.) Explain why $Z_4 \oplus Z_4$ has the same number of elements of order 4 as does $Z_{8000000} \oplus Z_{400000}$. Generalize to the case $Z_m \oplus Z_n$.
12. Give examples of four groups of order 12, no two of which are isomorphic. Give reasons why no two are isomorphic.
13. For each integer $n > 1$, give examples of two nonisomorphic groups of order n^2.
14. The dihedral group D_n of order $2n$ ($n \geq 3$) has a subgroup of n rotations and a subgroup of order 2. Explain why D_n cannot be isomorphic to the external direct product of two such groups.
15. Prove that the group of complex numbers under addition is isomorphic to $\mathbf{R} \oplus \mathbf{R}$.
16. Suppose that $G_1 \approx G_2$ and $H_1 \approx H_2$. Prove that $G_1 \oplus H_1 \approx G_2 \oplus H_2$. State the general case.
17. If $G \oplus H$ is cyclic, prove that G and H are cyclic. State the general case.
18. In $Z_{40} \oplus Z_{30}$, find two subgroups of order 12.

19. If r is a divisor of m and s is a divisor of n, find a subgroup of $Z_m \oplus Z_n$ that is isomorphic to $Z_r \oplus Z_s$.
20. Find a subgroup of $Z_{12} \oplus Z_{18}$ that is isomorphic to $Z_9 \oplus Z_4$.
21. Let G and H be finite groups and $(g, h) \in G \oplus H$. State a necessary and sufficient condition for $\langle(g, h)\rangle = \langle g\rangle \oplus \langle h\rangle$.
22. Determine the number of elements of order 15 and the number of cyclic subgroups of order 15 in $Z_{30} \oplus Z_{20}$.
23. What is the order of any nonidentity element of $Z_3 \oplus Z_3 \oplus Z_3$? Generalize.
24. Let $m > 2$ be an even integer and let $n > 2$ be an odd integer. Find a formula for the number of elements of order 2 in $D_m \oplus D_n$.
25. Let M be the group of all real 2×2 matrices under addition. Let $N = \mathbf{R} \oplus \mathbf{R} \oplus \mathbf{R} \oplus \mathbf{R}$ under componentwise addition. Prove that M and N are isomorphic. What is the corresponding theorem for the group of $m \times n$ matrices under addition?
26. The group $S_3 \oplus Z_2$ is isomorphic to one of the following groups: Z_{12}, $Z_6 \oplus Z_2$, A_4, D_6. Determine which one by elimination.
27. Let G be a group, and let $H = \{(g, g) \mid g \in G\}$. Show that H is a subgroup of $G \oplus G$. (This subgroup is called the *diagonal* of $G \oplus G$.) When G is the set of real numbers under addition, describe $G \oplus G$ and H geometrically.
28. Find a subgroup of $Z_4 \oplus Z_2$ that is not of the form $H \oplus K$, where H is a subgroup of Z_4 and K is a subgroup of Z_2.
29. Find all subgroups of order 3 in $Z_9 \oplus Z_3$.
30. Find all subgroups of order 4 in $Z_4 \oplus Z_4$.
31. What is the largest order of any element in $Z_{30} \oplus Z_{20}$?
32. What is the order of the largest cyclic subgroup of $Z_6 \oplus Z_{10} \oplus Z_{15}$? What is the order of the largest cyclic subgroup of $Z_{n_1} \oplus Z_{n_2} \oplus \cdots \oplus Z_{n_k}$?
33. Find three cyclic subgroups of maximum possible order in $Z_6 \oplus Z_{10} \oplus Z_{15}$ of the form $\langle a\rangle \oplus \langle b\rangle \oplus \langle c\rangle$, where $a \in Z_6$, $b \in Z_{10}$, and $c \in Z_{15}$.
34. How many elements of order 2 are in $Z_{2000000} \oplus Z_{4000000}$? Generalize.
35. Find a subgroup of $Z_{800} \oplus Z_{200}$ that is isomorphic to $Z_2 \oplus Z_4$.
36. Find a subgroup of $Z_{12} \oplus Z_4 \oplus Z_{15}$ that has order 9.
37. Prove that $\mathbf{R}^* \oplus \mathbf{R}^*$ is not isomorphic to $\mathbf{C}^*$. (Compare this with Exercise 15.)
38. Let

$$H = \left\{ \begin{bmatrix} 1 & a & b \\ 0 & 1 & 0 \\ 0 & 0 & 1 \end{bmatrix} \middle| \, a, b \in Z_3 \right\}.$$

(See Exercise 48 in Chapter 2 for the definition of multiplication.) Show that H is an Abelian group of order 9. Is H isomorphic to Z_9 or to $Z_3 \oplus Z_3$?

39. Let $G = \{3^m6^n \mid m, n \in Z\}$ under multiplication. Prove that G is isomorphic to $Z \oplus Z$. Does your proof remain valid if $G = \{3^m9^n \mid m, n \in Z\}$?

40. Let $(a_1, a_2, \ldots, a_n) \in G_1 \oplus G_2 \oplus \cdots \oplus G_n$. Give a necessary and sufficient condition for $|(a_1, a_2, \ldots, a_n)| = \infty$.

41. Prove that $D_3 \oplus D_4 \not\approx D_{12} \oplus Z_2$.

42. Determine the number of cyclic subgroups of order 15 in $Z_{90} \oplus Z_{36}$. Provide a generator for each of the subgroups of order 15.

43. List the elements in the groups $U_5(35)$ and $U_7(35)$.

44. Prove or disprove that $U(40) \oplus Z_6$ is isomorphic to $U(72) \oplus Z_4$.

45. Prove or disprove that C^* has a subgroup isomorphic to $Z_2 \oplus Z_2$.

46. Let G be a group isomorphic to $Z_{n_1} \oplus Z_{n_2} \oplus \cdots \oplus Z_{nk}$. Let x be the product of all elements in G. Describe all possibilities for x.

47. If a group has exactly 24 elements of order 6, how many cyclic subgroups of order 6 does it have?

48. For any Abelian group G and any positive integer n, let $G^n = \{g^n \mid g \in G\}$ (see Exercise 17, Supplementary Exercises for Chapters 1–4). If H and K are Abelian, show that $(H \oplus K)^n = H^n \oplus K^n$.

49. Express Aut($U(25)$) in the form $Z_m \oplus Z_n$.

50. Determine Aut($Z_2 \oplus Z_2$).

51. Suppose that $n_1, n_2, \ldots, n_k$ are positive even integers. How many elements of order 2 does $Z_{n_1} \oplus Z_{n_2} \oplus \cdots \oplus Z_{n_k}$ have ? How many are there if we drop the requirement that $n_1, n_2, \ldots, n_k$ must be even?

52. Is $Z_{10} \oplus Z_{12} \oplus Z_6 \approx Z_{60} \oplus Z_6 \oplus Z_2$?

53. Is $Z_{10} \oplus Z_{12} \oplus Z_6 \approx Z_{15} \oplus Z_4 \oplus Z_{12}$?

54. Find an isomorphism from Z_{12} to $Z_4 \oplus Z_3$.

55. How many isomorphisms are there from Z_{12} to $Z_4 \oplus Z_3$?

56. Suppose that ϕ is an isomorphism from $Z_3 \oplus Z_5$ to Z_{15} and $\phi(2, 3) = 2$. Find the element in $Z_3 \oplus Z_5$ that maps to 1.

57. If ϕ is an isomorphism from $Z_4 \oplus Z_3$ to Z_{12}, what is $\phi(2, 0)$? What are the possibilities for $\phi(1, 0)$? Give reasons for your answer.

58. Prove that $Z_5 \oplus Z_5$ has exactly six subgroups of order 5.

59. Let (a, b) belong to $Z_m \oplus Z_n$. Prove that $|(a, b)|$ divides lcm(m, n).

60. Let $G = \{ax^2 + bx + c \mid a, b, c \in Z_3\}$. Add elements of G as you would polynomials with integer coefficients, except use modulo 3 addition. Prove that G is isomorphic to $Z_3 \oplus Z_3 \oplus Z_3$. Generalize.

61. Determine all cyclic groups that have exactly two generators.

62. Explain a way that a string of length n of the four nitrogen bases A, T, G, and C could be modeled with the external direct product of n copies of $Z_2 \oplus Z_2$.

63. Let p be a prime. Prove that $Z_p \oplus Z_p$ has exactly $p + 1$ subgroups of order p.

64. Give an example of an infinite non-Abelian group that has exactly six elements of finite order.

65. Give an example to show that there exists a group with elements a and b such that $|a| = \infty$, $|b| = \infty$, and $|ab| = 2$.

66. Express $U(165)$ as an external direct product of cyclic groups of the form Z_n.

67. Express $U(165)$ as an external direct product of U-groups in four different ways.

68. Without doing any calculations in $\text{Aut}(Z_{20})$, determine how many elements of $\text{Aut}(Z_{20})$ have order 4. How many have order 2?

69. Without doing any calculations in $\text{Aut}(Z_{720})$, determine how many elements of $\text{Aut}(Z_{720})$ have order 6.

70. Without doing any calculations in $U(27)$, decide how many subgroups $U(27)$ has.

71. What is the largest order of any element in $U(900)$?

72. Let p and q be odd primes and let m and n be positive integers. Explain why $U(p^m) \oplus U(q^n)$ is not cyclic.

73. Use the results presented in this chapter to prove that $U(55)$ is isomorphic to $U(75)$.

74. Use the results presented in this chapter to prove that $U(144)$ is isomorphic to $U(140)$.

75. For every $n > 2$, prove that $U(n)^2 = \{x^2 \mid x \in U(n)\}$ is a proper subgroup of $U(n)$.

76. Show that $U(55)^3 = \{x^3 \mid x \in U(55)\}$ is $U(55)$.

77. Find an integer n such that $U(n)$ contains a subgroup isomorphic to $Z_5 \oplus Z_5$.

78. Find a subgroup of order 6 in $U(700)$.

79. Show that there is a U-group containing a subgroup isomorphic to $Z_3 \oplus Z_3$.

80. Find an integer n such that $U(n)$ is isomorphic to $Z_2 \oplus Z_4 \oplus Z_9$.

81. What is the smallest positive integer k such that $x^k = e$ for all x in $U(7 \cdot 17)$? Generalize to $U(pq)$ where p and q are distinct primes.

82. If k divides m and m divides n, how are $U_m(n)$ and $U_k(n)$ related?

83. Let $p_1, p_2, \ldots, p_k$ be distinct odd primes and $n_1, n_2, \ldots, n_k$ be positive integers. Determine the number of elements of order 2 in $U(p_1^{n_1} p_2^{n_2} \cdots p_k^{n_k})$. How many are there in $U(2^n p_1^{n_1} p_2^{n_2} \cdots p_k^{n_k})$ where n is at least 3?

84. Show that no U-group has order 14.

85. Show that there is a U-group containing a subgroup isomorphic to Z_{14}.

86. Show that no U-group is isomorphic to $Z_4 \oplus Z_4$.

87. Show that there is a U-group containing a subgroup isomorphic to $Z_4 \oplus Z_4$.

88. Using the RSA scheme with $p = 37$, $q = 73$, and $e = 5$, what number would be sent for the message "RM"?

89. Assuming that a message has been sent via the RSA scheme with $p = 37$, $q = 73$, and $e = 5$, decode the received message "34."

Computer Exercises

Computer exercises in this chapter are available at the website:

http://www.d.umn.edu/~jgallian

References

1. J. A. Gallian and D. Rusin, "Factoring Groups of Integers Modulo n," *Mathematics Magazine* 53 (1980): 33–36.
2. D. Shanks, *Solved and Unsolved Problems in Number Theory,* 2nd ed., New York: Chelsea, 1978.
3. S. Washburn, T. Marlowe, and C. Ryan, *Discrete Mathematics,* Reading, MA: Addison-Wesley, 1999.

Suggested Readings

Y. Cheng, "Decompositions of U-Groups," *Mathematics Magazine* 62 (1989): 271–273.

This article explores the decomposition of $U(st)$, where s and t are relatively prime, in greater detail than we have provided.

David J. Devries, "The Group of Units in Z_m," *Mathematics Magazine* 62 (1989): 340–342.

This article provides a simple proof that $U(n)$ is not cyclic when n is not of the form 1, 2, 4, p^k, or $2p^k$, where p is an odd prime.

David R. Guichard, "When Is $U(n)$ Cyclic? An Algebraic Approach," *Mathematics Magazine* 72 (1999): 139–142.

The author provides a group theoretic proof of the fact that $U(n)$ is cyclic if and only if n is 1, 2, 4, p^k, or $2p^k$, where p is an odd prime.

Markku Niemenmaa, "A Check Digit System for Hexadecimal Numbers," *Applicable Algebra in Engineering, Communication, and Computing* 22 (2011):109–112.

This article provides a new check-digit system for hexadecimal numbers that is based on the use of a suitable automorphism of the group $Z_2 \oplus Z_2 \oplus Z_2 \oplus Z_2$. It is able to detect all single errors, adjacent transpositions, twin errors, jump transpositions, and jump twin errors.

Leonard Adleman

"For their ingenious contribution for making public-key cryptography useful in practice."

Citation for the ACM A. M. Turing Award

LEONARD ADLEMAN was born on December 31, 1945 in San Francisco, California. He received a B.A. degree in mathematics in 1968 and a Ph.D. degree in computer science in 1976 from the University of California, Berkeley. He spent 1976–1980 as professor of mathematics at the Massachusetts Institute of Technology where he met Ronald Rivest and Adi Shamir. Rivest and Shamir were attempting to devise a secure public key cryptosystem and asked Adleman if he could break their codes. Eventually, they invented what is now known as the RSA code that was simple to implement yet secure.

In 1983, Adleman, Shamir, and Rivest formed the RSA Data Security company to license their algorithm. Their algorithm has become the primary cryptosystem used for security on the World Wide Web. They sold their company for $200 million in 1996.

In the early 1990s, Adleman became interested in trying to find out a way to use DNA as a computer. His pioneering work on this problem lead to the field now called "DNA computing."

Among his many honors are: the Association for Computing Machinery A. M. Turing Award, the Kanallakis Award for Theory and Practice, and election to the National Academy of Engineering, the American Academy of Arts and Sciences, and the National Academy of Sciences.

Adleman's current position is the Henry Salvatori Distinguished Chair in Computer Science and Professor of Computer Science and Biological Sciences at the University of Southern California, where he has been since 1980.

For more information on Adleman, visit:

http://www.wikipedia.com

and

http://www.nytimes.com/1994/12/13/science/scientist-at-work-leonard-adleman-hitting-the-high-spots-of-computer-theory.html?pagewanted=all&src=pm

Supplementary Exercises for Chapters 5–8

My mind rebels at stagnation. Give me problems, give me work, give me the most obstruse cryptogram, or the most intricate analysis, and I am in my own proper atmosphere.

SHERLOCK HOLMES, *The Sign of Four*

True/false questions for Chapters 5–8 are available on the Web at:

www.d.umn.edu/~jgallian/TF

1. A subgroup N of a group G is called a *characteristic subgroup* if $\phi(N) = N$ for all automorphisms ϕ of G. (The term *characteristic* was first applied by G. Frobenius in 1895.) Prove that every subgroup of a cyclic group is characteristic.
2. Prove that the center of a group is characteristic.
3. The *commutator subgroup* G' of a group G is the subgroup generated by the set $\{x^{-1}y^{-1}xy \mid x, y \in G\}$. (That is, every element of G' has the form $a_1^{i_1}a_2^{i_2} \cdots a_k^{i_k}$, where each a_j has the form $x^{-1}y^{-1}xy$, each $i_j = \pm 1$, and k is any positive integer.) Prove that G' is a characteristic subgroup of G. (This subgroup was first introduced by G. A. Miller in 1898.)
4. Prove that the property of being a characteristic subgroup is transitive. That is, if N is a characteristic subgroup of K and K is a characteristic subgroup of G, then N is a characteristic subgroup of G.
5. Let $G = Z_3 \oplus Z_3 \oplus Z_3$ and let H be the subgroup of $SL(3, Z_3)$ consisting of

$$H = \left\{ \begin{bmatrix} 1 & a & b \\ 0 & 1 & c \\ 0 & 0 & 1 \end{bmatrix} \middle| \, a, b, c \in Z_3 \right\}.$$

   (See Exercise 48 in Chapter 2 for the definition of multiplication.) Determine the number of elements of each order in G and H. Are G and H isomorphic? (This exercise shows that two groups with the same number of elements of each order need not be isomorphic.)
6. Let H and K be subgroups of a group G and let $HK = \{hk \mid h \in H, k \in K\}$ and $KH = \{kh \mid k \in K, h \in H\}$. Prove that HK is a group if and only if $HK = KH$.
7. Let G be a finite Abelian group in which every nonidentity element has order 2. If $|G| > 2$, prove that the product of all the elements in G is the identity.

8. Prove that S_4 is not isomorphic to $D_4 \oplus Z_3$.
9. Let G be a group. For any element g of G, define $gZ(G) = \{gh \mid h \in Z(G)\}$. If a is an element of G of order 4, prove that $H = Z(G) \cup aZ(G) \cup a^2Z(G) \cup a^3Z(G)$ is a subgroup of G. Generalize to the case that $|a| = k$.
10. The *exponent* of a group is the smallest positive integer n such that $x^n = e$ for all x in the group. Prove that every finite group has an exponent that divides the order of the group.
11. Determine all U-groups of exponent 2.
12. Suppose that H and K are subgroups of a group and that $|H|$ and $|K|$ are relatively prime. Show that $H \cap K = \{e\}$.
13. Let $\mathbf{R}^+$ denote the multiplicative group of positive real numbers and let $T = \{a + bi \in \mathbf{C}^* \mid a^2 + b^2 = 1\}$ be the multiplicative group of complex numbers on the unit circle. Show that every element of $\mathbf{C}^*$ can be uniquely expressed in the form rz, where $r \in \mathbf{R}^+$ and $z \in T$.
14. Prove that Q^* under multiplication is not isomorphic to $\mathbf{R}^*$ under multiplication.
15. Prove that Q under addition is not isomorphic to $\mathbf{R}$ under addition.
16. Prove that $\mathbf{R}$ under addition is not isomorphic to $\mathbf{R}^*$ under multiplication.
17. Show that Q^+ (the set of positive rational numbers) under multiplication is not isomorphic to Q under addition.
18. Suppose that $G = \{e, x, x^2, y, yx, yx^2\}$ is a non-Abelian group with $|x| = 3$ and $|y| = 2$. Show that $xy = yx^2$.
19. Let p be an odd prime. Show that 1 is the only solution of $x^{p-2} = 1$ in $U(p)$.
20. Let G be an Abelian group under addition. Let n be a fixed positive integer and let $H = \{(g, ng) \mid g \in G\}$. Show that H is a subgroup of $G \oplus G$. When G is the set of real numbers under addition, describe H geometrically.
21. Find a subgroup of $Z_{12} \oplus Z_{20}$ that is isomorphic to $Z_4 \oplus Z_5$.
22. Suppose that $G = G_1 \oplus G_2 \oplus \cdots \oplus G_n$. Prove that $Z(G) = Z(G_1) \oplus Z(G_2) \oplus \cdots \oplus Z(G_n)$.
23. Exhibit four nonisomorphic groups of order 18.
24. What is the order of the largest cyclic subgroup in Aut(Z_{720})? (*Hint:* It is not necessary to consider automorphisms of Z_{720}.)
25. Let G be the group of all permutations of the positive integers. Let H be the subset of elements of G that can be expressed as a product of a finite number of cycles. Prove that H is a subgroup of G.

26. Let G be a group and let $g \in G$. Show that $Z(G)\langle g \rangle$ is a subgroup of G.

27. Show that $D_{11} \oplus Z_3 \not\approx D_3 \oplus Z_{11}$. (This exercise is referred to in Chapter 24.)

28. Show that $D_{33} \not\approx D_{11} \oplus Z_3$. (This exercise is referred to in Chapter 24.)

29. Show that $D_{33} \not\approx D_3 \oplus Z_{11}$. (This exercise is referred to in Chapter 24.)

30. Exhibit four nonisomorphic groups of order 66. (This exercise is referred to in Chapter 24.)

31. Prove that $|\text{Inn}(G)| = 1$ if and only if G is Abelian.

32. Prove that $x^{100} = 1$ for all x in $U(1000)$.

33. Find a subgroup of order 6 in $U(450)$.

34. List four elements of $Z_{20} \oplus Z_5 \oplus Z_{60}$ that form a noncyclic subgroup.

35. In S_{10}, let $\beta = (13)(17)(265)(289)$. Find an element in S_{10} that commutes with β but is not a power of β.

36. Prove or disprove that $Z_4 \oplus Z_{15} \approx Z_6 \oplus Z_{10}$.

37. Prove or disprove that $D_{12} \approx Z_3 \oplus D_4$.

38. Describe a three-dimensional solid whose symmetry group is isomorphic to D_5.

39. Let $G = U(15) \oplus Z_{10} \oplus S_5$. Find the order of $(2, 3, (123)(15))$. Find the inverse of $(2, 3, (123)(15))$.

40. Let $G = Z \oplus Z_{10}$ and let $H = \{g \in G \mid |g| = \infty \text{ or } |g| = 1\}$. Prove or disprove that H is a subgroup of G.

41. Find a subgroup H of $Z_{p^2} \oplus Z_{p^2}$ such that $(Z_{p^2} \oplus Z_{p^2})/H$ is isomorphic to $Z_p \oplus Z_p$.

42. Find three subgroups H_1, H_2, and H_3 of $Z_{p^2} \oplus Z_{p^2}$ such that $(Z_{p^2} \oplus Z_{p^2})/H_i$ is isomorphic to Z_{p^2} for $i = 1, 2, 3$.

43. Find an element of order 10 in A_9.

44. In the left regular representation for D_4, write $T_{R_{90}}$ and T_H in matrix form and in cycle form.

45. How many elements of order 6 are in S_7?

46. Prove that $S_3 \oplus S_4$ is not isomorphic to a subgroup of S_6.

47. Find a permutation β such that $\beta^2 = (13579)(268)$.

48. In $\mathbf{R} \oplus \mathbf{R}$ under componentwise addition, let $H = \{(x, 3x) \mid x \in \mathbf{R}\}$. (Note that H is the subgroup of all points on the line $y = 3x$.) Show that $(2, 5) + H$ is a straight line passing through the point $(2, 5)$ and parallel to the line $y = 3x$.

49. In $\mathbf{R} \oplus \mathbf{R}$, suppose that H is the subgroup of all points lying on a line through the origin. Show that any left coset of H is a line parallel to H.

50. Let G be a group of permutations on the set $\{1, 2, \ldots, n\}$. Recall that $\text{stab}_G(1) = \{\alpha \in G \mid \alpha(1) = 1\}$. If γ sends 1 to k, prove that $\gamma\,\text{stab}_G(1) = \{\beta \in G \mid \beta(1) = k\}$.

51. Let H be a subgroup of G and let $a, b \in G$. Show that $aH = bH$ if and only if $Ha^{-1} = Hb^{-1}$.

52. Suppose that G is a finite Abelian group that does not contain a subgroup isomorphic to $Z_p \oplus Z_p$ for any prime p. Prove that G is cyclic.

53. Let p be a prime. Determine the number of elements of order p in $Z_{p^2} \oplus Z_{p^2}$.

54. Show that $Z_{p^2} \oplus Z_{p^2}$ has exactly one subgroup isomorphic to $Z_p \oplus Z_p$.

55. Let p be a prime. Determine the number of subgroups of $Z_{p^2} \oplus Z_{p^2}$ that are isomorphic to Z_{p^2}.

56. Find a group of order $3^2 \cdot 5^2 \cdot 7^2 \cdot 2^8$ that contains a subgroup isomorphic to A_8.

57. Let p and q be distinct odd primes. Let $n = \text{lcm}(p - 1, q - 1)$. Prove that $x^n = 1$ for all $x \in U(pq)$.

58. Give a simple characterization of all positive integers n for which $Z_n \approx H \oplus Z_n/H$ for every subgroup H of Z_n.

59. Prove that the permutations (12) and $(123 \ldots n)$ generate S_n. (That is, every member of S_n can be expressed as some combination of these elements.)

60. Suppose that n is even and σ is an $(n - 1)$-cycle in S_n. Show that σ does not commute with any element of order 2.

61. Suppose that n is odd and σ is an n-cycle in S_n. Prove that σ does not commute with any element of order 2.

62. Let $H = \{\alpha \in S_n \mid \alpha \text{ maps the set } \{1, 2\} \text{ to itself}\}$. Prove that $C((12)) = H$.

63. Let m be a positive integer. For any n-cycle σ, show that σ^m is the product of $\gcd(m, n)$ disjoint cycles, each of length $n/\gcd(m, n)$.

9 Normal Subgroups and Factor Groups

It is tribute to the genius of Galois that he recognized that those subgroups for which the left and right cosets coincide are distinguished ones. Very often in mathematics the crucial problem is to recognize and to discover what are the relevant concepts; once this is accomplished the job may be more than half done.

I. N. HERSTEIN, *Topics in Algebra*

Normal Subgroups

As we saw in Chapter 7, if G is a group and H is a subgroup of G, it is not always true that $aH = Ha$ for all a in G. There are certain situations where this does hold, however, and these cases turn out to be of critical importance in the theory of groups. It was Galois, about 180 years ago, who first recognized that such subgroups were worthy of special attention.

Definition Normal Subgroup

A subgroup H of a group G is called a *normal* subgroup of G if $aH = Ha$ for all a in G. We denote this by $H \triangleleft G$.

You should think of a normal subgroup in this way: You can switch the order of a product of an element a from the group and an element h from the normal subgroup H, but you must "fudge" a bit on the element from the normal subgroup H by using some h' from H rather than h. That is, there is an element h' in H such that $ah = h'a$. Likewise, there is some h'' in H such that $ha = ah''$. (It is possible that $h' = h$ or $h'' = h$, but we may not assume this.)

There are several equivalent formulations of the definition of normality. We have chosen the one that is the easiest to use in applications. However, to *verify* that a subgroup is normal, it is usually better to use Theorem 9.1, which is a weaker version of property 8 of the lemma in Chapter 7. It allows us to substitute a condition about two subgroups of G for a condition about two cosets of G.

Theorem 9.1 Normal Subgroup Test

A subgroup H of G is normal in G if and only if $xHx^{-1} \subseteq H$ *for all x in G.*

PROOF If H is normal in G, then for any $x \in G$ and $h \in H$ there is an h' in H such that $xh = h'x$. Thus, $xhx^{-1} = h'$, and therefore $xHx^{-1} \subseteq H$.

Conversely, if $xHx^{-1} \subseteq H$ for all x, then, letting $x = a$, we have $aHa^{-1} \subseteq H$ or $aH \subseteq Ha$. On the other hand, letting $x = a^{-1}$, we have $a^{-1}H(a^{-1})^{-1} = a^{-1}Ha \subseteq H$ or $Ha \subseteq aH$. ∎

EXAMPLE 1 Every subgroup of an Abelian group is normal. (In this case, $ah = ha$ for a in the group and h in the subgroup.) ∎

EXAMPLE 2 The center $Z(G)$ of a group is always normal. [Again, $ah = ha$ for any $a \in G$ and any $h \in Z(G)$.] ∎

EXAMPLE 3 The alternating group A_n of even permutations is a normal subgroup of S_n. [Note, for example, that for $(12) \in S_n$ and $(123) \in A_n$, we have $(12)(123) \neq (123)(12)$ but $(12)(123) = (132)(12)$ and $(132) \in A_n$.] ∎

EXAMPLE 4 Every subgroup of D_n consisting solely of rotations is normal in D_n. (For any rotation R and any reflection F, we have $FR = R^{-1}F$ and any two rotations commute.) ∎

The next example illustrates a way to use a normal subgroup to create new subgroups from existing ones.

EXAMPLE 5 Let H be a normal subgroup of a group G and K be any subgroup of G. Then $HK = \{hk \mid h \in H, k \in K\}$ is a subgroup of G. To verify this, note that $e = ee$ is in HK. Then for any $a = h_1k_1$ and $b = h_2k_2$, where h_1, h_2 are in H and k_1, k_2 are in K, there is an element h' in H such that $ab^{-1} = h_1k_1k_2^{-1}h_2^{-1} = h_1(k_1k_2^{-1})h_2^{-1} = (h_1h')(k_1k_2^{-1})$. So, ab^{-1} is in HK. ∎

Be careful not to assume that for any subgroups H and K of a group G, the set HK is a subgroup of G. See Exercise 57.

Combining Examples 4 and 5, we form a non-Abelian subgroup of D_8 of order 8.

EXAMPLE 6 In D_8, let $H = \{R_0, R_{90}, R_{180}, R_{270}\}$ and $K = \{R_0, F\}$, where F is any reflection. Then $HK = \{R_0, R_{90}, R_{180}, R_{270}, R_0F, R_{90}F, R_{180}F, R_{270}F\}$ is a subgroup of D_8. ∎

EXAMPLE 7 If a group G has a unique subgroup H of some finite order, then H is normal in G. To see that this is so, observe that for any $g \in G$, gHg^{-1} is a subgroup of G and $|gHg^{-1}| = |H|$. ∎

EXAMPLE 8 The group $SL(2, \mathbf{R})$ of 2×2 matrices with determinant 1 is a normal subgroup of $GL(2, \mathbf{R})$, the group of 2×2 matrices with nonzero determinant. To verify this, we use the Normal Subgroup Test given in Theorem 9.1. Let $x \in GL(2, \mathbf{R}) = G$, $h \in SL(2, \mathbf{R}) = H$, and note that $\det xhx^{-1} = (\det x)(\det h)(\det x)^{-1} = (\det x)(\det x)^{-1} = 1$. So, $xhx^{-1} \in H$, and, therefore, $xHx^{-1} \subseteq H$. ∎

EXAMPLE 9 Referring to the group table for A_4 given in Table 5.1 on page 111, we may observe that $H = \{\alpha_1, \alpha_2, \alpha_3, \alpha_4\}$ is a normal subgroup of A_4, whereas $K = \{\alpha_1, \alpha_5, \alpha_9\}$ is *not* a normal subgroup of A_4. To see that H is normal, simply note that for any β in A_4, $\beta H \beta^{-1}$ is a subgroup of order 4 and H is the only subgroup of A_4 of order 4 (see Table 5.1). Thus, $\beta H \beta^{-1} = H$. In contrast, $\alpha_2 \alpha_5 \alpha_2^{-1} = \alpha_7$, so that $\alpha_2 K \alpha_2^{-1} \nsubseteq K$. ∎

Factor Groups

We have yet to explain why normal subgroups are of special significance. The reason is simple. When the subgroup H of G is normal, then the set of left (or right) cosets of H in G is itself a group—called the *factor group of G by H* (or the *quotient group of G by H*). Quite often, one can obtain information about a group by studying one of its factor groups. This method will be illustrated in the next section of this chapter.

Theorem 9.2 Factor Groups (O. Hölder, 1889)

Let G be a group and let H be a normal subgroup of G. The set $G/H = \{aH \mid a \in G\}$ is a group under the operation $(aH)(bH) = abH$.[†]

[†]The notation G/H was first used by C. Jordan.

PROOF Our first task is to show that the operation is well-defined; that is, we must show that the correspondence defined above from $G/H \times G/H$ into G/H is actually a function. To do this, we assume that for some elements a, a', b, and b' from G, we have $aH = a'H$ and $bH = b'H$, and verify that $aHbH = a'Hb'H$. That is, verify that $abH = a'b'H$. (This shows that the definition of multiplication depends on only the cosets and not on the coset representatives.) From $aH = a'H$ and $bH = b'H$, we have $a' = ah_1$ and $b' = bh_2$ for some h_1, h_2 in H, and therefore $a'b'H = ah_1bh_2H = ah_1bH = ah_1Hb = aHb = abH$. Here we have made multiple use of associativity, property 2 of the lemma in Chapter 7, and the fact that $H \lhd G$. The rest is easy: $eH = H$ is the identity; $a^{-1}H$ is the inverse of aH; and $(aHbH)cH = (ab)HcH = (ab)cH = a(bc)H = aH(bc)H = aH(bHcH)$. This proves that G/H is a group. ■

Although it is merely a curiosity, we point out that the converse of Theorem 9.2 is also true; that is, if the correspondence $aHbH = abH$ defines a group operation on the set of left cosets of H in G, then H is normal in G.

The next few examples illustrate the factor group concept.

■ **EXAMPLE 10** Let $4Z = \{0, \pm 4, \pm 8, \ldots\}$. To construct $Z/4Z$, we first must determine the left cosets of $4Z$ in Z. Consider the following four cosets:

$$\begin{aligned}
0 + 4Z &= 4Z = \{0, \pm 4, \pm 8, \ldots\},\\
1 + 4Z &= \{1, 5, 9, \ldots; -3, -7, -11, \ldots\},\\
2 + 4Z &= \{2, 6, 10, \ldots; -2, -6, -10, \ldots\},\\
3 + 4Z &= \{3, 7, 11, \ldots; -1, -5, -9, \ldots\}.
\end{aligned}$$

We claim that there are no others. For if $k \in Z$, then $k = 4q + r$, where $0 \le r < 4$; and, therefore, $k + 4Z = r + 4q + 4Z = r + 4Z$. Now that we know the elements of the factor group, our next job is to determine the structure of $Z/4Z$. Its Cayley table is

	0 + 4Z	1 + 4Z	2 + 4Z	3 + 4Z
0 + 4Z	0 + 4Z	1 + 4Z	2 + 4Z	3 + 4Z
1 + 4Z	1 + 4Z	2 + 4Z	3 + 4Z	0 + 4Z
2 + 4Z	2 + 4Z	3 + 4Z	0 + 4Z	1 + 4Z
3 + 4Z	3 + 4Z	0 + 4Z	1 + 4Z	2 + 4Z

Clearly, then, $Z/4Z \approx Z_4$. More generally, if for any $n > 0$ we let $nZ = \{0, \pm n, \pm 2n, \pm 3n, \ldots\}$, then Z/nZ is isomorphic to Z_n. ■

EXAMPLE 11 Let $G = Z_{18}$ and let $H = \langle 6 \rangle = \{0, 6, 12\}$. Then $G/H = \{0 + H, 1 + H, 2 + H, 3 + H, 4 + H, 5 + H\}$. To illustrate how the group elements are combined, consider $(5 + H) + (4 + H)$. This should be one of the six elements listed in the set G/H. Well, $(5 + H) + (4 + H) = 5 + 4 + H = 9 + H = 3 + 6 + H = 3 + H$, since H absorbs all multiples of 6. ∎

A few words of caution about notation are warranted here. When H is a normal subgroup of G, the expression $|aH|$ has two possible interpretations. One could be thinking of aH as a *set* of elements and $|aH|$ as the size of the set; or, as is more often the case, one could be thinking of aH as a group element of the factor group G/H and $|aH|$ as the order of the *element* aH in G/H. In Example 11, for instance, the *set* $3 + H$ has size 3, since $3 + H = \{3, 9, 15\}$. But the *group element* $3 + H$ has order 2, since $(3 + H) + (3 + H) = 6 + H = 0 + H$. As is usually the case when one notation has more than one meaning, the appropriate interpretation will be clear from the context.

EXAMPLE 12 Let $\mathcal{H} = \{R_0, R_{180}\}$, and consider the factor group of the dihedral group D_4 (see the back inside cover for the multiplication table for D_4)

$$D_4/\mathcal{H} = \{\mathcal{H}, R_{90}\mathcal{H}, H\mathcal{H}, D\mathcal{H}\}.$$

The multiplication table for $D_4/\mathcal{H}$ is given in Table 9.1. (Notice that even though $R_{90}H = D'$, we have used $D\mathcal{H}$ in Table 9.1 for $R_{90}\mathcal{H}H\mathcal{H}$ because $D'\mathcal{H} = D\mathcal{H}$.)

Table 9.1

	$\mathcal{H}$	$R_{90}\mathcal{H}$	$H\mathcal{H}$	$D\mathcal{H}$
$\mathcal{H}$	$\mathcal{H}$	$R_{90}\mathcal{H}$	$H\mathcal{H}$	$D\mathcal{H}$
$R_{90}\mathcal{H}$	$R_{90}\mathcal{H}$	$\mathcal{H}$	$D\mathcal{H}$	$H\mathcal{H}$
$H\mathcal{H}$	$H\mathcal{H}$	$D\mathcal{H}$	$\mathcal{H}$	$R_{90}\mathcal{H}$
$D\mathcal{H}$	$D\mathcal{H}$	$H\mathcal{H}$	$R_{90}\mathcal{H}$	$\mathcal{H}$

$D_4/\mathcal{H}$ provides a good opportunity to demonstrate how a factor group of G is related to G itself. Suppose we arrange the heading of the Cayley table for D_4 in such a way that elements from the same coset of $\mathcal{H}$ are in adjacent columns (Table 9.2). Then, the multiplication table for D_4 can be blocked off into boxes that are cosets of $\mathcal{H}$, and the substitution that replaces a box containing the element x with the coset $x\mathcal{H}$ yields the Cayley table for $D_4/\mathcal{H}$.

For example, when we pass from D_4 to $D_4/\mathcal{H}$, the box

H	V
V	H

in Table 9.2 becomes the element $H\mathcal{H}$ in Table 9.1. Similarly, the box

D	D'
D'	D

becomes the element $D\mathcal{H}$, and so on.

Table 9.2

	R_0	R_{180}	R_{90}	R_{270}	H	V	D	D'
R_0	R_0	R_{180}	R_{90}	R_{270}	H	V	D	D'
R_{180}	R_{180}	R_0	R_{270}	R_{90}	V	H	D'	D
R_{90}	R_{90}	R_{270}	R_{180}	R_0	D'	D	H	V
R_{270}	R_{270}	R_{90}	R_0	R_{180}	D	D'	V	H
H	H	V	D	D'	R_0	R_{180}	R_{90}	R_{270}
V	V	H	D'	D	R_{180}	R_0	R_{270}	R_{90}
D	D	D'	V	H	R_{270}	R_{90}	R_0	R_{180}
D'	D'	D	H	V	R_{90}	R_{270}	R_{180}	R_0

■

In this way, one can see that the formation of a factor group G/H causes a systematic collapse of the elements of G. In particular, all the elements in the coset of H containing a collapse to the single group element aH in G/H.

■ **EXAMPLE 13** Consider the group A_4 as represented by Table 5.1 on page 111. (Here i denotes the permutation α_i.) Let $H = \{1, 2, 3, 4\}$. Then the three cosets of H are H, $5H = \{5, 6, 7, 8\}$, and $9H = \{9, 10, 11, 12\}$. (In this case, rearrangement of the headings is unnecessary.) Blocking off the table for A_4 into boxes that are cosets of H and replacing the boxes containing 1, 5, and 9 (see Table 9.3) with the cosets $1H$, $5H$, and $9H$, we obtain the Cayley table for G/H given in Table 9.4.

Table 9.3

	1	2	3	4	5	6	7	8	9	10	11	12
1	1	2	3	4	5	6	7	8	9	10	11	12
2	2	1	4	3	6	5	8	7	10	9	12	11
3	3	4	1	2	7	8	5	6	11	12	9	10
4	4	3	2	1	8	7	6	5	12	11	10	9
5	5	8	6	7	9	12	10	11	1	4	2	3
6	6	7	5	8	10	11	9	12	2	3	1	4
7	7	6	8	5	11	10	12	9	3	2	4	1
8	8	5	7	6	12	9	11	10	4	1	3	2
9	9	11	12	10	1	3	4	2	5	7	8	6
10	10	12	11	9	2	4	3	1	6	8	7	5
11	11	9	10	12	3	1	2	4	7	5	6	8
12	12	10	9	11	4	2	1	3	8	6	5	7

Table 9.4

	$1H$	$5H$	$9H$
$1H$	$1H$	$5H$	$9H$
$5H$	$5H$	$9H$	$1H$
$9H$	$9H$	$1H$	$5H$

This procedure can be illustrated more vividly with colors. Let's say we had printed the elements of H in green, the elements of $5H$ in red, and the elements of $9H$ in blue. Then, in Table 9.3, each box would consist of elements of a uniform color. We could then think of the factor group as consisting of the three colors that define a group table isomorphic to G/H.

	Green	**Red**	**Blue**
Green	Green	Red	Blue
Red	Red	Blue	Green
Blue	Blue	Green	Red

It is instructive to see what happens if we attempt the same procedure with a group G and a subgroup H that is not normal in G—that is, if we arrange the headings of the Cayley table so that the elements from the same coset of H are in adjacent columns and attempt to block off the table into boxes that are also cosets of H to produce a Cayley table for the set of cosets. Say, for instance, we were to take G to be A_4 and $H = \{1, 5, 9\}$. The cosets of H would be H, $2H = \{2, 6, 10\}$, $3H = \{3, 7, 11\}$, and $4H = \{4, 8, 12\}$. Then the first three rows of the rearranged Cayley table for A_4 would be the following.

	1	5	9	2	6	10	3	7	11	4	8	12
1	1	5	9	2	6	10	3	7	11	4	8	12
5	5	9	1	8	12	4	6	10	2	7	11	3
9	9	1	5	11	3	7	12	4	8	10	2	6

But already we are in trouble, for blocking these off into 3×3 boxes yields boxes that contain elements of different cosets. Hence, it is impossible to represent an entire box by a single element of the box in the same way we could for boxes made from the cosets of a normal subgroup. Had we printed the rearranged table in four colors with all members of the same coset having the same color, we would see multicolored boxes rather than the uniformly colored boxes produced by a normal subgroup. ■

In Chapter 11, we will prove that every finite Abelian group is isomorphic to a direct product of cyclic groups. In particular, an Abelian group of order 8 is isomorphic to one of Z_8, $Z_4 \oplus Z_2$, or $Z_2 \oplus Z_2 \oplus Z_2$. In the next two examples, we examine Abelian factor groups of order 8 and determine the isomorphism type of each.

■ **EXAMPLE 14** Let $G = U(32) = \{1, 3, 5, 7, 9, 11, 13, 15, 17, 19, 21, 23, 25, 27, 29, 31\}$ and $H = U_{16}(32) = \{1, 17\}$. Then G/H is an Abelian group of order $16/2 = 8$. Which of the three Abelian groups of order 8 is it—Z_8, $Z_4 \oplus Z_2$, or $Z_2 \oplus Z_2 \oplus Z_2$? To answer this question, we need only determine the elements of G/H and their orders. Observe that the eight cosets

$$\begin{array}{llll} 1H = \{1, 17\}, & 3H = \{3, 19\}, & 5H = \{5, 21\}, & 7H = \{7, 23\}, \\ 9H = \{9, 25\}, & 11H = \{11, 27\}, & 13H = \{13, 29\}, & 15H = \{15, 31\} \end{array}$$

are all distinct, so that they form the factor group G/H. Clearly, $(3H)^2 = 9H \neq H$, and so $3H$ has order at least 4. Thus, G/H is not $Z_2 \oplus Z_2 \oplus Z_2$. On the other hand, direct computations show that both $7H$ and $9H$ have order 2, so that G/H cannot be Z_8 either, since a cyclic group of even order has exactly one element of order 2 (Theorem 4.4). This proves that $U(32)/U_{16}(32) \approx Z_4 \oplus Z_2$, which (not so incidentally!) is isomorphic to $U(16)$. ■

EXAMPLE 15 Let $G = U(32)$ and $K = \{1, 15\}$. Then $|G/K| = 8$, and we ask, which of the three Abelian groups of order 8 is G/K? Since $(3K)^4 = 81K = 17K \neq K$, $|3K| = 8$. Thus, $G/K \approx Z_8$. ∎

It is crucial to understand that when we factor out by a normal subgroup H, what we are essentially doing is defining every element in H to be the *identity*. Thus, in Example 12, we are making $R_{180}\mathcal{K} = \mathcal{K}$ the identity. Likewise, $R_{270}\mathcal{K} = R_{90}R_{180}\mathcal{K} = R_{90}\mathcal{K}$. Similarly, in Example 10, we are declaring any multiple of 4 to be 0 in the factor group $Z/4Z$. This is why $5 + 4Z = 1 + 4 + 4Z = 1 + 4Z$, and so on. In Example 14, we have $3H = 19H$, since $19 = 3 \cdot 17$ in $U(32)$ and going to the factor group makes 17 the identity. Algebraists often refer to the process of creating the factor group G/H as "killing" H.

Applications of Factor Groups

Why are factor groups important? Well, when G is finite and $H \neq \{e\}$, G/H is smaller than G, and its structure is usually less complicated than that of G. At the same time, G/H simulates G in many ways. In fact, we may think of a factor group of G as a less complicated approximation of G (similar to using the rational number 3.14 for the irrational number π). What makes factor groups important is that one can often deduce properties of G by examining the less complicated group G/H instead. We illustrate this by giving another proof that A_4 has no subgroup of order 6.

EXAMPLE 16 A_4 has no subgroup of order 6.
The group A_4 of even permutations on the set $\{1, 2, 3, 4\}$ has no subgroup H of order 6. To see this, suppose that A_4 does have a subgroup H of order 6. By Exercise 9 in this chapter, we know that $H \triangleleft A_4$. Thus, the factor group A_4/H exists and has order 2. Since the order of an element divides the order of the group, we have for all $\alpha \in A_4$ that $\alpha^2 H = (\alpha H)^2 = H$. Thus, $\alpha^2 \in H$ for all α in A_4. Referring to the main diagonal of the group table for A_4 given in Table 5.1 on page 111, however, we observe that A_4 has nine different elements of the form α^2, all of which must belong to H, a subgroup of order 6. This is clearly impossible, so a subgroup of order 6 cannot exist in A_4.† ∎

The next three theorems illustrate how knowledge of a factor group of G reveals information about G itself.

†"How often have I said to you that when you have eliminated the impossible, whatever remains, however improbable, must be the truth." Sherlock Holmes, *The Sign of Four*

Theorem 9.3 *G/Z* Theorem

Let G be a group and let Z(G) be the center of G. If G/Z(G) is cyclic, then G is Abelian.

PROOF Since G is Abelian is equivalent to $Z(G) = G$, it suffices to show that the only element of $G/Z(G)$ is the identity coset $Z(G)$. To this end, let $G/Z(G) = \langle gZ(G)\rangle$ and let $a \in G$. Then there exists an integer i such that $aZ(G) = (gZ(G))^i = g^iZ(G)$. Thus, $a = g^iz$ for some z in $Z(G)$. Since both g^i and z belong to $C(g)$, so does a. Because a is an arbitrary element of G this means that every element of G commutes with g so $g \in Z(G)$. Thus, $gZ(G) = Z(G)$ is the only element of $G/Z(G)$. ∎

A few remarks about Theorem 9.3 are in order. First, our proof shows that a better result is possible: If G/H is cyclic, where H is a subgroup of $Z(G)$, then G is Abelian. Second, in practice, it is the contrapositive of the theorem that is most often used—that is, if G is non-Abelian, then $G/Z(G)$ is not cyclic. For example, it follows immediately from this statement and Lagrange's Theorem that a non-Abelian group of order pq, where p and q are primes, must have a trivial center. Third, if $G/Z(G)$ is cyclic, it must be trivial.

Theorem 9.4 $G/Z(G) \approx \text{Inn}(G)$

For any group G, G/Z(G) is isomorphic to Inn(G).

PROOF Consider the correspondence from $G/Z(G)$ to $\text{Inn}(G)$ given by $T: gZ(G) \to \phi_g$ [where, recall, $\phi_g(x) = gxg^{-1}$ for all x in G]. First, we show that T is well defined. To do this, we assume that $gZ(G) = hZ(G)$ and verify that $\phi_g = \phi_h$. (This shows that the image of a coset of $Z(G)$ depends only on the coset itself and not on the element representing the coset.) From $gZ(G) = hZ(G)$, we have that $h^{-1}g$ belongs to $Z(G)$. Then, for all x in G, $h^{-1}gx = xh^{-1}g$. Thus, $gxg^{-1} = hxh^{-1}$ for all x in G, and, therefore, $\phi_g = \phi_h$. Reversing this argument shows that T is one-to-one, as well. Clearly, T is onto.

That T is operation-preserving follows directly from the fact that $\phi_g\phi_h = \phi_{gh}$ for all g and h in G. ∎

As an application of Theorems 9.3 and 9.4, we may easily determine $\text{Inn}(D_6)$ without looking at $\text{Inn}(D_6)$!

EXAMPLE 17 We know from Example 14 in Chapter 3 that $|Z(D_6)| = 2$. Thus, $|D_6/Z(D_6)| = 6$. So, by our classification of groups of order 6 (Theorem 7.3), we know that $\text{Inn}(D_6)$ is isomorphic to D_3 or Z_6. Now, if $\text{Inn}(D_6)$ were cyclic, then, by Theorem 9.4, $D_6/Z(D_6)$ would be also. But then, Theorem 9.3 would tell us that D_6 is Abelian. So, $\text{Inn}(D_6)$ is isomorphic to D_3. ∎

The next theorem demonstrates one of the most powerful proof techniques available in the theory of finite groups—the combined use of factor groups and induction.

Theorem 9.5 Cauchy's Theorem for Abelian Groups

Let G be a finite Abelian group and let p be a prime that divides the order of G. Then G has an element of order p.

PROOF Clearly, this statement is true for the case in which G has order 2. We prove the theorem by using the Second Principle of Mathematical Induction on $|G|$. That is, we assume that the statement is true for all Abelian groups with fewer elements than G and use this assumption to show that the statement is true for G as well. Certainly, G has elements of prime order, for if $|x| = m$ and $m = qn$, where q is prime, then $|x^n| = q$. So let x be an element of G of some prime order q, say. If $q = p$, we are finished; so assume that $q \neq p$. Since every subgroup of an Abelian group is normal, we may construct the factor group $\overline{G} = G/\langle x\rangle$. Then $\overline{G}$ is Abelian and p divides $|\overline{G}|$, since $|\overline{G}| = |G|/q$. By induction, then, $\overline{G}$ has an element—call it $y\langle x\rangle$—of order p.

Then, $(y\langle x\rangle)^p = y^p\langle x\rangle = \langle x\rangle$ and therefore $y^p \in \langle x\rangle$. If $y^p = e$, we are done. If not, then y^p has order q and y^q has order p. ∎

Internal Direct Products

As we have seen, the external direct product provides a method of putting groups together to get a larger group in such a way that we can determine many properties of the larger group from the properties of the component pieces. For example: If $G = H \oplus K$, then $|G| = |H||K|$; every element of G has the form (h, k) where $h \in H$ and $k \in K$; if $|h|$ and $|k|$ are finite, then $|(h, k)| = \text{lcm}(|h|, |k|)$; if H and K are Abelian, then G is Abelian; if H and K are cyclic and $|H|$ and $|K|$ are relatively prime, then $H \oplus K$ is cyclic. It would be quite useful to be able to reverse this process—that is, to be able to start with a large group G and break it down into a product of subgroups in such a way that we could glean many properties of G from properties of the component pieces. It is occasionally possible to do this.

Definition Internal Direct Product of H and K

We say that G is the *internal direct product* of H and K and write $G = H \times K$ if H and K are normal subgroups of G and

$$G = HK \quad \text{and} \quad H \cap K = \{e\}.$$

The wording of the phrase "internal direct product" is easy to justify. We want to call G the internal direct product of H and K if H and K are subgroups of G, and if G is naturally isomorphic to the external direct product of H and K. One forms the internal direct product by *starting* with a group G and then proceeding to find two subgroups H and K within G such that G is *isomorphic* to the external direct product of H and K. (The definition ensures that this is the case—see Theorem 9.6.) On the other hand, one forms an external direct product by *starting* with any two groups H and K, related or not, and proceeding to produce the larger group $H \oplus K$. The difference between the two products is that the internal direct product can be formed within G itself, using subgroups of G and the operation of G, whereas the external direct product can be formed with totally unrelated groups by creating a new set and a new operation. (See Figures 9.1 and 9.2.)

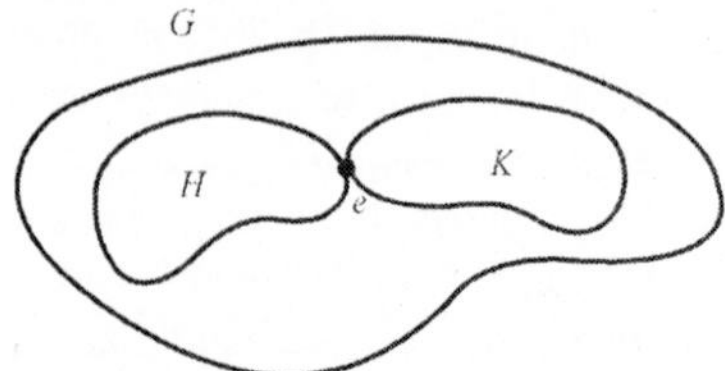

Figure 9.1 For the internal direct product, H and K must be subgroups of the same group.

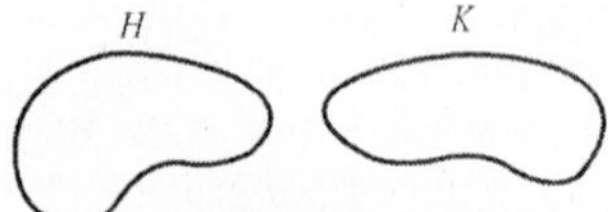

Figure 9.2 For the external direct product, H and K can be any groups.

Perhaps the following analogy with integers will be useful in clarifying the distinction between the two products of groups discussed in the preceding paragraph. Just as we may take any (finite) collection of integers and form their product, we may also take any collection of groups and form their external direct product. Conversely, just as we may start with a particular integer and express it as a product of certain of its divisors, we may be able to start with a particular group and factor it as an internal direct product of certain of its subgroups.

EXAMPLE 18 In D_6, the dihedral group of order 12, let F denote some reflection and let R_k denote a rotation of k degrees. Then,

$$D_6 = \{R_0, R_{120}, R_{240}, F, R_{120}F, R_{240}F\} \times \{R_0, R_{180}\}.$$

Students should be cautioned about the necessity of having all conditions of the definition of internal direct product satisfied to ensure that $HK \approx H \oplus K$. For example, if we take

$$G = S_3, \qquad H = \langle(123)\rangle, \qquad \text{and} \qquad K = \langle(12)\rangle,$$

then $G = HK$, and $H \cap K = \{(1)\}$. But G is *not* isomorphic to $H \oplus K$, since, by Theorem 8.2, $H \oplus K$ is cyclic, whereas S_3 is not. Note that K is not normal.

A group G can also be the internal direct product of a collection of subgroups.

Definition Internal Direct Product $H_1 \times H_2 \times \cdots \times H_n$

Let $H_1, H_2, \ldots, H_n$ be a finite collection of normal subgroups of G. We say that G is the *internal direct product* of $H_1, H_2, \ldots, H_n$ and write $G = H_1 \times H_2 \times \cdots \times H_n$, if

1. $G = H_1H_2 \cdots H_n = \{h_1h_2 \cdots h_n \mid h_i \in H_i\}$,
2. $(H_1H_2 \cdots H_i) \cap H_{i+1} = \{e\}$ for $i = 1, 2, \ldots, n - 1$.

This definition is somewhat more complicated than the one given for two subgroups. The student may wonder about the motivation for it—that is, why should we want the subgroups to be normal and why is it desirable for each subgroup to be disjoint from the product of all previous ones? The reason is quite simple. We want the internal direct product to be isomorphic to the external direct product. As the next theorem shows, the conditions in the definition of internal direct product were chosen to ensure that the two products are isomorphic.

Theorem 9.6 $H_1 \times H_2 \times \cdots \times H_n \approx H_1 \oplus H_2 \oplus \cdots \oplus H_n$

> ***If a group G is the internal direct product of a finite number of subgroups $H_1, H_2, \ldots, H_n$, then G is isomorphic to the external direct product of $H_1, H_2, \ldots, H_n$.***

PROOF We first show that the normality of the H's together with the second condition of the definition guarantees that h's from different H_i's commute. For if $h_i \in H_i$ and $h_j \in H_j$ with $i \neq j$, then

$$(h_i h_j h_i^{-1}) h_j^{-1} \in H_j h_j^{-1} = H_j$$

and

$$h_i (h_j h_i^{-1} h_j^{-1}) \in h_i H_i = H_i.$$

Thus, $h_i h_j h_i^{-1} h_j^{-1} \in H_i \cap H_j = \{e\}$ (see Exercise 5), and, therefore, $h_i h_j = h_j h_i$. We next claim that each member of G can be expressed uniquely in the form $h_1 h_2 \cdots h_n$, where $h_i \in H_i$. That there is at least one such representation is the content of condition 1 of the definition. To prove uniqueness, suppose that $g = h_1 h_2 \cdots h_n$ and $g = h_1' h_2' \cdots h_n'$, where h_i and h_i' belong to H_i for $i = 1, \ldots, n$. Then, using the fact that the h's from different H_i's commute, we can solve the equation

$$h_1 h_2 \cdots h_n = h_1' h_2' \cdots h_n' \qquad (1)$$

for $h_n' h_n^{-1}$ to obtain

$$h_n' h_n^{-1} = (h_1')^{-1} h_1 (h_2')^{-1} h_2 \cdots (h_{n-1}')^{-1} h_{n-1}.$$

But then

$$h_n' h_n^{-1} \in H_1 H_2 \cdots H_{n-1} \cap H_n = \{e\},$$

so that $h_n' h_n^{-1} = e$ and, therefore, $h_n' = h_n$. At this point, we can cancel h_n and h_n' from opposite sides of the equal sign in Equation (1) and repeat the preceding argument to obtain $h_{n-1} = h_{n-1}'$. Continuing in this fashion, we eventually have $h_i = h_i'$ for $i = 1, \ldots, n$. With our claim established, we may now define a function ϕ from G to $H_1 \oplus H_2 \oplus \cdots \oplus H_n$ by $\phi(h_1 h_2 \cdots h_n) = (h_1, h_2, \ldots, h_n)$. We leave to the reader the easy verification that ϕ is an isomorphism. ∎

The next theorem provides an important application of Theorem 9.6.

Theorem 9.7 Classification of Groups of Order p^2

Every group of order p^2, where p is a prime, is isomorphic to Z_{p^2} or $Z_p \oplus Z_p$.

PROOF Let G be a group of order p^2, where p is a prime. If G has an element of order p^2, then G is isomorphic to Z_{p^2}. So, by Corollary 2 of Lagrange's Theorem, we may assume that every nonidentity element of G has order p. First we show that for any element a, the subgroup $\langle a \rangle$ is normal in G. If this is not the case, then there is an element b in G such that bab^{-1} is not in $\langle a \rangle$. Then $\langle a \rangle$ and $\langle bab^{-1} \rangle$ are distinct subgroups of order p. Since $\langle a \rangle \cap \langle bab^{-1} \rangle$ is a subgroup of both $\langle a \rangle$ and $\langle bab^{-1} \rangle$, we have that $\langle a \rangle \cap \langle bab^{-1} \rangle = \{e\}$. From this it follows that the distinct left cosets of $\langle bab^{-1} \rangle$ are $\langle bab^{-1} \rangle$, $a\langle bab^{-1} \rangle$, $a^2\langle bab^{-1} \rangle, \ldots,$ $a^{p-1}\langle bab^{-1} \rangle$. Since b^{-1} must lie in one of these cosets, we may write b^{-1} in the form $b^{-1} = a^i(bab^{-1})^j = a^iba^jb^{-1}$ for some i and j. Canceling the b^{-1} terms, we obtain $e = a^iba^j$ and therefore $b = a^{-i-j} \in \langle a \rangle$. This contradiction verifies our assertion that every subgroup of the form $\langle a \rangle$ is normal in G. To complete the proof, let x be any nonidentity element in G and y be any element of G not in $\langle x \rangle$. Then, by comparing orders and using Theorem 9.6, we see that $G = \langle x \rangle \times \langle y \rangle \approx Z_p \oplus Z_p$. ∎

As an immediate corollary of Theorem 9.7, we have the following important fact.

Corollary

If G is a group of order p^2, where p is a prime, then G is Abelian.

We mention in passing that if $G = H_1 \oplus H_2 \oplus \cdots \oplus H_n$, then G can be expressed as the internal direct product of subgroups isomorphic to $H_1, H_2, \ldots, H_n$. For example, if $G = H_1 \oplus H_2$, then $G = \overline{H_1} \times \overline{H_2}$, where $\overline{H_1} = H_1 \oplus \{e\}$ and $\overline{H_2} = \{e\} \oplus H_2$.

The topic of direct products is one in which notation and terminology vary widely. Many authors use $H \times K$ to denote both the internal direct product and the external direct product of H and K, making no notational distinction between the two products. A few authors define only the external direct product. Many people reserve the notation $H \oplus K$ for the situation where H and K are Abelian groups under addition and call it the *direct sum* of H and K. In fact, we will adopt this terminology in the section on rings (Part 3), since rings are always Abelian groups under addition.

The U-groups provide a convenient way to illustrate the preceding ideas and to clarify the distinction between internal and external direct products. It follows directly from Theorem 8.3, its corollary, and Theorem 9.6 that if $m = n_1 n_2 \cdots n_k$, where $\gcd(n_i, n_j) = 1$ for $i \neq j$, then

$$\begin{aligned} U(m) &= U_{m/n_1}(m) \times U_{m/n_2}(m) \times \cdots \times U_{m/n_k}(m) \\ &\approx U(n_1) \oplus U(n_2) \oplus \cdots \oplus U(n_k). \end{aligned}$$

Let us return to the examples given following Theorem 8.3.

$$\begin{aligned} U(105) &= U(15 \cdot 7) = U_{15}(105) \times U_7(105) \\ &= \{1, 16, 31, 46, 61, 76\} \times \{1, 8, 22, 29, 43, 64, 71, 92\} \\ &\approx U(7) \oplus U(15), \\ U(105) &= U(5 \cdot 21) = U_5(105) \times U_{21}(105) \\ &= \{1, 11, 16, 26, 31, 41, 46, 61, 71, 76, 86, 101\} \\ &\quad \times \{1, 22, 43, 64\} \approx U(21) \oplus U(5), \\ U(105) &= U(3 \cdot 5 \cdot 7) = U_{35}(105) \times U_{21}(105) \times U_{15}(105) \\ &= \{1, 71\} \times \{1, 22, 43, 64\} \times \{1, 16, 31, 46, 61, 76\} \\ &\approx U(3) \oplus U(5) \oplus U(7). \end{aligned}$$

Exercises

The heart of mathematics is its problems.

Paul Halmos

1. Let $H = \{(1), (12)\}$. Is H normal in S_3?
2. Prove that A_n is normal in S_n.
3. In D_4, let $K = \{R_0, R_{90}, R_{180}, R_{270}\}$. Write HR_{90} in the form xH, where $x \in K$. Write DR_{270} in the form xD, where $x \in K$. Write $R_{90}V$ in the form Vx, where $x \in K$.
4. Write $(12)(13)(14)$ in the form $\alpha(12)$, where $\alpha \in A_4$. Write $(1234)(12)(23)$, in the form $\alpha(1234)$, where $\alpha \in A_4$.
5. Show that if G is the internal direct product of $H_1, H_2, \ldots, H_n$ and $i \neq j$ with $1 \leq i \leq n$, $1 \leq j \leq n$, then $H_i \cap H_j = \{e\}$. (This exercise is referred to in this chapter.)
6. Let $H = \left\{ \begin{bmatrix} a & b \\ 0 & d \end{bmatrix} \middle| \, a, b, d \in \mathbf{R}, ad \neq 0 \right\}$. Is H a normal subgroup of $GL(2, \mathbf{R})$?
7. Let $G = GL(2, \mathbf{R})$ and let K be a subgroup of $\mathbf{R}^*$. Prove that $H = \{A \in G \mid \det A \in K\}$ is a normal subgroup of G.
8. Viewing $\langle 3 \rangle$ and $\langle 12 \rangle$ as subgroups of Z, prove that $\langle 3 \rangle/\langle 12 \rangle$ is isomorphic to Z_4. Similarly, prove that $\langle 8 \rangle/\langle 48 \rangle$ is isomorphic to Z_6. Generalize to arbitrary integers k and n.

9. Prove that if H has index 2 in G, then H is normal in G. (This exercise is referred to in Chapters 24 and 25 and this chapter.)
10. Let $H = \{(1), (12)(34)\}$ in A_4.
    a. Show that H is not normal in A_4.
    b. Referring to the multiplication table for A_4 in Table 5.1 on page 111, show that, although $\alpha_6 H = \alpha_7 H$ and $\alpha_9 H = \alpha_{11} H$, it is not true that $\alpha_6 \alpha_9 H = \alpha_7 \alpha_{11} H$. Explain why this proves that the left cosets of H do not form a group under coset multiplication.
11. Let $G = Z_4 \oplus U(4)$, $H = \langle(2, 3)\rangle$, and $K = \langle(2, 1)\rangle$. Show that G/H is not isomorphic to G/K. (This shows that $H \approx K$ does not imply that $G/H \approx G/K$.)
12. Prove that a factor group of a cyclic group is cyclic.
13. Prove that a factor group of an Abelian group is Abelian.
14. What is the order of the element $14 + \langle 8 \rangle$ in the factor group $Z_{24}/\langle 8 \rangle$?
15. What is the order of the element $4U_5(105)$ in the factor group $U(105)/U_5(105)$?
16. Recall that $Z(D_6) = \{R_0, R_{180}\}$. What is the order of the element $R_{60}Z(D_6)$ in the factor group $D_6/Z(D_6)$?
17. Let $G = Z/\langle 20 \rangle$ and $H = \langle 4 \rangle/\langle 20 \rangle$. List the elements of H and G/H.
18. What is the order of the factor group $Z_{60}/\langle 15 \rangle$?
19. What is the order of the factor group $(Z_{10} \oplus U(10))/\langle(2, 9)\rangle$?
20. Construct the Cayley table for $U(20)/U_5(20)$.
21. Prove that an Abelian group of order 33 is cyclic.
22. Determine the order of $(Z \oplus Z)/\langle(2, 2)\rangle$. Is the group cyclic?
23. Determine the order of $(Z \oplus Z)/\langle(4, 2)\rangle$. Is the group cyclic?
24. The group $(Z_4 \oplus Z_{12})/\langle(2, 2)\rangle$ is isomorphic to one of Z_8, $Z_4 \oplus Z_2$, or $Z_2 \oplus Z_2 \oplus Z_2$. Determine which one by elimination.
25. Let $G = U(32)$ and $H = \{1, 31\}$. The group G/H is isomorphic to one of Z_8, $Z_4 \oplus Z_2$, or $Z_2 \oplus Z_2 \oplus Z_2$. Determine which one by elimination.
26. Let G be the group of quaternions given by the table in Exercise 4 of the Supplementary Exercises for Chapters 1–4, and let H be the subgroup $\{e, a^2\}$. Is G/H isomorphic to Z_4 or $Z_2 \oplus Z_2$?
27. Let $G = U(16)$, $H = \{1, 15\}$, and $K = \{1, 9\}$. Are H and K isomorphic? Are G/H and G/K isomorphic?
28. Let $G = Z_4 \oplus Z_4$, $H = \{(0, 0), (2, 0), (0, 2), (2, 2)\}$, and $K = \langle(1, 2)\rangle$. Is G/H isomorphic to Z_4 or $Z_2 \oplus Z_2$? Is G/K isomorphic to Z_4 or $Z_2 \oplus Z_2$?
29. Prove that $A_4 \oplus Z_3$ has no subgroup of order 18.

30. Express $U(165)$ as an internal direct product of proper subgroups in four different ways.
31. Let $\mathbf{R}^*$ denote the group of all nonzero real numbers under multiplication. Let $\mathbf{R}^+$ denote the group of positive real numbers under multiplication. Prove that $\mathbf{R}^*$ is the internal direct product of $\mathbf{R}^+$ and the subgroup $\{1, -1\}$.
32. Prove that D_4 cannot be expressed as an internal direct product of two proper subgroups.
33. Let H and K be subgroups of a group G. If $G = HK$ and $g = hk$, where $h \in H$ and $k \in K$, is there any relationship among $|g|$, $|h|$, and $|k|$? What if $G = H \times K$?
34. In Z, let $H = \langle 5 \rangle$ and $K = \langle 7 \rangle$. Prove that $Z = HK$. Does $Z = H \times K$?
35. Let $G = \{3^a 6^b 10^c \mid a, b, c \in Z\}$ under multiplication and $H = \{3^a 6^b 12^c \mid a, b, c \in Z\}$ under multiplication. Prove that $G = \langle 3 \rangle \times \langle 6 \rangle \times \langle 10 \rangle$, whereas $H \neq \langle 3 \rangle \times \langle 6 \rangle \times \langle 12 \rangle$.
36. Determine all subgroups of $\mathbf{R}^*$ (nonzero reals under multiplication) of index 2.
37. Let G be a finite group and let H be a normal subgroup of G. Prove that the order of the element gH in G/H must divide the order of g in G.
38. Let H be a normal subgroup of G and let a belong to G. If the element aH has order 3 in the group G/H and $|H| = 10$, what are the possibilities for the order of a?
39. If H is a normal subgroup of a group G, prove that $C(H)$, the centralizer of H in G, is a normal subgroup of G.
40. Let ϕ be an isomorphism from a group G onto a group $\overline{G}$. Prove that if H is a normal subgroup of G, then $\phi(H)$ is a normal subgroup of $\overline{G}$.
41. Show that Q, the group of rational numbers under addition, has no proper subgroup of finite index.
42. An element is called a *square* if it can be expressed in the form b^2 for some b. Suppose that G is an Abelian group and H is a subgroup of G. If every element of H is a square and every element of G/H is a square, prove that every element of G is a square. Does your proof remain valid when "square" is replaced by "nth power," where n is any integer?
43. Show, by example, that in a factor group G/H it can happen that $aH = bH$ but $|a| \neq |b|$.
44. Observe from the table for A_4 given in Table 5.1 on page 111 that the subgroup given in Example 9 of this chapter is the only subgroup of A_4 of order 4. Why does this imply that this subgroup must be normal in A_4? Generalize this to arbitrary finite groups.

45. Let p be a prime. Show that if H is a subgroup of a group of order $2p$ that is not normal, then H has order 2.
46. Show that D_{13} is isomorphic to $\text{Inn}(D_{13})$.
47. Suppose that N is a normal subgroup of a finite group G and H is a subgroup of G. If $|G/N|$ is prime, prove that H is contained in N or that $NH = G$.
48. If G is a group and $|G: Z(G)| = 4$, prove that $G/Z(G) \approx Z_2 \oplus Z_2$.
49. Suppose that G is a non-Abelian group of order p^3, where p is a prime, and $Z(G) \neq \{e\}$. Prove that $|Z(G)| = p$.
50. If $|G| = pq$, where p and q are primes that are not necessarily distinct, prove that $|Z(G)| = 1$ or pq.
51. Let N be a normal subgroup of G and let H be a subgroup of G. If N is a subgroup of H, prove that H/N is a normal subgroup of G/N if and only if H is a normal subgroup of G.
52. Let G be an Abelian group and let H be the subgroup consisting of all elements of G that have finite order. (See Exercise 20 in the Supplementary Exercises for Chapters 1–4.) Prove that every nonidentity element in G/H has infinite order.
53. Determine all subgroups of $\mathbf{R}^*$ that have finite index.
54. Let $G = \{\pm 1, \pm i, \pm j, \pm k\}$, where $i^2 = j^2 = k^2 = -1$, $-i = (-1)i$, $1^2 = (-1)^2 = 1$, $ij = -ji = k$, $jk = -kj = i$, and $ki = -ik = j$.
    a. Construct the Cayley table for G.
    b. Show that $H = \{1, -1\} \lhd G$.
    c. Construct the Cayley table for G/H. Is G/H isomorphic to Z_4 or $Z_2 \oplus Z_2$?

    (The rules involving i, j, and k can be remembered by using the circle below.

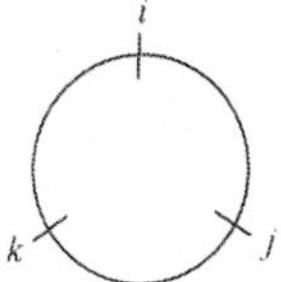

Going clockwise, the product of two consecutive elements is the third one. The same is true for going counterclockwise, except that we obtain the negative of the third element.) This is the group of quaternions that was given in another form in Exercise 4 in the Supplementary Exercises for Chapters 1–4. It was invented by William Hamilton in 1843. The quaternions are used to describe rotations in three-dimensional space, and they are used in physics. The quaternions can be used to extend the complex numbers in a natural way.

55. In D_4, let $K = \{R_0, D\}$ and let $L = \{R_0, D, D', R_{180}\}$. Show that $K \triangleleft L \triangleleft D_4$, but that K is not normal in D_4. (Normality is not transitive. Compare Exercise 4, Supplementary Exercises for Chapters 5–8.)

56. Show that the intersection of two normal subgroups of G is a normal subgroup of G. Generalize.

57. Give an example of subgroups H and K of a group G such that HK is not a subgroup of G.

58. If N and M are normal subgroups of G, prove that NM is also a normal subgroup of G.

59. Let N be a normal subgroup of a group G. If N is cyclic, prove that every subgroup of N is also normal in G. (This exercise is referred to in Chapter 24.)

60. Without looking at inner automorphisms of D_n, determine the number of such automorphisms.

61. Let H be a normal subgroup of a finite group G and let $x \in G$. If $\gcd(|x|, |G/H|) = 1$, show that $x \in H$. (This exercise is referred to in Chapter 25.)

62. Let G be a group and let G' be the subgroup of G generated by the set $S = \{x^{-1}y^{-1}xy \mid x, y \in G\}$. (See Exercise 3, Supplementary Exercises for Chapters 5–8, for a more complete description of G'.)

a. Prove that G' is normal in G.

b. Prove that G/G' is Abelian.

c. If G/N is Abelian, prove that $G' \leq N$.

d. Prove that if H is a subgroup of G and $G' \leq H$, then H is normal in G.

63. If N is a normal subgroup of G and $|G/N| = m$, show that $x^m \in N$ for all x in G.

64. Suppose that a group G has a subgroup of order n. Prove that the intersection of all subgroups of G of order n is a normal subgroup of G.

65. If G is non-Abelian, show that Aut(G) is not cyclic.

66. Let $|G| = p^n m$, where p is prime and $\gcd(p, m) = 1$. Suppose that H is a normal subgroup of G of order p^n. If K is a subgroup of G of order p^k, show that $K \subseteq H$.

67. Suppose that H is a normal subgroup of a finite group G. If G/H has an element of order n, show that G has an element of order n. Show, by example, that the assumption that G is finite is necessary.

68. Recall that a subgroup N of a group G is called characteristic if $\phi(N) = N$ for all automorphisms ϕ of G. If N is a characteristic subgroup of G, show that N is a normal subgroup of G.

69. In D_4, let $\mathcal{K} = \{R_0, H\}$. Form an operation table for the cosets $\mathcal{K}$, $D\mathcal{K}$, $V\mathcal{K}$, and $D'\mathcal{K}$. Is the result a group table? Does your answer contradict Theorem 9.2?

70. Prove that A_4 is the only subgroup of S_4 of order 12.

71. If $|G| = 30$ and $|Z(G)| = 5$, what is the structure of $G/Z(G)$?

72. If H is a normal subgroup of G and $|H| = 2$, prove that H is contained in the center of G.

73. Prove that A_5 cannot have a normal subgroup of order 2.

74. Let G be a finite group and let H be an odd-order subgroup of G of index 2. Show that the product of all the elements of G (taken in any order) cannot belong to H.

75. Let G be a group and p a prime. Suppose that $H = \{g^p \mid g \in G\}$ is a subgroup of G. Show that H is normal and that every nonidentity element of G/H has order p.

76. Suppose that H is a normal subgroup of G. If $|H| = 4$ and gH has order 3 in G/H, find a subgroup of order 12 in G.

77. Let G be a group and H an odd-order subgroup of G of index 2. Show that H contains every element of G of odd order.

78. A proper subgroup H of a group G is called *maximal* if there is no subgroup K such that $H \subset K \subset G$ (that is, there is no subgroup K properly contained between H and G). Show that $Z(G)$ is never a maximal subgroup of a group G.

79. Let G be a group of order 100 that has exactly one subgroup of order 5. Prove that it has a subgroup of order 10.

Suggested Readings

Michael Brennan and Des MacHale, "Variations on a Theme: A_4 Definitely Has No Subgroup of Order Six!," *Mathematics Magazine* 73 (2000): 36–40.

The authors offer 11 proofs that A_4 has no subgroup of order 6. These proofs provide a review of many of the ideas covered thus far in this text.

J. A. Gallian, R. S. Johnson, and S. Peng, "On Quotient Structures of Z^n," *Pi Mu Epsilon Journal* 9 (1993): 524–526.

The authors determine the structure of the group $(Z \oplus Z)/\langle(a, b)\rangle$ and related groups.

Tony Rothman, "Genius and Biographers: The Fictionalization of Évariste Galois," *The American Mathematical Monthly* 89 (1982): 84–106.

The author argues that many popular accounts of Galois's life have been greatly embroidered.

Paul F. Zweifel, "Generalized Diatonic and Pentatonic Scales: A Group-theoretic Approach," *Perspectives of New Music* 34 (1996): 140–161.

The author discusses how group theoretic notions such as subgroups, cosets, factor groups, and isomorphisms of Z_{12} and Z_{20} relate to musical scales, tuning, temperament, and structure.

Évariste Galois

Galois at seventeen was making discoveries of epochal significance in the theory of equations, discoveries whose consequences are not yet exhausted after more than a century.

E. T. BELL, *Men of Mathematics*

The Granger Collection, New York

This French stamp was issued as part of the 1984 "Celebrity Series" in support of the Red Cross Fund.

ÉVARISTE GALOIS (pronounced gal-WAH) was born on October 25, 1811, near Paris. Although he had mastered the works of Legendre and Lagrange at age 15, Galois twice failed his entrance examination to the École Polytechnique. He did not know some basic mathematics, and he did mathematics almost entirely in his head, to the annoyance of the examiner.

At 18, Galois wrote his important research on the theory of equations and submitted it to the French Academy of Sciences for publication. The paper was given to Cauchy for refereeing. Cauchy, impressed by the paper, agreed to present it to the academy, but he never did. At the age of 19, Galois entered a paper of the highest quality in the competition for the Grand Prize in Mathematics, given by the French Academy of Sciences. The paper was given to Fourier, who died shortly thereafter. Galois's paper was never seen again.

Galois spent most of the last year and a half of his life in prison for revolutionary political offenses. While in prison, he attempted suicide and prophesied that he would die in a duel. On May 30, 1832, Galois was shot in a duel; he died the next day at the age of 20.

Among the many concepts introduced by Galois are normal subgroups, isomorphisms, simple groups, finite fields, and Galois theory. His work provided a method for disposing of several famous constructability problems, such as trisecting an arbitrary angle and doubling a cube. Galois's entire collected works fill only 60 pages.

To find more information about Galois, visit:

http://www-groups.dcs.st-and.ac.uk/~history/

10 Group Homomorphisms

When it comes to laws, there is absolutely no doubt that symmetry and group theory are extremely *useful* concepts. Without the introduction of symmetry and the language of groups into particle physics the description of the elementary particles and their interactions would have been an intricate nightmare. Groups truly flesh out order and identify patterns like no other mathematical machinery.

MARIO LIVIO, *The Equation That Couldn't be Solved*

Definition and Examples

In this chapter, we consider one of the most fundamental ideas of algebra—homomorphisms. The term *homomorphism* comes from the Greek words *homo*, "like," and *morphe,* "form." We will see that a homomorphism is a natural generalization of an isomorphism and that there is an intimate connection between factor groups of a group and homomorphisms of a group. The concept of group homomorphisms was introduced by Camille Jordan in 1870, in his influential book *Traité des substitutions*.

Definition Group Homomorphism
A *homomorphism* ϕ from a group G to a group $\overline{G}$ is a mapping from G into $\overline{G}$ that preserves the group operation; that is, $\phi(ab) = \phi(a)\phi(b)$ for all a, b in G.

Before giving examples and stating numerous properties of homomorphisms, it is convenient to introduce an important subgroup that is intimately related to the image of a homomorphism. (See property 4 of Theorem 10.1.)

Definition Kernel of a Homomorphism
The *kernel* of a homomorphism ϕ from a group G to a group with identity e is the set $\{x \in G \mid \phi(x) = e\}$. The kernel of ϕ is denoted by Ker ϕ.

EXAMPLE 1 Any isomorphism is a homomorphism that is also onto and one-to-one. The kernel of an isomorphism is the trivial subgroup. ∎

EXAMPLE 2 Let $\mathbf{R}^*$ be the group of nonzero real numbers under multiplication. Then the determinant mapping $A \to \det A$ is a homomorphism from $GL(2, \mathbf{R})$ to $\mathbf{R}^*$. The kernel of the determinant mapping is $SL(2, \mathbf{R})$. ∎

EXAMPLE 3 The mapping ϕ from $\mathbf{R}^*$ to $\mathbf{R}^*$, defined by $\phi(x) = |x|$, is a homomorphism with Ker $\phi = \{1, -1\}$. ∎

EXAMPLE 4 Let $\mathbf{R}[x]$ denote the group of all polynomials with real coefficients under addition. For any f in $\mathbf{R}[x]$, let f' denote the derivative of f. Then the mapping $f \to f'$ is a homomorphism from $\mathbf{R}[x]$ to itself. The kernel of the derivative mapping is the set of all constant polynomials. ∎

EXAMPLE 5 The mapping ϕ from Z to Z_n, defined by $\phi(m) = m$ mod n, is a homomorphism (see Exercise 9 in Chapter 0). The kernel of this mapping is $\langle n \rangle$. ∎

EXAMPLE 6 The mapping $\phi(x) = x^2$ from $\mathbf{R}^*$, the nonzero real numbers under multiplication, to itself is a homomorphism, since $\phi(ab) = (ab)^2 = a^2b^2 = \phi(a)\phi(b)$ for all a and b in $\mathbf{R}^*$. (See Exercise 5.) The kernel is $\{1, -1\}$. ∎

EXAMPLE 7 The mapping $\phi(x) = x^2$ from $\mathbf{R}$, the real numbers under addition, to itself is not a homomorphism, since $\phi(a + b) = (a + b)^2 = a^2 + 2ab + b^2$, whereas $\phi(a) + \phi(b) = a^2 + b^2$. ∎

When defining a homomorphism from a group in which there are several ways to represent the elements, caution must be exercised to ensure that the correspondence is a function. (The term *well-defined* is often used in this context.) For example, since $3(x + y) = 3x + 3y$ in Z_6, one might believe that the correspondence $x + \langle 3 \rangle \to 3x$ from $Z/\langle 3 \rangle$ to Z_6 is a homomorphism. But it is not a function, since $0 + \langle 3 \rangle = 3 + \langle 3 \rangle$ in $Z/\langle 3 \rangle$ but $3 \cdot 0 \neq 3 \cdot 3$ in Z_6.

For students who have had linear algebra, we remark that every linear transformation is a group homomorphism and the null-space is the same as the kernel. An invertible linear transformation is a group isomorphism.

Properties of Homomorphisms

Theorem 10.1 Properties of Elements Under Homomorphisms

Let ϕ be a homomorphism from a group G to a group $\overline{G}$ and let g be an element of G. Then

1. *ϕ carries the identity of G to the identity of $\overline{G}$.*
2. *$\phi(g^n) = (\phi(g))^n$ for all n in Z.*
3. *If $|g|$ is finite, then $|\phi(g)|$ divides $|g|$.*
4. *Ker ϕ is a subgroup of G.*
5. *$\phi(a) = \phi(b)$ if and only if $a\text{Ker }\phi = b\text{Ker }\phi$.*
6. *If $\phi(g) = g'$, then $\phi^{-1}(g') = \{x \in G \mid \phi(x) = g'\} = g\text{Ker }\phi$.*

PROOF The proofs of properties 1 and 2 are identical to the proofs of properties 1 and 2 of isomorphisms in Theorem 6.2. To prove property 3, notice that properties 1 and 2 together with $g^n = e$ imply that $e = \phi(e) = \phi(g^n) = (\phi(g))^n$. So, by Corollary 2 to Theorem 4.1, we have $|\phi(g)|$ divides n.

By property 1 we know that Ker ϕ is not empty. So, to prove property 4, we assume that $a, b \in$ Ker ϕ and show that $ab^{-1} \in$ Ker ϕ. Since $\phi(a) = e$ and $\phi(b) = e$, we have $\phi(ab^{-1}) = \phi(a)\phi(b^{-1}) = \phi(a)(\phi(b))^{-1} = ee^{-1} = e$. So, $ab^{-1} \in$ Ker ϕ.

To prove property 5, first assume that $\phi(a) = \phi(b)$. Then $e = (\phi(b))^{-1}\phi(a) = \phi(b^{-1})\phi(a) = \phi(b^{-1}a)$, so that $b^{-1}a \in$ Ker ϕ. It now follows from property 6 of the lemma in Chapter 7 that $b\text{Ker }\phi = a\text{Ker }\phi$. Reversing this argument completes the proof.

To prove property 6, we must show that $\phi^{-1}(g') \subseteq g\text{Ker }\phi$ and that $g\text{Ker }\phi \subseteq \phi^{-1}(g')$. For the first inclusion, let $x \in \phi^{-1}(g')$, so that $\phi(x) = g'$. Then $\phi(g) = \phi(x)$ and by property 5 we have $g\text{Ker }\phi = x\text{Ker }\phi$ and therefore $x \in g\text{Ker }\phi$. This completes the proof that $\phi^{-1}(g') \subseteq g\text{Ker }\phi$. To prove that $g\text{Ker }\phi \subseteq \phi^{-1}(g')$, suppose that $k \in$ Ker ϕ. Then $\phi(gk) = \phi(g)\phi(k) = g'e = g'$. Thus, by definition, $gk \in \phi^{-1}(g')$. ∎

Since homomorphisms preserve the group operation, it should not be a surprise that they preserve many group properties.

Theorem 10.2 Properties of Subgroups Under Homomorphisms

Let ϕ be a homomorphism from a group G to a group $\overline{G}$ and let H be a subgroup of G. Then

1. *$\phi(H) = \{\phi(h) \mid h \in H\}$ is a subgroup of $\overline{G}$.*
2. *If H is cyclic, then $\phi(H)$ is cyclic.*
3. *If H is Abelian, then $\phi(H)$ is Abelian.*
4. *If H is normal in G, then $\phi(H)$ is normal in $\phi(G)$.*
5. *If $|\text{Ker } \phi| = n$, then ϕ is an n-to-1 mapping from G onto $\phi(G)$.*
6. *If $|H| = n$, then $|\phi(H)|$ divides n.*
7. *If $\overline{K}$ is a subgroup of $\overline{G}$, then $\phi^{-1}(\overline{K}) = \{k \in G \mid \phi(k) \in \overline{K}\}$ is a subgroup of G.*
8. *If $\overline{K}$ is a normal subgroup of $\overline{G}$, then $\phi^{-1}(\overline{K}) = \{k \in G \mid \phi(k) \in \overline{K}\}$ is a normal subgroup of G.*
9. *If ϕ is onto and $\text{Ker } \phi = \{e\}$, then ϕ is an isomorphism from G to $\overline{G}$.*

PROOF First note that the proofs of properties 1, 2, and 3 are identical to the proofs of properties 4, 3, and 2, respectively, of Theorem 6.3, since those proofs use only the fact that an isomorphism is an operation-preserving mapping.

To prove property 4, let $\phi(h) \in \phi(H)$ and $\phi(g) \in \phi(G)$. Then $\phi(g)\phi(h)\phi(g)^{-1} = \phi(ghg^{-1}) \in \phi(H)$, since H is normal in G.

Property 5 follows directly from property 6 of Theorem 10.1 and the fact that all cosets of $\text{Ker } \phi = \phi^{-1}(e)$ have the same number of elements.

To prove property 6, let ϕ_H denote the restriction of ϕ to the elements of H. Then ϕ_H is a homomorphism from H onto $\phi(H)$. Suppose $|\text{Ker } \phi_H| = t$. Then, by property 5, ϕ_H is a t-to-1 mapping. So, $|\phi(H)|t = |H|$.

To prove property 7, we use the One-Step Subgroup Test. Clearly, $e \in \phi^{-1}(\overline{K})$, so that $\phi^{-1}(\overline{K})$ is not empty. Let $k_1, k_2 \in \phi^{-1}(\overline{K})$. Then, by the definition of $\phi^{-1}(\overline{K})$, we know that $\phi(k_1), \phi(k_2) \in \overline{K}$. Thus, $\phi(k_2)^{-1} \in \overline{K}$ as well and $\phi(k_1k_2^{-1}) = \phi(k_1)\phi(k_2)^{-1} \in \overline{K}$. So, by the definition of $\phi^{-1}(\overline{K})$, we have $k_1k_2^{-1} \in \phi^{-1}(\overline{K})$.

To prove property 8, we use the normality test given in Theorem 9.1. Note that every element in $x\phi^{-1}(\overline{K})x^{-1}$ has the form xkx^{-1}, where $\phi(k) \in \overline{K}$. Thus, since $\overline{K}$ is normal in $\overline{G}$, $\phi(xkx^{-1}) = \phi(x)\phi(k)(\phi(x))^{-1} \in \overline{K}$, and, therefore, $xkx^{-1} \in \phi^{-1}(\overline{K})$.

Finally, property 9 follows directly from property 5. ■

A few remarks about Theorems 10.1 and 10.2 are in order. Students should remember the various properties of these theorems in words. For example, properties 2 and 3 of Theorem 10.2 say that the homomorphic image of a cyclic group is cyclic and the homomorphic image of an Abelian group is Abelian. Property 4 of Theorem 10.2 says that the homomorphic image of a normal subgroup of G is normal in the image of G. Property 5 of Theorem 10.2 says that if ϕ is a homomorphism from G to $\overline{G}$, then every element of $\overline{G}$ that gets "hit" by ϕ gets hit the same number of times as does the identity. The set $\phi^{-1}(g')$ defined in property 6 of Theorem 10.1 is called the *inverse image of* g' (or the *pullback* of g'). Note that the inverse image of an element is a coset of the kernel and that every element in that coset has the same image. Similarly, the set $\phi^{-1}(\overline{K})$ defined in property 7 of Theorem 10.2 is called the *inverse image* of $\overline{K}$ (or the *pullback* of $\overline{K}$).

Property 6 of Theorem 10.1 is reminiscent of something from linear algebra and differential equations. Recall that if x is a particular solution to a system of linear equations and S is the entire solution set of the corresponding homogeneous system of linear equations, then $x + S$ is the entire solution set of the nonhomogeneous system. In reality, this statement is just a special case of property 6. Properties 1 and 6 of Theorem 10.1 and property 5 of Theorem 10.2 are pictorially represented in Figure 10.1.

The special case of property 8 of Theorem 10.2, where $\overline{K} = \{e\}$, is of such importance that we single it out.

Corollary Kernels Are Normal

Let ϕ be a group homomorphism from G to $\overline{G}$. Then Ker ϕ is a normal subgroup of G.

The next two examples illustrate several properties of Theorems 10.1 and 10.2.

EXAMPLE 8 Consider the mapping ϕ from $\mathbf{C}^*$ to $\mathbf{C}^*$ given by $\phi(x) = x^4$. Since $(xy)^4 = x^4y^4$, ϕ is a homomorphism. Clearly, Ker $\phi = \{x \mid x^4 = 1\} = \{1, -1, i, -i\}$. So, by property 5 of Theorem 10.2, we know that ϕ is a 4-to-1 mapping. Now let's find all elements that map to, say, 2. Certainly, $\phi(\sqrt[4]{2}) = 2$. Then, by property 6 of Theorem 10.1, the set of all elements that map to 2 is $\sqrt[4]{2}$ Ker $\phi = \{\sqrt[4]{2}, -\sqrt[4]{2}, \sqrt[4]{2}\,i, -\sqrt[4]{2}\,i\}$.

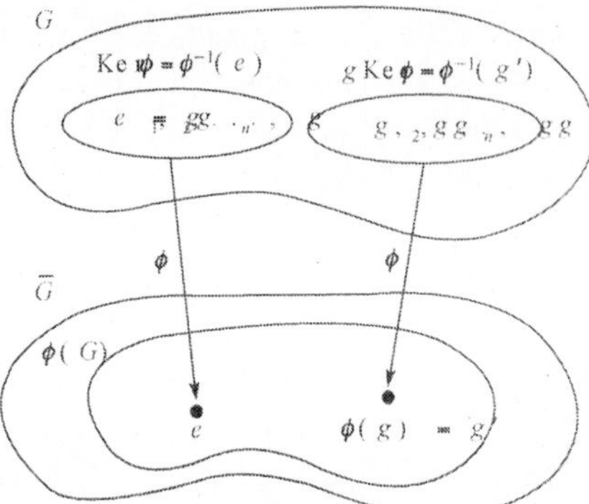

Figure 10.1

Finally, we verify a specific instance of property 3 of Theorem 10.1 and of properties 2 and 6 of Theorem 10.2. Let $H = \langle \cos 30° + i \sin 30° \rangle$. It follows from DeMoivre's Theorem (Example 10 in Chapter 0) that $|H| = 12$, $\phi(H) = \langle \cos 120° + i \sin 120° \rangle$, and $|\phi(H)| = 3$. ▮

▮ **EXAMPLE 9** Define ϕ: $Z_{12} \to Z_{12}$ by $\phi(x) = 3x$. To verify that ϕ is a homomorphism, we observe that in Z_{12}, $3(a + b) = 3a + 3b$ (since the group operation is addition modulo 12). Direct calculations show that $\text{Ker}\,\phi = \{0, 4, 8\}$. Thus, we know from property 5 of Theorem 10.2 that ϕ is a 3-to-1 mapping. Since $\phi(2) = 6$, we have by property 6 of Theorem 10.1 that $\phi^{-1}(6) = 2 + \text{Ker}\,\phi = \{2, 6, 10\}$. Notice also that $\langle 2 \rangle$ is cyclic and $\phi(\langle 2 \rangle) = \{0, 6\}$ is cyclic. Moreover, $|2| = 6$ and $|\phi(2)| = |6| = 2$, so $|\phi(2)|$ divides $|2|$ in agreement with property 3 of Theorem 10.1. Letting $\overline{K} = \{0, 6\}$, we see that the subgroup $\phi^{-1}(\overline{K}) = \{0, 2, 4, 6, 8, 10\}$. This verifies property 7 of Theorem 10.2 in this particular case. ▮

The next example illustrates how one can easily determine all homomorphisms from a cyclic group to a cyclic group.

▮ **EXAMPLE 10** We determine all homomorphisms from Z_{12} to Z_{30}. By property 2 of Theorem 10.1, such a homomorphism is completely specified by the image of 1. That is, if 1 maps to a, then x maps to xa. Lagrange's Theorem and property 3 of Theorem 10.1 require that $|a|$ divide both 12 and 30. So, $|a| = 1, 2, 3,$ or 6. Thus, $a = 0, 15, 10, 20, 5,$ or 25. This gives us a list of candidates for the homomorphisms. That each of these six possibilities yields an operation-preserving, well-defined function can now be verified by direct calculations. [Note that $\gcd(12, 30) = 6$. This is not a coincidence!] ▮

■ **EXAMPLE 11** The mapping from S_n to Z_2 that takes an even permutation to 0 and an odd permutation to 1 is a homomorphism. Figure 10.2 illustrates the telescoping nature of the mapping. ■

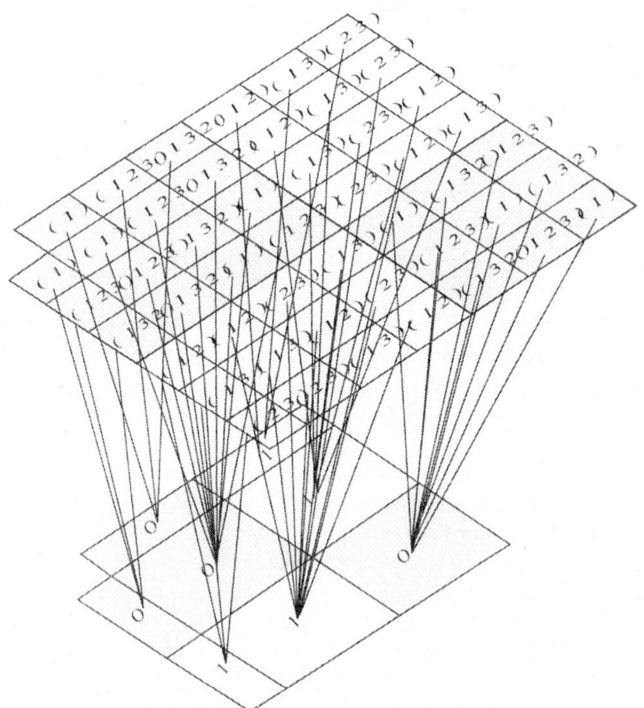

Figure 10.2 Homomorphism from S_3 to Z_2.

The First Isomorphism Theorem

In Chapter 9, we showed that for a group G and a normal subgroup H, we could arrange the Cayley table of G into boxes that represented the cosets of H in G, and that these boxes then became a Cayley table for G/H. The next theorem shows that for any homomorphism ϕ of G and the normal subgroup Ker ϕ, the same process produces a Cayley table isomorphic to the homomorphic image of G. Thus, homomorphisms, like factor groups, cause a *systematic* collapse of a group to a simpler but closely related group. This can be likened to viewing a group through the reverse end of a telescope—the general features of the group are present, but the apparent size is diminished. The important relationship between homomorphisms and factor groups given below is often called the Fundamental Theorem of Group Homomorphisms.

Theorem 10.3 First Isomorphism Theorem (Jordan, 1870)

Let ϕ be a group homomorphism from G to $\overline{G}$. Then the mapping from $G/\text{Ker } \phi$ to $\phi(G)$, given by $g\text{Ker } \phi \to \phi(g)$, is an isomorphism. In symbols, $G/\text{Ker } \phi \approx \phi(G)$.

PROOF Let us use ψ to denote the correspondence $g\text{Ker}\phi \to \phi(g)$. That ψ is well-defined (that is, the correspondence is independent of the particular coset representative chosen) and one-to-one follows directly from property 5 of Theorem 10.1. To show that ψ is operation-preserving, observe that $\psi(x\text{Ker } \phi\ y\text{Ker } \phi) = \psi(xy\text{Ker } \phi) = \phi(xy) = \phi(x)\,\phi(y) = \psi(x\text{Ker } \phi)\psi(y\text{Ker } \phi)$. ■

The next corollary follows directly from Theorem 10.3, property 1 of Theorem 10.2, and Lagrange's Theorem.

Corollary

If ϕ is a homomorphism from a finite group G to $\overline{G}$, then $|\phi(G)|$ divides $|G|$ and $|\overline{G}|$.

EXAMPLE 12 To illustrate Theorem 10.3 and its proof, consider the homomorphism ϕ from D_4 to itself given by the following.

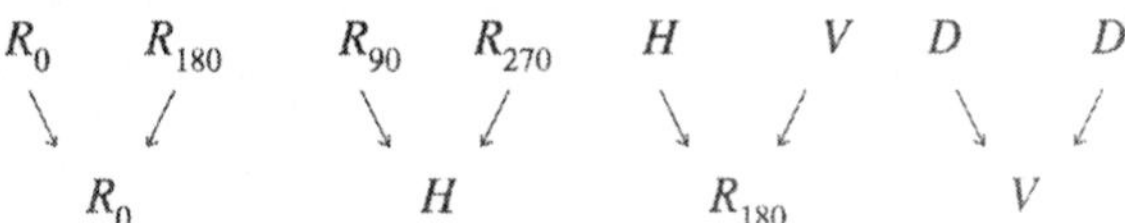

Then $\text{Ker } \phi = \{R_0, R_{180}\}$, and the mapping ψ in Theorem 10.3 is $R_0\text{Ker } \phi \to R_0$, $R_{90}\text{Ker } \phi \to H$, $H\text{Ker } \phi \to R_{180}$, $D\text{Ker } \phi \to V$. It is straightforward to verify that the mapping ψ is an isomorphism. ■

Mathematicians often give a pictorial representation of Theorem 10.3, as follows:

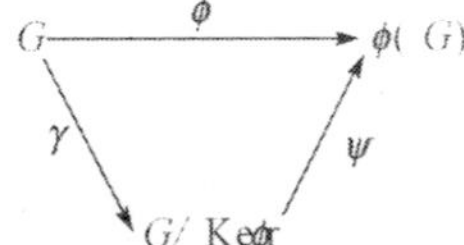

where γ: $G \to G/\text{Ker } \phi$ is defined as $\gamma(g) = g\text{Ker } \phi$. The mapping γ is called the *natural mapping* from G to $G/\text{Ker } \phi$. Our proof of Theorem 10.3 shows that $\psi\gamma = \phi$. In this case, one says that the preceding diagram is *commutative.*

As a consequence of Theorem 10.3, we see that all homomorphic images of G can be determined using G. We may simply consider the various factor groups of G. For example, we know that the homomorphic image of an Abelian group is Abelian because the factor group of an Abelian group is Abelian. We know that the number of homomorphic images of a cyclic group G of order n is the number of divisors of n, since there is exactly one subgroup of G (and therefore one factor group of G) for each divisor of n. (Be careful: The number of homomorphisms of a cyclic group of order n need not be the same as the number of divisors of n, since different homomorphisms can have the same image.)

An appreciation for Theorem 10.3 can be gained by looking at a few examples.

EXAMPLE 13 $Z/\langle n\rangle \approx Z_n$

Consider the mapping from Z to Z_n defined in Example 5. Clearly, its kernel is $\langle n\rangle$. So, by Theorem 10.3, $Z/\langle n\rangle \approx Z_n$.

EXAMPLE 14 Wrapping Function

Recall the wrapping function W from trigonometry. The real number line is wrapped around a unit circle in the plane centered at (0, 0) with the number 0 on the number line at the point (1, 0), the positive reals in the counterclockwise direction and the negative reals in the clockwise direction (see Figure 10.3). The function W assigns to each real number a the point a radians from (1, 0) on the circle. This mapping is a homomorphism from the group **R** under addition onto the circle group (the group of complex numbers of magnitude 1 under multiplication). Indeed, it follows from elementary facts of trigonometry that $W(x) = \cos x + i \sin x$ and $W(x + y) = W(x)W(y)$. Since W is periodic of period 2π, Ker $W = \langle 2\pi\rangle$. So, from the First Isomorphism Theorem, we see that $\mathbf{R}/\langle 2\pi\rangle$ is isomorphic to the circle group.

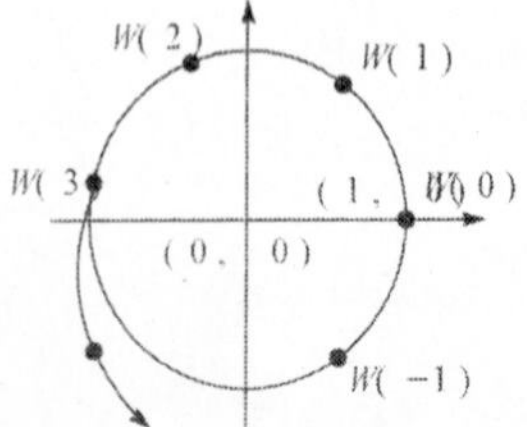

Figure 10.3

Our next example is a theorem that is used repeatedly in Chapters 24 and 25.

▮ EXAMPLE 15 *N/C* Theorem

Let H be a subgroup of a group G. Recall that the normalizer of H in G is $N(H) = \{x \in G \mid xHx^{-1} = H\}$ and the centralizer of H in G is $C(H) = \{x \in G \mid xhx^{-1} = h$ for all h in $H\}$. Consider the mapping from $N(H)$ to Aut(H) given by $g \to \phi_g$, where ϕ_g is the inner automorphism of H induced by g [that is, $\phi_g(h) = ghg^{-1}$ for all h in H]. This mapping is a homomorphism with kernel $C(H)$. So, by Theorem 10.3, $N(H)/C(H)$ is isomorphic to a subgroup of Aut(H). ▮

As an application of the *N/C* Theorem, we will show that every group of order 35 is cyclic.

▮ EXAMPLE 16 Let G be a group of order 35. By Lagrange's Theorem, every nonidentity element of G has order 5, 7, or 35. If some element has order 35, G is cyclic. So we may assume that all nonidentity elements have order 5 or 7. However, not all such elements can have order 5, since elements of order 5 come 4 at a time (if $|x| = 5$, then $|x^2| = |x^3| = |x^4| = 5$) and 4 does not divide 34. Similarly, since 6 does not divide 34, not all nonidentity elements can have order 7. So, G has elements of order 7 and order 5. Since G has an element of order 7, it has a subgroup of order 7. Let us call it H. In fact, H is the only subgroup of G of order 7, for if K is another subgroup of G of order 7, we have by Theorem 7.2 that $|HK| = |H||K|/|H \cap K| = 7 \cdot 7/1 = 49$. But, of course, this is impossible in a group of order 35. Since for every a in G, aHa^{-1} is also a subgroup of G of order 7 (see Exercise 1 of the Supplementary Exercises for Chapters 1–4), we must have $aHa^{-1} = H$. So, $N(H) = G$. Since H has prime order, it is cyclic and therefore Abelian. In particular, $C(H)$ contains H. So, 7 divides $|C(H)|$ and $|C(H)|$ divides 35. It follows, then, that $C(H) = G$ or $C(H) = H$. If $C(H) = G$, then we may obtain an element x of order 35 by letting $x = hk$, where h is a nonidentity element of H and k has order 5. On the other hand, if $C(H) = H$, then $|C(H)| = 7$ and $|N(H)/C(H)| = 35/7 = 5$. However, 5 does not divide $|\text{Aut}(H)| = |\text{Aut}(Z_7)| = 6$. This contradiction shows that G is cyclic. ▮

The corollary of Theorem 10.2 says that the kernel of every homomorphism of a group is a normal subgroup of the group. We conclude this chapter by verifying that the converse of this statement is also true.

Theorem 10.4 Normal Subgroups Are Kernels

Every normal subgroup of a group G is the kernel of a homomorphism of G. In particular, a normal subgroup N is the kernel of the mapping $g \to gN$ from G to G/N.

PROOF Define $\gamma: G \to G/N$ by $\gamma(g) = gN$. (This mapping is called the *natural homomorphism* from G to G/N.) Then, $\gamma(xy) = (xy)N = xNyN = \gamma(x)\gamma(y)$. Moreover, $g \in \text{Ker } \gamma$ if and only if $gN = \gamma(g) = N$, which is true if and only if $g \in N$ (see property 2 of the lemma in Chapter 7). ▮

Examples 13, 14, and 15 illustrate the utility of the First Isomorphism Theorem. But what about homomorphisms in general? Why would one care to study a homomorphism of a group? The answer is that, just as was the case with factor groups of a group, homomorphic images of a group tell us *some* of the properties of the original group. One measure of the likeness of a group and its homomorphic image is the size of the kernel. If the kernel of the homomorphism of group G is the identity, then the image of G tells us everything (group theoretically) about G (the two being isomorphic). On the other hand, if the kernel of the homomorphism is G itself, then the image tells us nothing about G. Between these two extremes, some information about G is preserved and some is lost. The utility of a particular homomorphism lies in its ability to preserve the group properties we want, while losing some inessential ones. In this way, we have replaced G by a group less complicated (and therefore easier to study) than G; but, in the process, we have saved enough information to answer questions that we have about G itself. For example, if G is a group of order 60 and G has a homomorphic image of order 12 that is cyclic, then we know from properties 5, 7, and 8 of Theorem 10.2 that G has normal subgroups of orders 5, 10, 15, 20, 30, and 60. To illustrate further, suppose we are asked to find an infinite group that is the union of three proper subgroups. Instead of attempting to do this directly, we first make the problem easier by finding a finite group that is the union of three proper subgroups. Observing that $Z_2 \oplus Z_2$ is the union of $H_1 = \langle 1, 0\rangle$, $H_2 = \langle 0, 1\rangle$, and $H_3 = \langle 1, 1\rangle$, we have found our finite group. Now all we need do is think of an infinite group that has $Z_2 \oplus Z_2$ as a homomorphic image and pull back H_1, H_2, and H_3, and our original problem is solved. Clearly, the mapping from $Z_2 \oplus Z_2 \oplus Z$ onto $Z_2 \oplus Z_2$ given by $\phi(a, b, c) = (a, b)$ is such a mapping, and therefore $Z_2 \oplus Z_2 \oplus Z$ is the union of $\phi^{-1}(H_1) = \{(a, 0, c,) \mid a \in Z_2, c \in Z\}$, $\phi^{-1}(H_2) = \{(0, b, c) \mid b \in Z_2, c \in Z\}$, and $\phi^{-1}(H_3) = \{(a, a, c) \mid a \in Z_2, c \in Z\}$.

Although an isomorphism is a special case of a homomorphism, the two concepts have entirely different roles. Whereas isomorphisms allow us to look at a group in an alternative way, homomorphisms act as investigative tools. The following analogy between homomorphisms and photography may be instructive.† A photograph of a person cannot tell us the person's exact height, weight, or age. Nevertheless, we *may* be able to decide from a photograph whether the person is tall or short, heavy or thin, old or young, male or female. In the same way, a homomorphic image of a group gives us *some* information about the group.

In certain branches of group theory, and especially in physics and chemistry, one often wants to know all homomorphic images of a group that are matrix groups over the complex numbers (these are called *group representations*). Here, we may carry our analogy with photography one step further by saying that this is like wanting photographs of a person from many different angles (front view, profile, head-to-toe view, close-up, etc.), as well as x-rays! Just as this composite information from the photographs reveals much about the person, several homomorphic images of a group reveal much about the group.

Exercises

The greater the difficulty, the more glory in surmounting it. Skillful pilots gain their reputation from storms and tempests.

EPICURUS

1. Prove that the mapping given in Example 2 is a homomorphism.
2. Prove that the mapping given in Example 3 is a homomorphism.
3. Prove that the mapping given in Example 4 is a homomorphism.
4. Prove that the mapping given in Example 11 is a homomorphism.
5. Let $\mathbf{R}^*$ be the group of nonzero real numbers under multiplication, and let r be a positive integer. Show that the mapping that takes x to x^r is a homomorphism from $\mathbf{R}^*$ to $\mathbf{R}^*$ and determine the kernel. Which values of r yield an isomorphism?
6. Let G be the group of all polynomials with real coefficients under addition. For each f in G, let $\int f$ denote the antiderivative of f that passes through the point (0, 0). Show that the mapping $f \rightarrow \int f$ from G to G is a homomorphism. What is the kernel of this mapping? Is this mapping a homomorphism if $\int f$ denotes the antiderivative of f that passes through (0, 1)?

†"All perception of truth is the detection of an analogy." Henry David Thoreau, *Journal*.

7. If ϕ is a homomorphism from G to H and σ is a homomorphism from H to K, show that $\sigma\phi$ is a homomorphism from G to K. How are Ker ϕ and Ker $\sigma\phi$ related? If ϕ and σ are onto and G is finite, describe [Ker $\sigma\phi$:Ker ϕ] in terms of $|H|$ and $|K|$.
8. Let G be a group of permutations. For each σ in G, define
$$\operatorname{sgn}(\sigma) = \begin{cases} +1 & \text{if } \sigma \text{ is an even permutation,} \\ -1 & \text{if } \sigma \text{ is an odd permutation.} \end{cases}$$
Prove that sgn is a homomorphism from G to the multiplicative group $\{+1, -1\}$. What is the kernel? Why does this homomorphism allow you to conclude that A_n is a normal subgroup of S_n of index 2? Why does this prove Exercise 23 of Chapter 5?
9. Prove that the mapping from $G \oplus H$ to G given by $(g, h) \to g$ is a homomorphism. What is the kernel? This mapping is called the *projection* of $G \oplus H$ onto G.
10. Let G be a subgroup of some dihedral group. For each x in G, define
$$\phi(x) = \begin{cases} +1 & \text{if } x \text{ is a rotation,} \\ -1 & \text{if } x \text{ is a reflection.} \end{cases}$$
Prove that ϕ is a homomorphism from G to the multiplicative group $\{+1, -1\}$. What is the kernel? Why does this prove Exercise 25 of Chapter 3?
11. Prove that $(Z \oplus Z)/(\langle(a, 0)\rangle \times \langle(0, b)\rangle)$ is isomorphic to $Z_a \oplus Z_b$.
12. Suppose that k is a divisor of n. Prove that $Z_n/\langle k\rangle \approx Z_k$.
13. Prove that $(A \oplus B)/(A \oplus \{e\}) \approx B$.
14. Explain why the correspondence $x \to 3x$ from Z_{12} to Z_{10} is not a homomorphism.
15. Suppose that ϕ is a homomorphism from Z_{30} to Z_{30} and Ker $\phi = \{0, 10, 20\}$. If $\phi(23) = 9$, determine all elements that map to 9.
16. Prove that there is no homomorphism from $Z_8 \oplus Z_2$ onto $Z_4 \oplus Z_4$.
17. Prove that there is no homomorphism from $Z_{16} \oplus Z_2$ onto $Z_4 \oplus Z_4$.
18. Can there be a homomorphism from $Z_4 \oplus Z_4$ onto Z_8? Can there be a homomorphism from Z_{16} onto $Z_2 \oplus Z_2$? Explain your answers.
19. Suppose that there is a homomorphism ϕ from Z_{17} to some group and that ϕ is not one-to-one. Determine ϕ.
20. How many homomorphisms are there from Z_{20} onto Z_8? How many are there to Z_8?
21. If ϕ is a homomorphism from Z_{30} onto a group of order 5, determine the kernel of ϕ.

22. Suppose that ϕ is a homomorphism from a finite group G onto $\overline{G}$ and that $\overline{G}$ has an element of order 8. Prove that G has an element of order 8. Generalize.

23. Suppose that ϕ is a homomorphism from Z_{36} to a group of order 24.

a. Determine the possible homomorphic images.

b. For each image in part a, determine the corresponding kernel of ϕ.

24. Suppose that ϕ: $Z_{50} \to Z_{15}$ is a group homomorphism with $\phi(7) = 6$.

a. Determine $\phi(x)$.

b. Determine the image of ϕ.

c. Determine the kernel of ϕ.

d. Determine $\phi^{-1}(3)$. That is, determine the set of all elements that map to 3.

25. How many homomorphisms are there from Z_{20} onto Z_{10}? How many are there to Z_{10}?

26. Determine all homomorphisms from Z_4 to $Z_2 \oplus Z_2$.

27. Determine all homomorphisms from Z_n to itself.

28. Suppose that ϕ is a homomorphism from S_4 onto Z_2. Determine Ker ϕ. Determine all homomorphisms from S_4 to Z_2.

29. Suppose that there is a homomorphism from a finite group G onto Z_{10}. Prove that G has normal subgroups of indexes 2 and 5.

30. Suppose that ϕ is a homomorphism from a group G onto $Z_6 \oplus Z_2$ and that the kernel of ϕ has order 5. Explain why G must have normal subgroups of orders 5, 10, 15, 20, 30, and 60.

31. Suppose that ϕ is a homomorphism from $U(30)$ to $U(30)$ and that Ker $\phi = \{1, 11\}$. If $\phi(7) = 7$, find all elements of $U(30)$ that map to 7.

32. Find a homomorphism ϕ from $U(30)$ to $U(30)$ with kernel $\{1, 11\}$ and $\phi(7) = 7$.

33. Suppose that ϕ is a homomorphism from $U(40)$ to $U(40)$ and that Ker $\phi = \{1, 9, 17, 33\}$. If $\phi(11) = 11$, find all elements of $U(40)$ that map to 11.

34. Find a homomorphism ϕ from $U(40)$ to $U(40)$ with kernel $\{1, 9, 17, 33\}$ and $\phi(11) = 11$.

35. Prove that the mapping ϕ: $Z \oplus Z \to Z$ given by $(a, b) \to a - b$ is a homomorphism. What is the kernel of ϕ? Describe the set $\phi^{-1}(3)$ (that is, all elements that map to 3).

36. Suppose that there is a homomorphism ϕ from $Z \oplus Z$ to a group G such that $\phi((3, 2)) = a$ and $\phi((2, 1)) = b$. Determine $\phi((4, 4))$ in terms of a and b. Assume that the operation of G is addition.

37. Let $H = \{z \in C^* \mid |z| = 1\}$. Prove that C^*/H is isomorphic to $\mathbf{R}^+$, the group of positive real numbers under multiplication.
38. Let α be a homomorphism from G_1 to H_1 and β be a homomorphism from G_2 to H_2. Determine the kernel of the homomorphism γ from $G_1 \oplus G_2$ to $H_1 \oplus H_2$ defined by $\gamma(g_1, g_2) = (\alpha(g_1), \beta(g_2))$.
39. Prove that the mapping $x \rightarrow x^6$ from $\mathbf{C}^*$ to $\mathbf{C}^*$ is a homomorphism. What is the kernel?
40. For each pair of positive integers m and n, we can define a homomorphism from Z to $Z_m \oplus Z_n$ by $x \rightarrow (x \bmod m, x \bmod n)$. What is the kernel when $(m, n) = (3, 4)$? What is the kernel when $(m, n) = (6, 4)$? Generalize.
41. (Second Isomorphism Theorem) If K is a subgroup of G and N is a normal subgroup of G, prove that $K/(K \cap N)$ is isomorphic to KN/N.
42. (Third Isomorphism Theorem) If M and N are normal subgroups of G and $N \leq M$, prove that $(G/N)/(M/N) \approx G/M$.
43. Let $\phi(d)$ denote the Euler phi function of d (see page 85). Show that the number of homomorphisms from Z_n to Z_k is $\Sigma\phi(d)$, where the sum runs over all common divisors d of n and k. [It follows from number theory that this sum is actually $\gcd(n, k)$.]
44. Let k be a divisor of n. Consider the homomorphism from $U(n)$ to $U(k)$ given by $x \rightarrow x \bmod k$. What is the relationship between this homomorphism and the subgroup $U_k(n)$ of $U(n)$?
45. Determine all homomorphic images of D_4 (up to isomorphism).
46. Let N be a normal subgroup of a finite group G. Use the theorems of this chapter to prove that the order of the group element gN in G/N divides the order of g.
47. Suppose that G is a finite group and that Z_{10} is a homomorphic image of G. What can we say about $|G|$? Generalize.
48. Suppose that Z_{10} and Z_{15} are both homomorphic images of a finite group G. What can be said about $|G|$? Generalize.
49. Suppose that for each prime p, Z_p is the homomorphic image of a group G. What can we say about $|G|$? Give an example of such a group.
50. (For students who have had linear algebra.) Suppose that x is a particular solution to a system of linear equations and that S is the entire solution set of the corresponding homogeneous system of linear equations. Explain why property 6 of Theorem 10.1 guarantees that $x + S$ is the entire solution set of the nonhomogeneous system. In particular, describe the relevant groups and the homomorphism between them.

51. Let N be a normal subgroup of a group G. Use property 7 of Theorem 10.2 to prove that every subgroup of G/N has the form H/N, where H is a subgroup of G. (This exercise is referred to in Chapter 24.)

52. Show that a homomorphism defined on a cyclic group is completely determined by its action on a generator of the group.

53. Use the First Isomorphism Theorem to prove Theorem 9.4.

54. Let α and β be group homomorphisms from G to $\overline{G}$ and let $H = \{g \in G \mid \alpha(g) = \beta(g)\}$. Prove or disprove that H is a subgroup of G.

55. Let $Z[x]$ be the group of polynomials in x with integer coefficients under addition. Prove that the mapping from $Z[x]$ into Z given by $f(x) \to f(3)$ is a homomorphism. Give a geometric description of the kernel of this homomorphism. Generalize.

56. Prove that the mapping from **R** under addition to $GL(2, \mathbf{R})$ that takes x to

$$\begin{bmatrix} \cos x & \sin x \\ -\sin x & \cos x \end{bmatrix}$$

is a group homomorphism. What is the kernel of the homomorphism?

57. Suppose there is a homomorphism ϕ from G onto $Z_2 \oplus Z_2$. Prove that G is the union of three proper normal subgroups.

58. If H and K are normal subgroups of G and $H \cap K = \{e\}$, prove that G is isomorphic to a subgroup of $G/H \oplus G/K$.

59. Suppose that H and K are distinct subgroups of G of index 2. Prove that $H \cap K$ is a normal subgroup of G of index 4 and that $G/(H \cap K)$ is not cyclic.

60. Suppose that the number of homomorphisms from G to H is n. How many homomorphisms are there from G to $H \oplus H \oplus \cdots \oplus H$ (s terms)? When H is Abelian, how many homomorphisms are there from $G \oplus G \oplus \cdots \oplus G$ (s terms) to H?

61. Prove that every group of order 77 is cyclic.

62. Determine all homomorphisms from Z onto S_3. Determine all homomorphisms from Z to S_3.

63. Let G be an Abelian group. Determine all homomorphisms from S_3 to G.

64. If ϕ is an isomorphism from a group G under addition to a group $\overline{G}$ under addition, prove that for any integer n, the mapping from G to $\overline{G}$ defined by $\gamma(x) = n\phi(x)$ is a homomorphism from G to $\overline{G}$.

65. Prove that the mapping from $\mathbf{C}^*$ to $\mathbf{C}^*$ given by $\phi(z) = z^2$ is a homomorphism and that $\mathbf{C}^*/\{1, -1\}$ is isomorphic to $\mathbf{C}^*$.

66. Let p be a prime. Determine the number of homomorphisms from $Z_p \oplus Z_p$ into Z_p.

67. Suppose G is an Abelian group under addition with the property that for every positive integer n, the set $nG = \{ng \mid g \in G\} = G$. Show that every proper subgroup of G is properly contained in a proper subgroup of G. Name two familiar groups that satisfy the hypothesis.

Computer Exercise

A computer exercise for this chapter is available at the website:

http://www.d.umn.edu/~jgallian

Camille Jordan

Although these contributions [to analysis and topology] would have been enough to rank Jordan very high among his mathematical contemporaries, it is chiefly as an algebraist that he reached celebrity when he was barely thirty; and during the next forty years he was universally regarded as the undisputed master of group theory.

J. DIEUDONNÉ, *Dictionary of Scientific Biography*

CAMILLE JORDAN was born into a well-to-do family on January 5, 1838, in Lyons, France. Like his father, he graduated from the École Polytechnique and became an engineer. Nearly all of his 120 research papers in mathematics were written before his retirement from engineering in 1885. From 1873 until 1912, Jordan taught simultaneously at the École Polytechnique and at the College of France.

In the great French tradition, Jordan was a universal mathematician who published in nearly every branch of mathematics. Among the concepts named after him are the Jordan canonical form in matrix theory, the Jordan curve theorem from topology, and the Jordan–Hölder Theorem from group theory. His classic book *Traité des substitutions,* published in 1870, was the first to be devoted solely to group theory and its applications to other branches of mathematics.

Another book that had great influence and set a new standard for rigor was his *Cours d'analyse.* This book gave the first clear definitions of the notions of volume and multiple integral. Nearly 100 years after this book appeared, the distinguished mathematician and mathematical historian B. L. van der Waerden wrote, "For me, every single chapter of the *Cours d'analyse* is a pleasure to read." Jordan died in Paris on January 22, 1922.

To find more information about Jordan, visit:

http://www-groups.dcs.st-and.ac.uk/~history/

11 Fundamental Theorem of Finite Abelian Groups

By a small sample we may judge of the whole piece.

MIGUEL DE CERVANTES, *Don Quixote*

The Fundamental Theorem

In this chapter, we present a theorem that describes to an algebraist's eye (that is, up to isomorphism) all finite Abelian groups in a standardized way. Before giving the proof, which is long and difficult, we discuss some consequences of the theorem and its proof. The first proof of the theorem was given by Leopold Kronecker in 1858.

Theorem 11.1 Fundamental Theorem of Finite Abelian Groups

Every finite Abelian group is a direct product of cyclic groups of prime-power order. Moreover, the number of terms in the product and the orders of the cyclic groups are uniquely determined by the group.

Since a cyclic group of order n is isomorphic to Z_n, Theorem 11.1 shows that every finite Abelian group G is isomorphic to a group of the form

$$Z_{p_1^{n_1}} \oplus Z_{p_2^{n_2}} \oplus \cdots \oplus Z_{p_k^{n_k}},$$

where the p_i's are not necessarily distinct primes and the prime powers $p_1^{n_1}, p_2^{n_2}, \ldots, p_k^{n_k}$ are uniquely determined by G. Writing a group in this form is called *determining the isomorphism class of G.*

The Isomorphism Classes of Abelian Groups

The Fundamental Theorem is extremely powerful. As an application, we can use it as an algorithm for constructing all Abelian groups of any order. Let's look at groups whose orders have the form p^k, where p is prime and $k \leq 4$. In general, there is one group of order p^k for each set of positive integers whose sum is k (such a set is called a *partition* of k); that is, if k can be written as

$$k = n_1 + n_2 + \cdots + n_t,$$

where each n_i is a positive integer, then

$$Z_{p^{n_1}} \oplus Z_{p^{n_2}} \oplus \cdots \oplus Z_{p^{n_t}}$$

is an Abelian group of order p^k.

Order of G	Partitions of k	Possible direct products for G
p	1	Z_p
p^2	2	Z_{p^2}
	1 + 1	$Z_p \oplus Z_p$
p^3	3	Z_{p^3}
	2 + 1	$Z_{p^2} \oplus Z_p$
	1 + 1 + 1	$Z_p \oplus Z_p \oplus Z_p$
p^4	4	Z_{p^4}
	3 + 1	$Z_{p^3} \oplus Z_p$
	2 + 2	$Z_{p^2} \oplus Z_{p^2}$
	2 + 1 + 1	$Z_{p^2} \oplus Z_p \oplus Z_p$
	1 + 1 + 1 + 1	$Z_p \oplus Z_p \oplus Z_p \oplus Z_p$

Furthermore, the uniqueness portion of the Fundamental Theorem guarantees that distinct partitions of k yield distinct isomorphism classes. Thus, for example, $Z_9 \oplus Z_3$ is not isomorphic to $Z_3 \oplus Z_3 \oplus Z_3$. A reliable mnemonic for comparing external direct products is the cancellation property: If A is *finite*, then

$$A \oplus B \approx A \oplus C \quad \text{if and only if} \quad B \approx C \quad \text{(see [1])}.$$

Thus, $Z_4 \oplus Z_4$ is not isomorphic to $Z_4 \oplus Z_2 \oplus Z_2$, because Z_4 is not isomorphic to $Z_2 \oplus Z_2$.

To appreciate fully the potency of the Fundamental Theorem, contrast the ease with which the Abelian groups of order p^k, $k \leq 4$, were determined with the corresponding problem for non-Abelian groups. Even a description of the two non-Abelian groups of order 8 is a challenge (see Theorem 26.4), and a description of the nine non-Abelian groups of order 16 is well beyond the scope of this text.

Now that we know how to construct all the Abelian groups of prime-power order, we move to the problem of constructing all Abelian groups of a certain order n, where n has two or more distinct prime divisors. We begin by writing n in prime-power decomposition form $n = p_1^{n_1}p_2^{n_2} \cdots p_k^{n_k}$. Next, we individually form all Abelian groups of order $p_1^{n_1}$, then $p_2^{n_2}$, and so on, as described earlier. Finally, we form all possible external direct products of these groups. For example, let $n = 1176 = 2^3 \cdot 3 \cdot 7^2$. Then, the complete list of the distinct isomorphism classes of Abelian groups of order 1176 is

$$\begin{aligned}
&Z_8 \oplus Z_3 \oplus Z_{49},\\
&Z_4 \oplus Z_2 \oplus Z_3 \oplus Z_{49},\\
&Z_2 \oplus Z_2 \oplus Z_2 \oplus Z_3 \oplus Z_{49},\\
&Z_8 \oplus Z_3 \oplus Z_7 \oplus Z_7,\\
&Z_4 \oplus Z_2 \oplus Z_3 \oplus Z_7 \oplus Z_7,\\
&Z_2 \oplus Z_2 \oplus Z_2 \oplus Z_3 \oplus Z_7 \oplus Z_7.
\end{aligned}$$

If we are given any particular Abelian group G of order 1176, the question we want to answer about G is: Which of the preceding six isomorphism classes represents the structure of G? We can answer this question by comparing the orders of the elements of G with the orders of the elements in the six direct products, since it can be shown that two finite Abelian groups are isomorphic if and only if they have the same number of elements of each order. For instance, we could determine whether G has any elements of order 8. If so, then G must be isomorphic to the first or fourth group above, since these are the only ones with elements of order 8. To narrow G down to a single choice, we now need only check whether or not G has an element of order 49, since the first product above has such an element, whereas the fourth one does not.

What if we have some specific Abelian group G of order $p_1^{n_1}p_2^{n_2} \cdots p_k^{n_k}$, where the p_i's are distinct primes? How can G be expressed as an *internal* direct product of cyclic groups of prime-power order? For simplicity, let us say that the group has 2^n elements. First, we must compute the orders of the elements. After this is done, pick an element of maximum order 2^r, call it a_1. Then $\langle a_1 \rangle$ is one of the factors in the desired internal direct product. If $G \neq \langle a_1 \rangle$, choose an element a_2 of maximum order 2^s such that $s \leq n - r$ and none of $a_2, a_2^2, a_2^4, \ldots, a_2^{2^{s-1}}$ is in $\langle a_1 \rangle$. Then $\langle a_2 \rangle$ is a second direct factor. If $n \neq r + s$, select an element a_3 of maximum order 2^t such that $t \leq n - r - s$ and none of $a_3, a_3^2, a_3^4, \ldots, a_3^{2^{t-1}}$ is in $\langle a_1 \rangle \times \langle a_2 \rangle = \{a_1^i a_2^j \mid 0 \leq i < 2^r, 0 \leq j < 2^s\}$. Then $\langle a_3 \rangle$ is another direct factor. We continue in this fashion until our direct product has the same order as G.

A formal presentation of this algorithm for any Abelian group G of prime-power order p^n is as follows.

Greedy Algorithm for an Abelian Group of Order p^n

1. Compute the orders of the elements of the group G.
2. Select an element a_1 of maximum order and define $G_1 = \langle a_1 \rangle$. Set $i = 1$.
3. If $|G| = |G_i|$, stop. Otherwise, replace i by $i + 1$.
4. Select an element a_i of maximum order p^k such that $p^k \leq |G|/|G_{i-1}|$ and none of $a_i, a_i^p, a_i^{p^2}, \ldots, a_i^{p^{k-1}}$ is in G_{i-1}, and define $G_i = G_{i-1} \times \langle a_i \rangle$.
5. Return to step 3.

In the general case where $|G| = p_1^{n_1} p_2^{n_2} \cdots p_k^{n_k}$, we simply use the algorithm to build up a direct product of order $p_1^{n_1}$, then another of order $p_2^{n_2}$, and so on. The direct product of all of these pieces is the desired factorization of G. The following example is small enough that we can compute the appropriate internal and external direct products by hand.

EXAMPLE 1 Let $G = \{1, 8, 12, 14, 18, 21, 27, 31, 34, 38, 44, 47, 51, 53, 57, 64\}$ under multiplication modulo 65. Since G has order 16, we know it is isomorphic to one of

$$Z_{16},$$
$$Z_8 \oplus Z_2,$$
$$Z_4 \oplus Z_4,$$
$$Z_4 \oplus Z_2 \oplus Z_2,$$
$$Z_2 \oplus Z_2 \oplus Z_2 \oplus Z_2.$$

To decide which one, we dirty our hands to calculate the orders of the elements of G.

Element	1	8	12	14	18	21	27	31	34	38	44	47	51	53	57	64
Order	1	4	4	2	4	4	4	4	4	4	4	4	2	4	4	2

From the table of orders, we can instantly rule out all but $Z_4 \oplus Z_4$ and $Z_4 \oplus Z_2 \oplus Z_2$ as possibilities. Finally, we observe that since this latter group has a subgroup isomorphic to $Z_2 \oplus Z_2 \oplus Z_2$, it has more than three elements of order 2, and therefore we must have $G \approx Z_4 \oplus Z_4$.

Expressing G as an internal direct product is even easier. Pick an element of maximum order, say the element 8. Then $\langle 8 \rangle$ is a factor in the product. Next, choose a second element, say a, so that a has order 4 and a and a^2 are not in $\langle 8 \rangle = \{1, 8, 64, 57\}$. Since 12 has this property, we have $G = \langle 8 \rangle \times \langle 12 \rangle$. ■

Example 1 illustrates how quickly and easily one can write an Abelian group as a direct product given the orders of the elements of the group. But calculating all those orders is certainly not an appealing prospect! The good news is that, in practice, a combination of theory and calculation of the orders of a few elements will usually suffice.

EXAMPLE 2 Let $G = \{1, 8, 17, 19, 26, 28, 37, 44, 46, 53, 62, 64, 71, 73, 82, 89, 91, 98, 107, 109, 116, 118, 127, 134\}$ under multiplication modulo 135. Since G has order 24, it is isomorphic to one of

$$Z_8 \oplus Z_3 \approx Z_{24},$$
$$Z_4 \oplus Z_2 \oplus Z_3 \approx Z_{12} \oplus Z_2,$$
$$Z_2 \oplus Z_2 \oplus Z_2 \oplus Z_3 \approx Z_6 \oplus Z_2 \oplus Z_2.$$

Consider the element 8. Direct calculations show that $8^6 = 109$ and $8^{12} = 1$. (Be sure to mod as you go. For example, $8^3 \bmod 135 = 512 \bmod 135 = 107$, so compute 8^4 as $8 \cdot 107$ rather than $8 \cdot 512$.) But now we know G. Why? Clearly, $|8| = 12$ rules out the third group in the list. At the same time, $|109| = 2 = |134|$ (remember, $134 = -1 \bmod 135$) implies that G is not Z_{24} (see Theorem 4.4). Thus, $G \approx Z_{12} \oplus Z_2$, and $G = \langle 8 \rangle \times \langle 134 \rangle$. ∎

Rather than express an Abelian group as a direct product of cyclic groups of prime-power orders, it is often more convenient to combine the cyclic factors of relatively prime order, as we did in Example 2, to obtain a direct product of the form $Z_{n_1} \oplus Z_{n_2} \oplus \cdots \oplus Z_{n_k}$, where n_i divides n_{i-1}. For example, $Z_4 \oplus Z_4 \oplus Z_2 \oplus Z_9 \oplus Z_3 \oplus Z_5$ would be written as $Z_{180} \oplus Z_{12} \oplus Z_2$ (see Exercise 11). The algorithm above is easily adapted to accomplish this by replacing step 4 by 4′: Select an element a_i of maximum order m such that $m \leq |G|/|G_{i-1}|$ and none of $a_i, a_i^2, \ldots, a_i^{m-1}$ is in G_{i-1}, and define $G_i = G_{i-1} \times \langle a_i \rangle$.

As a consequence of the Fundamental Theorem of Finite Abelian Groups, we have the following corollary, which shows that the converse of Lagrange's Theorem is true for finite Abelian groups.

Corollary Existence of Subgroups of Abelian Groups

If m divides the order of a finite Abelian group G, then G has a subgroup of order m.

It is instructive to verify this corollary for a specific case. Let us say that G is an Abelian group of order 72 and we wish to produce a subgroup of order 12. According to the Fundamental Theorem, G is isomorphic to one of the following six groups:

$$Z_8 \oplus Z_9, \qquad Z_8 \oplus Z_3 \oplus Z_3,$$
$$Z_4 \oplus Z_2 \oplus Z_9, \qquad Z_4 \oplus Z_2 \oplus Z_3 \oplus Z_3,$$
$$Z_2 \oplus Z_2 \oplus Z_2 \oplus Z_9, \qquad Z_2 \oplus Z_2 \oplus Z_2 \oplus Z_3 \oplus Z_3.$$

Obviously, $Z_8 \oplus Z_9 \approx Z_{72}$ and $Z_4 \oplus Z_2 \oplus Z_3 \oplus Z_3 \approx Z_{12} \oplus Z_6$ both have a subgroup of order 12. To construct a subgroup of order 12 in $Z_4 \oplus Z_2 \oplus Z_9$, we simply piece together all of Z_4 and the subgroup of order 3 in Z_9; that is, $\{(a, 0, b) \mid a \in Z_4, b \in \{0, 3, 6\}\}$. A subgroup of order 12 in $Z_8 \oplus Z_3 \oplus Z_3$ is given by $\{(a, b, 0) \mid a \in \{0, 2, 4, 6\}, b \in Z_3\}$. An analogous procedure applies to the remaining cases and indeed to any finite Abelian group.

Proof of the Fundamental Theorem

Because of the length and complexity of the proof of the Fundamental Theorem of Finite Abelian Groups, we will break it up into a series of lemmas.

Lemma 1

Let G be a finite Abelian group of order $p^n m$, where p is a prime that does not divide m. Then $G = H \times K$, where $H = \{x \in G \mid x^{p^n} = e\}$ and $K = \{x \in G \mid x^m = e\}$. Moreover, $|H| = p^n$.

PROOF It is an easy exercise to prove that H and K are subgroups of G (see Exercise 45 in Chapter 3). Because G is Abelian, to prove that $G = H \times K$ we need only prove that $G = HK$ and $H \cap K = \{e\}$. Since we have $\gcd(m, p^n) = 1$, there are integers s and t such that $1 = sm + tp^n$. For any x in G, we have $x = x^1 = x^{sm+tp^n} = x^{sm}x^{tp^n}$ and, by Corollary 4 of Lagrange's Theorem (Theorem 7.1), $x^{sm} \in H$ and $x^{tp^n} \in K$. Thus, $G = HK$. Now suppose that some $x \in H \cap K$. Then $x^{p^n} = e = x^m$ and, by Corollary 2 of Theorem 4.1, $|x|$ divides both p^n and m. Since p does not divide m, we have $|x| = 1$ and, therefore, $x = e$.

To prove the second assertion of the lemma, note that $p^n m = |HK| = |H||K|/|H \cap K| = |H||K|$ (Theorem 7.2). It follows from Theorem 9.5 and Corollary 2 to Theorem 4.1 that p does not divide $|K|$ and therefore $|H| = p^n$. ∎

Given an Abelian group G with $|G| = p_1^{n_1} p_2^{n_2} \cdots p_k^{n_k}$, where the p's are distinct primes, we let $G(p_i)$ denote the set $\{x \in G \mid x^{p_i^{n_i}} = e\}$.

It then follows immediately from Lemma 1 and induction that $G = G(p_1) \times G(p_2) \times \cdots \times G(p_k)$ and $|G(p_i)| = p_i^{n_i}$. Hence, we turn our attention to groups of prime-power order.

Lemma 2

Let G be an Abelian group of prime-power order and let a be an element of maximum order in G. Then G can be written in the form $\langle a \rangle \times K$.

PROOF We denote $|G|$ by p^n and induct on n. If $n = 1$, then $G = \langle a \rangle \times \langle e \rangle$. Now assume that the statement is true for all Abelian groups of order p^k, where $k < n$. Among all the elements of G, choose a of maximum order p^m. Then $x^{p^m} = e$ for all x in G. We may assume that $G \neq \langle a \rangle$, for otherwise there is nothing to prove. Now, among all the elements of G, choose b of smallest order such that $b \notin \langle a \rangle$. We claim that $\langle a \rangle \cap \langle b \rangle = \{e\}$. Since $|b^p| = |b|/p$, we know that $b^p \in \langle a \rangle$ by the manner in which b was chosen. Say $b^p = a^i$. Notice that $e = b^{p^m} = (b^p)^{p^{m-1}} = (a^i)^{p^{m-1}}$, so $|a^i| \leq p^{m-1}$. Thus, a^i is not a generator of $\langle a \rangle$ and, therefore, by Corollary 3 to Theorem 4.2, $\gcd(p^m, i) \neq 1$. This proves that p divides i, so that we can write $i = pj$. Then $b^p = a^i = a^{pj}$. Consider the element $c = a^{-j}b$. Certainly, c is not in $\langle a \rangle$, for if it were, b would be, too. Also, $c^p = a^{-jp}b^p = a^{-i}b^p = b^{-p}b^p = e$. Thus, we have found an element c of order p such that $c \notin \langle a \rangle$. Since b was chosen to have smallest order such that $b \notin \langle a \rangle$, we conclude that b also has order p. It now follows that $\langle a \rangle \cap \langle b \rangle = \{e\}$, because any nonidentity element of the intersection would generate $\langle b \rangle$ and thus contradict $b \notin \langle a \rangle$.

Now consider the factor group $\overline{G} = G/\langle b \rangle$. To simplify the notation, we let $\overline{x}$ denote the coset $x\langle b \rangle$ in $\overline{G}$. If $|\overline{a}| < |a| = p^m$, then $\overline{a}^{p^{m-1}} = \overline{e}$. This means that $(a\langle b \rangle)^{p^{m-1}} = a^{p^{m-1}}\langle b \rangle = \langle b \rangle$, so that $a^{p^{m-1}} \in \langle a \rangle \cap \langle b \rangle = \{e\}$, contradicting the fact that $|a| = p^m$. Thus, $|\overline{a}| = |a| = p^m$, and therefore $\overline{a}$ is an element of maximum order in $\overline{G}$. By induction, we know that $\overline{G}$ can be written in the form $\langle \overline{a} \rangle \times \overline{K}$ for some subgroup $\overline{K}$ of $\overline{G}$. Let K be the pullback of $\overline{K}$ under the natural homomorphism from G to $\overline{G}$ (that is, $K = \{x \in G \mid \overline{x} \in \overline{K}\}$). We claim that $\langle a \rangle \cap K = \{e\}$. For if $x \in \langle a \rangle \cap K$, then $\overline{x} \in \langle \overline{a} \rangle \cap \overline{K} = \{\overline{e}\} = \langle b \rangle$ and $x \in \langle a \rangle \cap \langle b \rangle = \{e\}$. It now follows from an order argument (see Exercise 35) that $G = \langle a \rangle K$, and therefore $G = \langle a \rangle \times K$. ∎

Lemma 2 and induction on the order of the group now give the following.

Lemma 3

A finite Abelian group of prime-power order is an internal direct product of cyclic groups.

Let us pause to determine where we are in our effort to prove the Fundamental Theorem of Finite Abelian Groups. The remark following Lemma 1 shows that $G = G(p_1) \times G(p_2) \times \cdots \times G(p_n)$, where each $G(p_i)$ is a group of prime-power order, and Lemma 3 shows that each of these factors is an internal direct product of cyclic groups. Thus, we have proved that G is an internal direct product of cyclic groups of prime-power order. All that remains to be proved is the uniqueness of the factors. Certainly the groups $G(p_i)$ are uniquely determined by G, since they comprise the elements of G whose orders are powers of p_i. So we must prove that there is only one way (up to isomorphism and rearrangement of factors) to write each $G(p_i)$ as an internal direct product of cyclic groups.

Lemma 4

Suppose that G is a finite Abelian group of prime-power order. If $G = H_1 \times H_2 \times \cdots \times H_m$ and $G = K_1 \times K_2 \times \cdots \times K_n$, where the H's and K's are nontrivial cyclic subgroups with $|H_1| \geq |H_2| \geq \cdots \geq |H_m|$ and $|K_1| \geq |K_2| \geq \cdots \geq |K_n|$, then $m = n$ and $|H_i| = |K_i|$ for all i.

PROOF We proceed by induction on $|G|$. Clearly, the case where $|G| = p$ is true. Now suppose that the statement is true for all Abelian groups of order less than $|G|$. For any Abelian group L, the set $L^p = \{x^p \mid x \in L\}$ is a subgroup of L (see Exercise 17 in the Supplementary Exercises for Chapters 1–4) and, by Theorem 9.5, is a proper subgroup if p divides $|L|$. It follows that $G^p = H_1{}^p \times H_2{}^p \times \cdots \times H_{m'}{}^p$, and $G^p = K_1{}^p \times K_2{}^p \times \cdots \times K_{n'}{}^p$ where m' is the largest integer i such that $|H_i| > p$, and n' is the largest integer j such that $|K_j| > p$. (This ensures that our two direct products for G^p do not have trivial factors.) Since $|G^p| < |G|$, we have, by induction, $m' = n'$ and $|H_i{}^p| = |K_i{}^p|$ for $i = 1, \ldots, m'$. Since $|H_i| = p|H_i{}^p|$, this proves that $|H_i| = |K_i|$ for all $i = 1, \ldots, m'$. All that remains to be proved is that the number of H_i of order p equals the number of K_i of order p; that is, we must prove that $m - m' = n - n'$ (since $n' = m'$). This follows directly from the facts that $|H_1||H_2| \cdots |H_{m'}|p^{m-m'} = |G| = |K_1||K_2| \cdots |K_{n'}|p^{n-n'}$, $|H_i| = |K_i|$, and $m' = n'$. ■

Exercises

You know it ain't easy, you know how hard it can be.

JOHN LENNON AND PAUL McCARTNEY,
"The Ballad of John and Yoko"*

1. What is the smallest positive integer n such that there are two nonisomorphic groups of order n? Name the two groups.
2. What is the smallest positive integer n such that there are three nonisomorphic Abelian groups of order n? Name the three groups.
3. What is the smallest positive integer n such that there are exactly four nonisomorphic Abelian groups of order n? Name the four groups.
4. Calculate the number of elements of order 2 in each of Z_{16}, $Z_8 \oplus Z_2$, $Z_4 \oplus Z_4$, and $Z_4 \oplus Z_2 \oplus Z_2$. Do the same for the elements of order 4.
5. Prove that any Abelian group of order 45 has an element of order 15. Does every Abelian group of order 45 have an element of order 9?
6. Show that there are two Abelian groups of order 108 that have exactly one subgroup of order 3.
7. Show that there are two Abelian groups of order 108 that have exactly four subgroups of order 3.
8. Show that there are two Abelian groups of order 108 that have exactly 13 subgroups of order 3.
9. Suppose that G is an Abelian group of order 120 and that G has exactly three elements of order 2. Determine the isomorphism class of G.
10. Find all Abelian groups (up to isomorphism) of order 360.
11. Prove that every finite Abelian group can be expressed as the (external) direct product of cyclic groups of orders $n_1, n_2, \ldots, n_t$, where n_{i+1} divides n_i for $i = 1, 2, \ldots, t - 1$. (This exercise is referred to in this chapter.)
12. Suppose that the order of some finite Abelian group is divisible by 10. Prove that the group has a cyclic subgroup of order 10.
13. Show, by example, that if the order of a finite Abelian group is divisible by 4, the group need not have a cyclic subgroup of order 4.
14. On the basis of Exercises 12 and 13, draw a general conclusion about the existence of cyclic subgroups of a finite Abelian group.

15. How many Abelian groups (up to isomorphism) are there
    a. of order 6?
    b. of order 15?
    c. of order 42?
    d. of order pq, where p and q are distinct primes?
    e. of order pqr, where p, q, and r are distinct primes?
    f. Generalize parts d and e.
16. How does the number (up to isomorphism) of Abelian groups of order n compare with the number (up to isomorphism) of Abelian groups of order m where
    a. $n = 3^2$ and $m = 5^2$?
    b. $n = 2^4$ and $m = 5^4$?
    c. $n = p^r$ and $m = q^r$, where p and q are prime?
    d. $n = p^r$ and $m = p^r q$, where p and q are distinct primes?
    e. $n = p^r$ and $m = p^r q^2$, where p and q are distinct primes?
17. Up to isomorphism, how many additive Abelian groups of order 16 have the property that $x + x + x + x = 0$ for all x in the group?
18. Let $p_1, p_2, \ldots, p_n$ be distinct primes. Up to isomorphism, how many Abelian groups are there of order $p_1^4 p_2^4 \ldots p_n^4$?
19. The symmetry group of a nonsquare rectangle is an Abelian group of order 4. Is it isomorphic to Z_4 or $Z_2 \oplus Z_2$?
20. Verify the corollary to the Fundamental Theorem of Finite Abelian Groups in the case that the group has order 1080 and the divisor is 180.
21. The set $\{1, 9, 16, 22, 29, 53, 74, 79, 81\}$ is a group under multiplication modulo 91. Determine the isomorphism class of this group.
22. Suppose that G is a finite Abelian group that has exactly one subgroup for each divisor of $|G|$. Show that G is cyclic.
23. Characterize those integers n such that the only Abelian groups of order n are cyclic.
24. Characterize those integers n such that any Abelian group of order n belongs to one of exactly four isomorphism classes.
25. Refer to Example 1 in this chapter and explain why it is unnecessary to compute the orders of the last five elements listed to determine the isomorphism class of G.
26. Let $G = \{1, 7, 17, 23, 49, 55, 65, 71\}$ under multiplication modulo 96. Express G as an external and an internal direct product of cyclic groups.

27. Let $G = \{1, 7, 43, 49, 51, 57, 93, 99, 101, 107, 143, 149, 151, 157, 193, 199\}$ under multiplication modulo 200. Express G as an external and an internal direct product of cyclic groups.
28. The set $G = \{1, 4, 11, 14, 16, 19, 26, 29, 31, 34, 41, 44\}$ is a group under multiplication modulo 45. Write G as an external and an internal direct product of cyclic groups of prime-power order.
29. Suppose that G is an Abelian group of order 9. What is the maximum number of elements (excluding the identity) of which one needs to compute the order to determine the isomorphism class of G? What if G has order 18? What about 16?
30. Suppose that G is an Abelian group of order 16, and in computing the orders of its elements, you come across an element of order 8 and two elements of order 2. Explain why no further computations are needed to determine the isomorphism class of G.
31. Let G be an Abelian group of order 16. Suppose that there are elements a and b in G such that $|a| = |b| = 4$ and $a^2 \neq b^2$. Determine the isomorphism class of G.
32. Prove that an Abelian group of order 2^n ($n \geq 1$) must have an odd number of elements of order 2.
33. Without using Lagrange's Theorem, show that an Abelian group of odd order cannot have an element of even order.
34. Let G be the group of all $n \times n$ diagonal matrices with ± 1 diagonal entries. What is the isomorphism class of G?
35. Prove the assertion made in the proof of Lemma 2 that $G = \langle a \rangle K$.
36. Suppose that G is a finite Abelian group. Prove that G has order p^n, where p is prime, if and only if the order of every element of G is a power of p.
37. Dirichlet's Theorem says that, for every pair of relatively prime integers a and b, there are infinitely many primes of the form $at + b$. Use Dirichlet's Theorem to prove that every finite Abelian group is isomorphic to a subgroup of a U-group.
38. Determine the isomorphism class of $\text{Aut}(Z_2 \oplus Z_3 \oplus Z_5)$.
39. Give an example to show that Lemma 2 is false if G is non-Abelian.

Computer Exercises

Computer exercises for this chapter are available at the website:

http://www.d.umn.edu/~jgallian

Reference

1. R. Hirshon, "On Cancellation in Groups," *American Mathematical Monthly* 76 (1969): 1037–1039.

Suggested Readings

J. A. Gallian, "Computers in Group Theory," *Mathematics Magazine* 49 (1976): 69–73.

This paper discusses several computer-related projects in group theory done by undergraduate students.

J. Kane, "Distribution of Orders of Abelian Groups," *Mathematics Magazine* 49 (1976): 132–135.

In this article, the author determines the percentages of integers k between 1 and n, for sufficiently large n, that have exactly one isomorphism class of Abelian groups of order k, exactly two isomorphism classes of Abelian groups of order k, and so on, up to 13 isomorphism classes.

G. Mackiw, "Computing in Abstract Algebra," *The College Mathematics Journal* 27 (1996): 136–142.

This article explains how one can use computer software to implement the algorithm given in this chapter for expressing an Abelian group as an internal direct product.

Suggested Website

To find more information about the development of group theory, visit:

http://www-groups.dcs.st-and.ac.uk/~history/

Supplementary Exercises for Chapters 9–11

Every prospector drills many a dry hole, pulls out his rig, and moves on.

JOHN L. HESS

True/false questions for Chapters 9–11 are available on the Web at:

http://www.d.umn.edu/~jgallian/TF

1. Suppose that H is a subgroup of G and that each left coset of H in G is some right coset of H in G. Prove that H is normal in G.
2. Use a factor group-induction argument to prove that a finite Abelian group of order n has a subgroup of order m for every positive divisor m of n.
3. Let $\text{diag}(G) = \{(g, g) \mid g \in G\}$. Prove that $\text{diag}(G) \triangleleft G \oplus G$ if and only if G is Abelian. When G is finite, what is the index of $\text{diag}(G)$ in $G \oplus G$?
4. Let H be any group of rotations in D_n. Prove that H is normal in D_n.
5. Prove that $\text{Inn}(G) \triangleleft \text{Aut}(G)$.
6. Let H be a subgroup of G. Prove that H is a normal subgroup if and only if, for all a and b in G, $ab \in H$ implies $ba \in H$.
7. The factor group $GL(2, \mathbf{R})/SL(2, \mathbf{R})$ is isomorphic to some very familiar group. What is the group?
8. Let k be a divisor of n. The factor group $(Z/\langle n\rangle)/(\langle k\rangle/\langle n\rangle)$ is isomorphic to some very familiar group. What is the group?
9. Let

$$H = \left\{ \begin{bmatrix} 1 & a & b \\ 0 & 1 & c \\ 0 & 0 & 1 \end{bmatrix} \middle| \; a, b, c \in Q \right\}$$

under matrix multiplication.

   a. Find $Z(H)$.
   b. Prove that $Z(H)$ is isomorphic to Q under addition.
   c. Prove that $H/Z(H)$ is isomorphic to $Q \oplus Q$.
   d. Are your proofs for parts a and b valid when Q is replaced by $\mathbf{R}$? Are they valid when Q is replaced by Z_p, where p is prime?

10. Prove that $D_4/Z(D_4)$ is isomorphic to $Z_2 \oplus Z_2$.
11. Prove that Q/Z under addition is an infinite group in which every element has finite order.
12. Show that the intersection of any collection of normal subgroups of a group is a normal subgroup.

13. Let $n > 1$ be a fixed integer and let G be a group. If the set $H = \{x \in G \mid |x| = n\}$ together with the identity forms a subgroup of G, prove that it is a normal subgroup of G. In the case where such a subgroup exists, what can be said about n? Give an example of a non-Abelian group that has such a subgroup. Give an example of a group G and a prime n for which the set H together with the identity is not a subgroup.

14. Show that Q/Z has a unique subgroup of order n for each positive integer n.

15. If H and K are normal Abelian subgroups of a group and $H \cap K = \{e\}$, prove that HK is Abelian.

16. Let G be a group of odd order. Prove that the mapping $x \to x^2$ from G to itself is one-to-one.

17. Suppose that G is a group of permutations on some set. If $|G| = 60$ and $\text{orb}_G(5) = \{1, 5\}$, prove that $\text{stab}_G(5)$ is normal in G.

18. Suppose that $G = H \times K$ and that N is a normal subgroup of H. Prove that N is normal in G.

19. Show that there is no homomorphism from $Z_8 \oplus Z_2 \oplus Z_2$ onto $Z_4 \oplus Z_4$.

20. Show that there is no homomorphism from A_4 onto a group of order 2, 4, or 6, but that there is a homomorphism from A_4 onto a group of order 3.

21. Let H be a normal subgroup of S_4 of order 4. Prove that S_4/H is isomorphic to S_3.

22. Suppose that ϕ is a homomorphism of $U(36)$, $\text{Ker}\,\phi = \{1, 13, 25\}$, and $\phi(5) = 17$. Determine all elements that map to 17.

23. Let $n = 2m$, where m is odd. How many elements of order 2 does $D_n/Z(D_n)$ have? How many elements are in the subgroup $\langle R_{360/n}\rangle/Z(D_n)$? How do these numbers compare with the number of elements of order 2 in D_m?

24. Suppose that H is a normal subgroup of a group G of odd order and that $|H| = 5$. Show that $H \subseteq Z(G)$.

25. Let G be an Abelian group and let n be a positive integer. Let $G_n = \{g \mid g^n = e\}$ and $G^n = \{g^n \mid g \in G\}$. Prove that G/G_n is isomorphic to G^n.

26. Let $\mathbf{R}^+$ denote the multiplicative group of positive reals and let $T = \{a + bi \in \mathbf{C} \mid a^2 + b^2 = 1\}$ be the multiplicative group of complex numbers of norm 1. Show that $\mathbf{C}^*$ is the internal direct product of $\mathbf{R}^+$ and T.

27. Let G be a finite group and let p be a prime. If $p^2 > |G|$, show that any subgroup of order p is normal in G.

28. Let $G = Z \oplus Z$ and $H = \{(x, y) \mid x$ and y are even integers$\}$. Show that H is a subgroup of G. Determine the order of G/H. To which familiar group is G/H isomorphic?

29. Let n be a positive integer. Prove that every element of order n in Q/Z is contained in $\langle 1/n + Z \rangle$.

30. (1997 Putnam Competition) Let G be a group and let $\phi: G \to G$ be a function such that

$$\phi(g_1)\phi(g_2)\phi(g_3) = \phi(h_1)\phi(h_2)\phi(h_3)$$

whenever $g_1g_2g_3 = e = h_1h_2h_3$. Prove that there exists an element a in G such that $\psi(x) = a\phi(x)$ is a homomorphism.

31. Prove that every homomorphism from $Z \oplus Z$ into Z has the form $(x, y) \to ax + by$, where a and b are integers.

32. Prove that every homomorphism from $Z \oplus Z$ into $Z \oplus Z$ has the form $(x, y) \to (ax + by, cx + dy)$, where a, b, c, and d are integers.

33. Prove that Q/Z is not isomorphic to a proper subgroup of itself.

34. Prove that for each positive integer n, the group Q/Z has exactly $\phi(n)$ elements of order n (ϕ is the Euler phi function).

35. Show that any group with more than two elements has an automorphism other than the identity mapping.

36. A proper subgroup H of a group G is called *maximal* if there is no subgroup K such that $H \subset K \subset G$. Prove that Q under addition has no maximal subgroups.

37. Let G be the group of quaternions as given in Exercise 4 of the Supplementary Exercises for Chapters 1–4 and let $H = \langle a^2 \rangle$. Determine whether G/H is isomorphic to Z_4 or $Z_2 \oplus Z_2$. Is G/H isomorphic to a subgroup of G?

38. Write the dihedral group D_8 as $\{R_0, R_{45}, R_{90}, R_{135}, R_{180}, R_{225}, R_{270}, R_{315}, F_1, F_2, F_3, F_4, F_5, F_6, F_7, F_8\}$ and let $N = \{R_0, R_{90}, R_{180}, R_{270}\}$. Prove that N is normal in D_8. Given that $F_1N = \{F_1, F_4, F_3, F_2\}$, determine whether D_8/N is cyclic.

39. Let G be the group $\left\{\begin{bmatrix} 1 & a \\ 0 & b \end{bmatrix} \middle| \text{ where } a, b \in \mathbf{R}, b \neq 0\right\}$ and $H = \left\{\begin{bmatrix} 1 & x \\ 0 & 1 \end{bmatrix} \middle| \text{ where } x \in \mathbf{R}\right\}$. Show that H is a subgroup of G. Is H a normal subgroup of G? Justify your answer.

40. Find a subgroup H of $Z_{p^2} \oplus Z_{p^2}$ such that $(Z_{p^2} \oplus Z_{p^2})/H$ is isomorphic to $Z_p \oplus Z_p$.

41. Recall that H is a characteristic subgroup of K if $\phi(H) = H$ for every automorphism ϕ of K. Prove that if H is a characteristic subgroup of K, and K is a normal subgroup of G, then H is a normal subgroup of G.

PART 3

Rings

For online student resources, visit this textbook's website at
www.CengageBrain.com 

12 Introduction to Rings

Example is the school of mankind, and they will learn at no other.

EDMUND BURKE, *On a Regicide Peace*

Motivation and Definition

Many sets are naturally endowed with two binary operations: addition and multiplication. Examples that quickly come to mind are the integers, the integers modulo n, the real numbers, matrices, and polynomials. When considering these sets as groups, we simply used addition and ignored multiplication. In many instances, however, one wishes to take into account both addition and multiplication. One abstract concept that does this is the concept of a ring.[†] This notion was originated in the mid-19th century by Richard Dedekind, although its first formal abstract definition was not given until Abraham Fraenkel presented it in 1914.

Definition Ring

A *ring* R is a set with two binary operations, addition (denoted by $a + b$) and multiplication (denoted by ab), such that for all a, b, c in R:

1. $a + b = b + a$.
2. $(a + b) + c = a + (b + c)$.
3. **There is an additive identity 0. That is, there is an element 0 in R such that $a + 0 = a$ for all a in R.**
4. **There is an element $-a$ in R such that $a + (-a) = 0$.**
5. $a(bc) = (ab)c$.
6. $a(b + c) = ab + ac$ **and** $(b + c)a = ba + ca$.

So, a ring is an Abelian group under addition, also having an associative multiplication that is left and right distributive over addition. Note that multiplication need not be commutative. When it is, we say that the ring is *commutative*. Also, a ring need not have an identity

[†]The term *ring* was first applied in 1897 by the German mathematician David Hilbert (1862–1943).

under multiplication. A *unity* (or *identity*) in a ring is a nonzero element that is an identity under multiplication. A nonzero element of a commutative ring with unity need not have a multiplicative inverse. When it does, we say that it is a *unit* of the ring. Thus, a is a unit if a^{-1} exists.

The following terminology and notation are convenient. If a and b belong to a commutative ring R and a is nonzero, we say that *a divides b* (or that a is a *factor* of b) and write $a \mid b$, if there exists an element c in R such that $b = ac$. If a does not divide b, we write $a \nmid b$.

Recall that if a is an element from a group under the operation of addition and n is a positive integer, na means $a + a + \cdots + a$, where there are n summands. When dealing with rings, this notation can cause confusion, since we also use juxtaposition for the ring multiplication. When there is the potential for confusion, we will use $n \cdot a$ to mean $a + a + \cdots + a$ (n summands).

For an abstraction to be worthy of study, it must have many diverse concrete realizations. The following list of examples shows that the ring concept is pervasive.

Examples of Rings

EXAMPLE 1 The set Z of integers under ordinary addition and multiplication is a commutative ring with unity 1. The units of Z are 1 and -1. ∎

EXAMPLE 2 The set $Z_n = \{0, 1, \ldots, n - 1\}$ under addition and multiplication modulo n is a commutative ring with unity 1. The set of units is $U(n)$. ∎

EXAMPLE 3 The set $Z[x]$ of all polynomials in the variable x with integer coefficients under ordinary addition and multiplication is a commutative ring with unity $f(x) = 1$. ∎

EXAMPLE 4 The set $M_2(Z)$ of 2×2 matrices with integer entries is a noncommutative ring with unity $\begin{bmatrix} 1 & 0 \\ 0 & 1 \end{bmatrix}$. ∎

EXAMPLE 5 The set $2Z$ of even integers under ordinary addition and multiplication is a commutative ring without unity. ∎

EXAMPLE 6 The set of all continuous real-valued functions of a real variable whose graphs pass through the point (1, 0) is a commutative ring without unity under the operations of pointwise addition and multiplication [that is, the operations $(f + g)(a) = f(a) + g(a)$ and $(fg)(a) = f(a)g(a)$]. ▮

▮ **EXAMPLE 7** Let $R_1, R_2, \ldots, R_n$ be rings. We can use these to construct a new ring as follows. Let

$$R_1 \oplus R_2 \oplus \cdots \oplus R_n = \{(a_1, a_2, \ldots, a_n) \mid a_i \in R_i\}$$

and perform componentwise addition and multiplication; that is, define

$$(a_1, a_2, \ldots, a_n) + (b_1, b_2, \ldots, b_n) = (a_1 + b_1, a_2 + b_2, \ldots, a_n + b_n)$$

and

$$(a_1, a_2, \ldots, a_n)(b_1, b_2, \ldots, b_n) = (a_1b_1, a_2b_2, \ldots, a_nb_n).$$

This ring is called the *direct sum* of $R_1, R_2, \ldots, R_n$. ▮

Properties of Rings

Our first theorem shows how the operations of addition and multiplication intertwine. We use $b - c$ to denote $b + (-c)$.

▮ Theorem 12.1 Rules of Multiplication

Let a, b, and c belong to a ring R. Then

1. $a0 = 0a = 0$.
2. $a(-b) = (-a)b = -(ab)$.
3. $(-a)(-b) = ab$.[†]
4. $a(b - c) = ab - ac$ **and** $(b - c)a = ba - ca$.

Furthermore, if R has a unity element 1, then

5. $(-1)a = -a$.
6. $(-1)(-1) = 1$.

PROOF We will prove rules 1 and 2 and leave the rest as easy exercises (see Exercise 11). To prove statements such as those in Theorem 12.1, we need only "play off" the distributive property against the fact that R is a group under addition with additive identity 0. Consider rule 1. Clearly,

$$0 + a0 = a0 = a(0 + 0) = a0 + a0.$$

So, by cancellation, $0 = a0$. Similarly, $0a = 0$.

[†]"Minus times minus equals plus.
The reason for this we need not discuss."
W. H. Auden, *A Certain World: A Commonplace Book*

To prove rule 2, we observe that $a(-b) + ab = a(-b + b) = a0 = 0$. So, adding $-(ab)$ to both sides yields $a(-b) = -(ab)$. The remainder of rule 2 is done analogously. ∎

Recall that in the case of groups, the identity and inverses are unique. The same is true for rings, provided that these elements exist. The proofs are identical to the ones given for groups and therefore are omitted.

Theorem 12.2 Uniqueness of the Unity and Inverses

If a ring has a unity, it is unique. If a ring element has a multiplicative inverse, it is unique.

Many students have the mistaken tendency to treat a ring as if it were a group under *multiplication*. It is not. The two most common errors are the assumptions that ring elements have multiplicative inverses—they need not—and that a ring has a multiplicative identity—it need not. For example, if a, b, and c belong to a ring, $a \neq 0$ and $ab = ac$, we *cannot* conclude that $b = c$. Similarly, if $a^2 = a$, we *cannot* conclude that $a = 0$ or 1 (as is the case with real numbers). In the first place, the ring need not have multiplicative cancellation, and in the second place, the ring need not have a multiplicative identity. There is an important class of rings that contains Z and $Z[x]$ wherein multiplicative identities exist and for which multiplicative cancellation holds. This class is taken up in the next chapter.

Subrings

In our study of groups, subgroups played a crucial role. Subrings, the analogous structures in ring theory, play a much less prominent role than their counterparts in group theory. Nevertheless, subrings are important.

> Definition **Subring**
> **A subset S of a ring R is a *subring of R* if S is itself a ring with the operations of R.**

Just as was the case for subgroups, there is a simple test for subrings.

Theorem 12.3 Subring Test

A nonempty subset S of a ring R is a subring if S is closed under subtraction and multiplication—that is, if $a - b$ and ab are in S whenever a and b are in S.

PROOF Since addition in R is commutative and S is closed under subtraction, we know by the One-Step Subgroup Test (Theorem 3.1) that S is an Abelian group under addition. Also, since multiplication in R is associative as well as distributive over addition, the same is true for multiplication in S. Thus, the only condition remaining to be checked is that multiplication is a binary operation on S. But this is exactly what closure means. ■

We leave it to the student to confirm that each of the following examples is a subring.

■ **EXAMPLE 8** $\{0\}$ and R are subrings of any ring R. $\{0\}$ is called the *trivial* subring of R. ■

■ **EXAMPLE 9** $\{0, 2, 4\}$ is a subring of the ring Z_6, the integers modulo 6. Note that although 1 is the unity in Z_6, 4 is the unity in $\{0, 2, 4\}$. ■

■ **EXAMPLE 10** For each positive integer n, the set

$$nZ = \{0, \pm n, \pm 2n, \pm 3n, \ldots\}$$

is a subring of the integers Z. ■

■ **EXAMPLE 11** The set of Gaussian integers

$$Z[i] = \{a + bi \mid a, b \in Z\}$$

is a subring of the complex numbers **C**. ■

■ **EXAMPLE 12** Let R be the ring of all real-valued functions of a single real variable under pointwise addition and multiplication. The subset S of R of functions whose graphs pass through the origin forms a subring of R. ■

■ **EXAMPLE 13** The set

$$\left\{ \begin{bmatrix} a & 0 \\ 0 & b \end{bmatrix} \middle| \, a, b \in Z \right\}$$

of diagonal matrices is a subring of the ring of all 2×2 matrices over Z. ■

We can picture the relationship between a ring and its various subrings by way of a subring lattice diagram. In such a diagram, any ring is a subring of all the rings that it is connected to by one or more upward lines. Figure 12.1 shows the relationships among some of the rings we have already discussed.

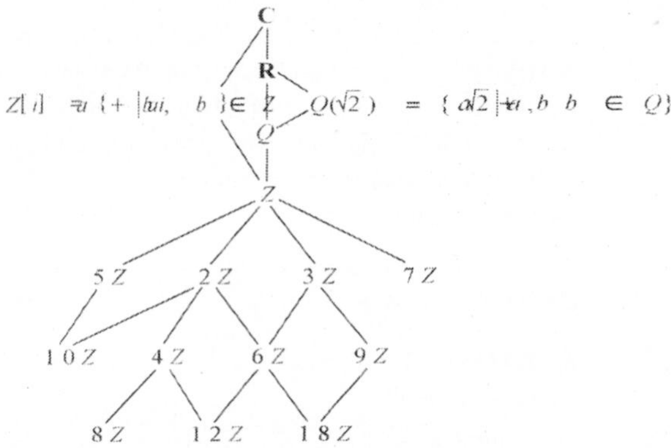

Figure 12.1 Partial subring lattice diagram of C.

In the next several chapters, we will see that many of the fundamental concepts of group theory can be naturally extended to rings. In particular, we will introduce ring homomorphisms and factor rings.

Exercises

There is no substitute for hard work.

THOMAS ALVA EDISON, *Life*

1. Give an example of a finite noncommutative ring. Give an example of an infinite noncommutative ring that does not have a unity.
2. The ring $\{0, 2, 4, 6, 8\}$ under addition and multiplication modulo 10 has a unity. Find it.
3. Give an example of a subset of a ring that is a subgroup under addition but not a subring.
4. Show, by example, that for fixed nonzero elements a and b in a ring, the equation $ax = b$ can have more than one solution. How does this compare with groups?
5. Prove Theorem 12.2.
6. Find an integer n that shows that the rings Z_n need not have the following properties that the ring of integers has.
   a. $a^2 = a$ implies $a = 0$ or $a = 1$.
   b. $ab = 0$ implies $a = 0$ or $b = 0$.
   c. $ab = ac$ and $a \neq 0$ imply $b = c$.
   Is the n you found prime?

7. Show that the three properties listed in Exercise 6 are valid for Z_p, where p is prime.
8. Show that a ring is commutative if it has the property that $ab = ca$ implies $b = c$ when $a \neq 0$.
9. Prove that the intersection of any collection of subrings of a ring R is a subring of R.
10. Verify that Examples 8 through 13 in this chapter are as stated.
11. Prove rules 3 through 6 of Theorem 12.1.
12. Let a, b, and c be elements of a commutative ring, and suppose that a is a unit. Prove that b divides c if and only if ab divides c.
13. Describe all the subrings of the ring of integers.
14. Let a and b belong to a ring R and let m be an integer. Prove that $m \cdot (ab) = (m \cdot a)b = a(m \cdot b)$.
15. Show that if m and n are integers and a and b are elements from a ring, then $(m \cdot a)(n \cdot b) = (mn) \cdot (ab)$. (This exercise is referred to in Chapters 13 and 15.)
16. Show that if n is an integer and a is an element from a ring, then $n \cdot (-a) = -(n \cdot a)$.
17. Show that a ring that is cyclic under addition is commutative.
18. Let a belong to a ring R. Let $S = \{x \in R \mid ax = 0\}$. Show that S is a subring of R.
19. Let R be a ring. The *center of R* is the set $\{x \in R \mid ax = xa$ for all a in $R\}$. Prove that the center of a ring is a subring.
20. Describe the elements of $M_2(Z)$ (see Example 4) that have multiplicative inverses.
21. Suppose that $R_1, R_2, \ldots, R_n$ are rings that contain nonzero elements. Show that $R_1 \oplus R_2 \oplus \cdots \oplus R_n$ has a unity if and only if each R_i has a unity.
22. Let R be a commutative ring with unity and let $U(R)$ denote the set of units of R. Prove that $U(R)$ is a group under the multiplication of R. (This group is called the *group of units of R*.)
23. Determine $U(Z[i])$ (see Example 11).
24. If $R_1, R_2, \ldots, R_n$ are commutative rings with unity, show that $U(R_1 \oplus R_2 \oplus \cdots \oplus R_n) = U(R_1) \oplus U(R_2) \oplus \cdots \oplus U(R_n)$.
25. Determine $U(Z[x])$. (This exercise is referred to in Chapter 17.)
26. Determine $U(\mathbf{R}[x])$.
27. Show that a unit of a ring divides every element of the ring.
28. In Z_6, show that $4 \mid 2$; in Z_8, show that $3 \mid 7$; in Z_{15}, show that $9 \mid 12$.

29. Suppose that a and b belong to a commutative ring R with unity. If a is a unit of R and $b^2 = 0$, show that $a + b$ is a unit of R.

30. Suppose that there is an integer $n > 1$ such that $x^n = x$ for all elements x of some ring. If m is a positive integer and $a^m = 0$ for some a, show that $a = 0$.

31. Give an example of ring elements a and b with the properties that $ab = 0$ but $ba \neq 0$.

32. Let n be an integer greater than 1. In a ring in which $x^n = x$ for all x, show that $ab = 0$ implies $ba = 0$.

33. Suppose that R is a ring such that $x^3 = x$ for all x in R. Prove that $6x = 0$ for all x in R.

34. Suppose that a belongs to a ring and $a^4 = a^2$. Prove that $a^{2n} = a^2$ for all $n \geq 1$.

35. Find an integer $n > 1$ such that $a^n = a$ for all a in Z_6. Do the same for Z_{10}. Show that no such n exists for Z_m when m is divisible by the square of some prime.

36. Let m and n be positive integers and let k be the least common multiple of m and n. Show that $mZ \cap nZ = kZ$.

37. Explain why every subgroup of Z_n under addition is also a subring of Z_n.

38. Is Z_6 a subring of Z_{12}?

39. Suppose that R is a ring with unity 1 and a is an element of R such that $a^2 = 1$. Let $S = \{ara \mid r \in R\}$. Prove that S is a subring of R. Does S contain 1?

40. Let $M_2(Z)$ be the ring of all 2×2 matrices over the integers and let $R = \left\{ \begin{bmatrix} a & a+b \\ a+b & b \end{bmatrix} \middle| \; a, b \in Z \right\}$. Prove or disprove that R is a subring of $M_2(Z)$.

41. Let $M_2(Z)$ be the ring of all 2×2 matrices over the integers and let $R = \left\{ \begin{bmatrix} a & a-b \\ a-b & b \end{bmatrix} \middle| \; a, b \in Z \right\}$. Prove or disprove that R is a subring of $M_2(Z)$.

42. Let $R = \left\{ \begin{bmatrix} a & a \\ b & b \end{bmatrix} \middle| \; a, b \in Z \right\}$. Prove or disprove that R is a subring of $M_2(Z)$.

43. Let $R = Z \oplus Z \oplus Z$ and $S = \{(a, b, c) \in R \mid a + b = c\}$. Prove or disprove that S is a subring of R.

44. Suppose that there is a positive even integer n such that $a^n = a$ for all elements a of some ring. Show that $-a = a$ for all a in the ring.

45. Let R be a ring with unity 1. Show that $S = \{n \cdot 1 \mid n \in Z\}$ is a subring of R.

46. Show that $2Z \cup 3Z$ is not a subring of Z.

47. Determine the smallest subring of Q that contains 1/2. (That is, find the subring S with the property that S contains 1/2 and, if T is any subring containing 1/2, then T contains S.)

48. Determine the smallest subring of Q that contains 2/3.

49. Let R be a ring. Prove that $a^2 - b^2 = (a + b)(a - b)$ for all a, b in R if and only if R is commutative.

50. Suppose that R is a ring and that $a^2 = a$ for all a in R. Show that R is commutative. [A ring in which $a^2 = a$ for all a is called a *Boolean* ring, in honor of the English mathematician George Boole (1815–1864).]

51. Give an example of a Boolean ring with four elements. Give an example of an infinite Boolean ring.

52. If a, b, and c are elements of a ring, does the equation $ax + b = c$ always have a solution x? If it does, must the solution be unique? Answer the same questions given that a is a unit.

Computer Exercises

Software for the computer exercises in this chapter is available at the website:

http://www.d.umn.edu/~jgallian

Suggested Reading

D. B. Erickson, "Orders for Finite Noncommutative Rings," *American Mathematical Monthly* 73 (1966): 376–377.

In this elementary paper, it is shown that there exists a noncommutative ring of order $m > 1$ if and only if m is divisible by the square of a prime.

I. N. Herstein

A whole generation of textbooks and an entire generation of mathematicians, myself included, have been profoundly influenced by that text [Herstein's *Topics in Algebra*].

GEORGIA BENKART

I. N. HERSTEIN was born on March 28, 1923, in Poland. His family moved to Canada when he was seven. He grew up in a poor and tough environment, on which he commented that in his neighborhood you became either a gangster or a college professor. During his school years he played football, hockey, golf, tennis, and pool. During this time he worked as a steeplejack and as a barber at a fair. Herstein received a B.S. degree from the University of Manitoba, an M.A. from the University of Toronto, and, in 1948, a Ph.D. degree from Indiana University under the supervision of Max Zorn. Before permanently settling at the University of Chicago in 1962, he held positions at the University of Kansas, the Ohio State University, the University of Pennsylvania, and Cornell University.

Herstein wrote more than 100 research papers and a dozen books. Although his principal interest was noncommutative ring theory, he also wrote papers on finite groups, linear algebra, and mathematical economics. His textbook *Topics in Algebra*, first published in 1964, dominated the field for 20 years and has become a classic. Herstein had great influence through his teaching and his collaboration with colleagues. He had 30 Ph.D. students, and traveled and lectured widely. His nonmathematical interests included languages and art. He spoke Italian, Hebrew, Polish, and Portuguese. Herstein died on February 9, 1988, after a long battle with cancer.

To find more information about Herstein, visit:

http://www-groups.dcs.st-and.ac.uk/~history/

13 Integral Domains

Don't just read it! Ask your own questions, look for your own examples, discover your own proofs. Is the hypothesis necessary? Is the converse true? What happens in the classical special case? Where does the proof use the hypothesis?

PAUL HALMOS

Definition and Examples

To a certain degree, the notion of a ring was invented in an attempt to put the algebraic properties of the integers into an abstract setting. A ring is not the appropriate abstraction of the integers, however, for too much is lost in the process. Besides the two obvious properties of commutativity and existence of a unity, there is one other essential feature of the integers that rings in general do not enjoy—the cancellation property. In this chapter, we introduce integral domains—a particular class of rings that have all three of these properties. Integral domains play a prominent role in number theory and algebraic geometry.

Definition **Zero-Divisors**

A *zero-divisor* is a nonzero element a of a commutative ring R such that there is a nonzero element $b \in R$ with $ab = 0$.

Definition **Integral Domain**

An *integral domain* is a commutative ring with unity and no zero-divisors.

Thus, in an integral domain, a product is 0 only when one of the factors is 0; that is, $ab = 0$ only when $a = 0$ or $b = 0$. The following examples show that many familiar rings are integral domains and some familiar rings are not. For each example, the student should verify the assertion made.

EXAMPLE 1 The ring of integers is an integral domain. ∎

EXAMPLE 2 The ring of Gaussian integers $Z[i] = \{a + bi \mid a, b \in Z\}$ is an integral domain. ■

EXAMPLE 3 The ring $Z[x]$ of polynomials with integer coefficients is an integral domain. ■

EXAMPLE 4 The ring $Z[\sqrt{2}] = \{a + b\sqrt{2} \mid a, b \in Z\}$ is an integral domain. ■

EXAMPLE 5 The ring Z_p of integers modulo a prime p is an integral domain. ■

EXAMPLE 6 The ring Z_n of integers modulo n is *not* an integral domain when n is not prime. ■

EXAMPLE 7 The ring $M_2(Z)$ of 2×2 matrices over the integers is *not* an integral domain. ■

EXAMPLE 8 $Z \oplus Z$ is *not* an integral domain. ■

What makes integral domains particularly appealing is that they have an important multiplicative group theoretic property, in spite of the fact that the nonzero elements need not form a group under multiplication. This property is cancellation.

Theorem 13.1 Cancellation

Let a, b, and c belong to an integral domain. If $a \neq 0$ and $ab = ac$, then $b = c$.

PROOF From $ab = ac$, we have $a(b - c) = 0$. Since $a \neq 0$, we must have $b - c = 0$. ■

Many authors prefer to define integral domains by the cancellation property—that is, as commutative rings with unity in which the cancellation property holds. This definition is equivalent to ours.

Fields

In many applications, a particular kind of integral domain called a *field* is necessary.

Definition Field
A *field* is a commutative ring with unity in which every nonzero element is a unit.

To verify that every field is an integral domain, observe that if a and b belong to a field with $a \neq 0$ and $ab = 0$, we can multiply both sides of the last expression by a^{-1} to obtain $b = 0$.

It is often helpful to think of ab^{-1} as a divided by b. With this in mind, a field can be thought of as simply an algebraic system that is closed under addition, subtraction, multiplication, and division (except by 0). We have had numerous examples of fields: the complex numbers, the real numbers, the rational numbers. The abstract theory of fields was initiated by Heinrich Weber in 1893. Groups, rings, and fields are the three main branches of abstract algebra. Theorem 13.2 says that, in the finite case, fields and integral domains are the same.

Theorem 13.2 Finite Integral Domains Are Fields

A finite integral domain is a field.

PROOF Let D be a finite integral domain with unity 1. Let a be any nonzero element of D. We must show that a is a unit. If $a = 1$, a is its own inverse, so we may assume that $a \neq 1$. Now consider the following sequence of elements of D: $a, a^2, a^3, \ldots$. Since D is finite, there must be two positive integers i and j such that $i > j$ and $a^i = a^j$. Then, by cancellation, $a^{i-j} = 1$. Since $a \neq 1$, we know that $i - j > 1$, and we have shown that a^{i-j-1} is the inverse of a. ∎

Corollary Z_p Is a Field

For every prime p, Z_p, the ring of integers modulo p is a field.

PROOF According to Theorem 13.2, we need only prove that Z_p has no zero-divisors. So, suppose that $a, b \in Z_p$ and $ab = 0$. Then $ab = pk$ for some integer k. But then, by Euclid's Lemma (see Chapter 0), p divides a or p divides b. Thus, in Z_p, $a = 0$ or $b = 0$. ∎

Putting the preceding corollary together with Example 6, we see that Z_n is a field if and only if n is prime. In Chapter 22, we will describe how all finite fields can be constructed. For now, we give one example of a finite field that is not of the form Z_p.

EXAMPLE 9 Field with Nine Elements

Let
$$\begin{aligned} Z_3[i] &= \{a + bi \mid a, b \in Z_3\} \\ &= \{0, 1, 2, i, 1 + i, 2 + i, 2i, 1 + 2i, 2 + 2i\}, \end{aligned}$$

where $i^2 = -1$. This is the ring of Gaussian integers modulo 3. Elements are added and multiplied as in the complex numbers, except that the coefficients are reduced modulo 3. In particular, $-1 = 2$. Table 13.1 is the multiplication table for the nonzero elements of $Z_3[i]$. ∎

Table 13.1 Multiplication Table for $Z_3[i]^*$

	1	2	i	$1+i$	$2+i$	$2i$	$1+2i$	$2+2i$
1	1	2	i	$1+i$	$2+i$	$2i$	$1+2i$	$2+2i$
2	2	1	$2i$	$2+2i$	$1+2i$	i	$2+i$	$1+i$
i	i	$2i$	2	$2+i$	$2+2i$	1	$1+i$	$1+2i$
$1+i$	$1+i$	$2+2i$	$2+i$	$2i$	1	$1+2i$	2	i
$2+i$	$2+i$	$1+2i$	$2+2i$	1	i	$1+i$	$2i$	2
$2i$	$2i$	i	1	$1+2i$	$1+i$	2	$2+2i$	$2+i$
$1+2i$	$1+2i$	$2+i$	$1+i$	2	$2i$	$2+2i$	i	1
$2+2i$	$2+2i$	$1+i$	$1+2i$	i	2	$2+i$	1	$2i$

∎ **EXAMPLE 10** Let $Q[\sqrt{2}] = \{a + b\sqrt{2} \mid a, b \in Q\}$. It is easy to see that $Q[\sqrt{2}]$ is a ring. Viewed as an element of **R**, the multiplicative inverse of any nonzero element of the form $a + b\sqrt{2}$ is simply $1/(a + b\sqrt{2})$. To verify that $Q[\sqrt{2}]$ is a field, we must show that $1/(a + b\sqrt{2})$ can be written in the form $c + d\sqrt{2}$. In high school algebra, this process is called "rationalizing the denominator." Specifically,

$$\frac{1}{a + b\sqrt{2}} = \frac{1}{a + b\sqrt{2}}\,\frac{a - b\sqrt{2}}{a - b\sqrt{2}} = \frac{a}{a^2 - 2b^2} - \frac{b}{a^2 - 2b^2}\sqrt{2}.$$

(Note that $a + b\sqrt{2} \neq 0$ guarantees that $a - b\sqrt{2} \neq 0$.) ∎

Characteristic of a Ring

Note that for any element x in $Z_3[i]$, we have $3x = x + x + x = 0$, since addition is done modulo 3. Similarly, in the subring $\{0, 3, 6, 9\}$ of Z_{12}, we have $4x = x + x + x + x = 0$ for all x. This observation motivates the following definition.

Definition Characteristic of a Ring

The *characteristic* of a ring R is the least positive integer n such that $nx = 0$ for all x in R. If no such integer exists, we say that R has characteristic 0. The characteristic of R is denoted by char R.

Thus, the ring of integers has characteristic 0, and Z_n has characteristic n. An infinite ring can have a nonzero characteristic. Indeed, the ring $Z_2[x]$ of all polynomials with coefficients in Z_2 has characteristic 2. (Addition and multiplication are done as for polynomials with ordinary integer coefficients except that the coefficients are reduced modulo 2.) When a ring has a unity, the task of determining the characteristic is simplified by Theorem 13.3.

Theorem 13.3 Characteristic of a Ring with Unity

Let R be a ring with unity 1. If 1 has infinite order under addition, then the characteristic of R is 0. If 1 has order n under addition, then the characteristic of R is n.

PROOF If 1 has infinite order, then there is no positive integer n such that $n \cdot 1 = 0$, so R has characteristic 0. Now suppose that 1 has additive order n. Then $n \cdot 1 = 0$, and n is the least positive integer with this property. So, for any x in R, we have

$$\begin{aligned} n \cdot x &= x + x + \cdots + x \ (n \text{ summands}) \\ &= 1x + 1x + \cdots + 1x \ (n \text{ summands}) \\ &= (1 + 1 + \cdots + 1)x \ (n \text{ summands}) \\ &= (n \cdot 1)x = 0x = 0. \end{aligned}$$

Thus, R has characteristic n. ∎

In the case of an integral domain, the possibilities for the characteristic are severely limited.

Theorem 13.4 Characteristic of an Integral Domain

The characteristic of an integral domain is 0 or prime.

PROOF By Theorem 13.3, it suffices to show that if the additive order of 1 is finite, it must be prime. Suppose that 1 has order n and that $n = st$, where $1 \leq s, t \leq n$. Then, by Exercise 15 in Chapter 12,

$$0 = n \cdot 1 = (st) \cdot 1 = (s \cdot 1)(t \cdot 1).$$

So, $s \cdot 1 = 0$ or $t \cdot 1 = 0$. Since n is the least positive integer with the property that $n \cdot 1 = 0$, we must have $s = n$ or $t = n$. Thus, n is prime. ∎

We conclude this chapter with a brief discussion of polynomials with coefficients from a ring—a topic we will consider in detail in later chapters. The existence of zero-divisors in a ring causes unusual results when one is finding zeros of polynomials with coefficients in the ring. Consider, for example, the equation $x^2 - 4x + 3 = 0$. In the integers, we could find all solutions by factoring

$$x^2 - 4x + 3 = (x - 3)(x - 1) = 0$$

and setting each factor equal to 0. But notice that when we say we can find *all* solutions in this manner, we are using the fact that the only way for a product to equal 0 is for one of the factors to be 0—that is, we are using the fact that Z is an integral domain. In Z_{12}, there are many pairs of nonzero elements whose products are 0: $2 \cdot 6 = 0$, $3 \cdot 4 = 0$, $4 \cdot 6 = 0$, $6 \cdot 8 = 0$, and so on. So, how do we find *all* solutions of $x^2 - 4x + 3 = 0$ in Z_{12}? The easiest way is simply to try every element! Upon doing so, we find four solutions: $x = 1$, $x = 3$, $x = 7$, and $x = 9$. Observe that we can find all solutions of $x^2 - 4x + 3 = 0$ over Z_{11} or Z_{13}, say, by setting the two factors $x - 3$ and $x - 1$ equal to 0. Of course, the reason this works for these rings is that they are integral domains. Perhaps this will convince you that integral domains are particularly advantageous rings. Table 13.2 gives a summary of some of the rings we have introduced and their properties.

Table 13.2 Summary of Rings and Their Properties

Ring	Form of Element	Unity	Commutative	Integral Domain	Field	Characteristic
Z	k	1	Yes	Yes	No	0
Z_n, n composite	k	1	Yes	No	No	n
Z_p, p prime	k	1	Yes	Yes	Yes	p
$Z[x]$	$a_nx^n + \cdots + a_1x + a_0$	$f(x) = 1$	Yes	Yes	No	0
nZ, $n > 1$	nk	None	Yes	No	No	0
$M_2(Z)$	$\begin{bmatrix} a & b \\ c & d \end{bmatrix}$	$\begin{bmatrix} 1 & 0 \\ 0 & 1 \end{bmatrix}$	No	No	No	0
$M_2(2Z)$	$\begin{bmatrix} 2a & 2b \\ 2c & 2d \end{bmatrix}$	None	No	No	No	0
$Z[i]$	$a + bi$	1	Yes	Yes	No	0
$Z_3[i]$	$a + bi$; $a, b \in Z_3$	1	Yes	Yes	Yes	3
$Z[\sqrt{2}]$	$a + b\sqrt{2}$; $a, b \in Z$	1	Yes	Yes	No	0
$Q[\sqrt{2}]$	$a + b\sqrt{2}$; $a, b \in Q$	1	Yes	Yes	Yes	0
$Z \oplus Z$	(a, b)	$(1, 1)$	Yes	No	No	0

Exercises

It looked absolutely impossible. But it so happens that you go on worrying away at a problem in science and it seems to get tired, and lies down and lets you catch it.

WILLIAM LAWRENCE BRAGG[†]

1. Verify that Examples 1 through 8 are as claimed.
2. Which of Examples 1 through 5 are fields?
3. Show that a commutative ring with the cancellation property (under multiplication) has no zero-divisors.
4. List all zero-divisors in Z_{20}. Can you see a relationship between the zero-divisors of Z_{20} and the units of Z_{20}?
5. Show that every nonzero element of Z_n is a unit or a zero-divisor.
6. Find a nonzero element in a ring that is neither a zero-divisor nor a unit.
7. Let R be a finite commutative ring with unity. Prove that every nonzero element of R is either a zero-divisor or a unit. What happens if we drop the "finite" condition on R?
8. Let $a \neq 0$ belong to a commutative ring. Prove that a is a zero-divisor if and only if $a^2b = 0$ for some $b \neq 0$.
9. Find elements a, b, and c in the ring $Z \oplus Z \oplus Z$ such that ab, ac, and bc are zero-divisors but abc is not a zero-divisor.
10. Describe all zero-divisors and units of $Z \oplus Q \oplus Z$.
11. Let d be an integer. Prove that $Z[\sqrt{d}] = \{a + b\sqrt{d} \mid a, b \in Z\}$ is an integral domain. (This exercise is referred to in Chapter 18.)
12. In Z_7, give a reasonable interpretation for the expressions $1/2$, $-2/3$, $\sqrt{-3}$, and $-1/6$.
13. Give an example of a commutative ring without zero-divisors that is not an integral domain.
14. Find two elements a and b in a ring such that both a and b are zero-divisors, $a + b \neq 0$, and $a + b$ is not a zero-divisor.
15. Let a belong to a ring R with unity and suppose that $a^n = 0$ for some positive integer n. (Such an element is called *nilpotent.*) Prove that $1 - a$ has a multiplicative inverse in R. [*Hint:* Consider $(1 - a)(1 + a + a^2 + \cdots + a^{n-1})$.]

[†]Bragg, at age 24, won the Nobel Prize for the invention of x-ray crystallography. He remains the youngest person ever to receive the Nobel Prize.

16. Show that the nilpotent elements of a commutative ring form a subring.

17. Show that 0 is the only nilpotent element in an integral domain.

18. A ring element a is called an *idempotent* if $a^2 = a$. Prove that the only idempotents in an integral domain are 0 and 1.

19. Let a and b be idempotents in a commutative ring. Show that each of the following is also an idempotent: ab, $a - ab$, $a + b - ab$, $a + b - 2ab$.

20. Show that Z_n has a nonzero nilpotent element if and only if n is divisible by the square of some prime.

21. Let R be the ring of real-valued continuous functions on $[-1, 1]$. Show that R has zero-divisors.

22. Prove that if a is a ring idempotent, then $a^n = a$ for all positive integers n.

23. Determine all ring elements that are both nilpotent elements and idempotents.

24. Find a zero-divisor in $Z_5[i] = \{a + bi \mid a, b \in Z_5\}$.

25. Find an idempotent in $Z_5[i] = \{a + bi \mid a, b \in Z_5\}$.

26. Find all units, zero-divisors, idempotents, and nilpotent elements in $Z_3 \oplus Z_6$.

27. Determine all elements of a ring that are both units and idempotents.

28. Let R be the set of all real-valued functions defined for all real numbers under function addition and multiplication.

a. Determine all zero-divisors of R.

b. Determine all nilpotent elements of R.

c. Show that every nonzero element is a zero-divisor or a unit.

29. (Subfield Test) Let F be a field and let K be a subset of F with at least two elements. Prove that K is a subfield of F if, for any a, b ($b \neq 0$) in K, $a - b$ and ab^{-1} belong to K.

30. Let d be a positive integer. Prove that $Q[\sqrt{d}] = \{a + b\sqrt{d} \mid a, b \in Q\}$ is a field.

31. Let R be a ring with unity 1. If the product of any pair of nonzero elements of R is nonzero, prove that $ab = 1$ implies $ba = 1$.

32. Let $R = \{0, 2, 4, 6, 8\}$ under addition and multiplication modulo 10. Prove that R is a field.

33. Formulate the appropriate definition of a subdomain (that is, a "sub" integral domain). Let D be an integral domain with unity 1. Show that $P = \{n \cdot 1 \mid n \in Z\}$ (that is, all integral multiples of 1) is a subdomain of D. Show that P is contained in every subdomain of D. What can we say about the order of P?

34. Prove that there is no integral domain with exactly six elements. Can your argument be adapted to show that there is no integral domain with exactly four elements? What about 15 elements? Use these observations to guess a general result about the number of elements in a finite integral domain.

35. Let F be a field of order 2^n. Prove that char $F = 2$.

36. Determine all elements of an integral domain that are their own inverses under multiplication.

37. Characterize those integral domains for which 1 is the only element that is its own multiplicative inverse.

38. Determine all integers $n > 1$ for which $(n - 1)!$ is a zero-divisor in Z_n.

39. Suppose that a and b belong to an integral domain.

a. If $a^5 = b^5$ and $a^3 = b^3$, prove that $a = b$.

b. If $a^m = b^m$ and $a^n = b^n$, where m and n are positive integers that are relatively prime, prove that $a = b$.

40. Find an example of an integral domain and distinct positive integers m and n such that $a^m = b^m$ and $a^n = b^n$, but $a \neq b$.

41. If a is an idempotent in a commutative ring, show that $1 - a$ is also an idempotent.

42. Construct a multiplication table for $Z_2[i]$, the ring of Gaussian integers modulo 2. Is this ring a field? Is it an integral domain?

43. The nonzero elements of $Z_3[i]$ form an Abelian group of order 8 under multiplication. Is it isomorphic to Z_8, $Z_4 \oplus Z_2$, or $Z_2 \oplus Z_2 \oplus Z_2$?

44. Show that $Z_7[\sqrt{3}] = \{a + b\sqrt{3} \mid a, b \in Z_7\}$ is a field. For any positive integer k and any prime p, determine a necessary and sufficient condition for $Z_p[\sqrt{k}] = \{a + b\sqrt{k} \mid a, b \in Z_p\}$ to be a field.

45. Show that a finite commutative ring with no zero-divisors and at least two elements has a unity.

46. Suppose that a and b belong to a commutative ring and ab is a zero-divisor. Show that either a or b is a zero-divisor.

47. Suppose that R is a commutative ring without zero-divisors. Show that all the nonzero elements of R have the same additive order.

48. Suppose that R is a commutative ring without zero-divisors. Show that the characteristic of R is 0 or prime.

49. Let x and y belong to a commutative ring R with prime characteristic p.

a. Show that $(x + y)^p = x^p + y^p$.

b. Show that, for all positive integers n, $(x + y)^{p^n} = x^{p^n} + y^{p^n}$.

c. Find elements x and y in a ring of characteristic 4 such that $(x + y)^4 \neq x^4 + y^4$. (This exercise is referred to in Chapter 20.)

50. Let R be a commutative ring with unity 1 and prime characteristic. If $a \in R$ is nilpotent, prove that there is a positive integer k such that $(1 + a)^k = 1$.

51. Show that any finite field has order p^n, where p is a prime. *Hint:* Use facts about finite Abelian groups. (This exercise is referred to in Chapter 22.)

52. Give an example of an infinite integral domain that has characteristic 3.

53. Let R be a ring and let $M_2(R)$ be the ring of 2×2 matrices with entries from R. Explain why these two rings have the same characteristic.

54. Let R be a ring with m elements. Show that the characteristic of R divides m.

55. Explain why a finite ring must have a nonzero characteristic.

56. Find all solutions of $x^2 - x + 2 = 0$ over $Z_3[i]$. (See Example 9.)

57. Consider the equation $x^2 - 5x + 6 = 0$.
a. How many solutions does this equation have in Z_7?
b. Find all solutions of this equation in Z_8.
c. Find all solutions of this equation in Z_{12}.
d. Find all solutions of this equation in Z_{14}.

58. Find the characteristic of $Z_4 \oplus 4Z$.

59. Suppose that R is an integral domain in which $20 \cdot 1 = 0$ and $12 \cdot 1 = 0$. (Recall that $n \cdot 1$ means the sum $1 + 1 + \cdots + 1$ with n terms.) What is the characteristic of R?

60. In a commutative ring of characteristic 2, prove that the idempotents form a subring.

61. Describe the smallest subfield of the field of real numbers that contains $\sqrt{2}$. (That is, describe the subfield K with the property that K contains $\sqrt{2}$ and if F is any subfield containing $\sqrt{2}$, then F contains K.)

62. Let F be a finite field with n elements. Prove that $x^{n-1} = 1$ for all nonzero x in F.

63. Let F be a field of prime characteristic p. Prove that $K = \{x \in F \mid x^p = x\}$ is a subfield of F.

64. Suppose that a and b belong to a field of order 8 and that $a^2 + ab + b^2 = 0$. Prove that $a = 0$ and $b = 0$. Do the same when the field has order 2^n with n odd.

65. Let F be a field of characteristic 2 with more than two elements. Show that $(x + y)^3 \neq x^3 + y^3$ for some x and y in F.

66. Suppose that F is a field with characteristic not 2, and that the nonzero elements of F form a cyclic group under multiplication. Prove that F is finite.

67. Suppose that D is an integral domain and that ϕ is a nonconstant function from D to the nonnegative integers such that $\phi(xy) = \phi(x)\phi(y)$. If x is a unit in D, show that $\phi(x) = 1$.

68. Let F be a field of order 32. Show that the only subfields of F are F itself and $\{0, 1\}$.

69. Suppose that F is a field with 27 elements. Show that for every element $a \in F$, $5a = -a$.

70. Let

$$R = \left\{ \begin{bmatrix} a & -b \\ b & a \end{bmatrix} \middle| \, a, b \in Z_7 \right\}$$

with the usual matrix addition and multiplication and mod 7 addition and multiplication of the entries. Prove that R is a commutative ring. How many elements are in R? Is R a field? What happens when Z_7 is replaced by Z_5?

Computer Exercises

Computer exercises for this chapter are available at the website:

http://www.d.umn.edu/~jgallian

Suggested Readings

Eric Berg, "A Family of Fields," *Pi Mu Epsilon* 9 (1990): 154–155.

In this article, the author uses properties of logarithms and exponents to define recursively an infinite family of fields starting with the real numbers.

N. A. Khan, "The Characteristic of a Ring," *American Mathematical Monthly* 70 (1963): 736–738.

Here it is shown that a ring has nonzero characteristic n if and only if n is the maximum of the orders of the elements of R.

K. Robin McLean, "Groups in Modular Arithmetic," *The Mathematical Gazette* 62 (1978): 94–104.

This article explores the interplay between various groups of integers under multiplication modulo n and the ring Z_n. It shows how to construct groups of integers in which the identity is not obvious; for example, 1977 is the identity of the group $\{1977, 5931\}$ under multiplication modulo 7908.

Nathan Jacobson

American Mathematical Society

Few mathematicians have been as productive over such a long career or have had as much influence on the profession as has Professor Jacobson.

Citation for the Steele Prize for Lifetime Achievement

NATHAN JACOBSON was born on September 8, 1910, in Warsaw, Poland. After arriving in the United States in 1917, Jacobson grew up in Alabama, Mississippi, and Georgia, where his father owned small clothing stores. He received a B.A. degree from the University of Alabama in 1930 and a Ph.D. from Princeton in 1934. After brief periods as a professor at Bryn Mawr, the University of Chicago, the University of North Carolina, and Johns Hopkins, Jacobson accepted a position at Yale, where he remained until his retirement in 1981.

Jacobson's principal contributions to algebra were in the areas of rings, Lie algebras, and Jordan algebras. In particular, he developed structure theories for these systems. He was the author of nine books and numerous articles, and he had 33 Ph.D. students.

Jacobson held visiting positions in France, India, Italy, Israel, China, Australia, and Switzerland. Among his many honors were the presidency of the American Mathematical Society, memberships in the National Academy of Sciences and the American Academy of Arts and Sciences, a Guggenheim Fellowship, and an honorary degree from the University of Chicago. Jacobson died on December 5, 1999, at the age of 89.

To find more information about Jacobson, visit:

http://www-groups.dcs.st-and.ac.uk/~history/

14 Ideals and Factor Rings

> The secret of science is to ask the right questions, and it is the choice of problem more than anything else that marks the man of genius in the scientific world.
>
> SIR HENRY TIZARD IN C. P. SNOW,
> A postscript to *Science and Government*

Ideals

Normal subgroups play a special role in group theory—they permit us to construct factor groups. In this chapter, we introduce the analogous concepts for rings—ideals and factor rings.

Definition Ideal
A subring A of a ring R is called a (two-sided) *ideal* of R if for every $r \in R$ and every $a \in A$ both ra and ar are in A.

So, a subring A of a ring R is an ideal of R if A "absorbs" elements from R—that is, if $rA = \{ra \mid a \in A\} \subseteq A$ and $Ar = \{ar \mid a \in A\} \subseteq A$ for all $r \in R$.

An ideal A of R is called a *proper* ideal of R if A is a proper subset of R. In practice, one identifies ideals with the following test, which is an immediate consequence of the definition of ideal and the subring test given in Theorem 12.3.

Theorem 14.1 Ideal Test

A nonempty subset A of a ring R is an ideal of R if

1. *$a - b \in A$ whenever $a, b \in A$.*
2. *ra and ar are in A whenever $a \in A$ and $r \in R$.*

EXAMPLE 1 For any ring R, $\{0\}$ and R are ideals of R. The ideal $\{0\}$ is called the *trivial* ideal.

EXAMPLE 2 For any positive integer n, the set $nZ = \{0, \pm n, \pm 2n, \ldots\}$ is an ideal of Z.

EXAMPLE 3 Let R be a commutative ring with unity and let $a \in R$. The set $\langle a \rangle = \{ra \mid r \in R\}$ is an ideal of R called the *principal ideal generated by a*. (Notice that $\langle a \rangle$ is also the notation we used for the cyclic subgroup generated by a. However, the intended meaning will always be clear from the context.) The assumption that R is commutative is necessary in this example (see Exercise 31 in the Supplementary Exercises for Chapters 12–14).

EXAMPLE 4 Let $\mathbf{R}[x]$ denote the set of all polynomials with real coefficients and let A denote the subset of all polynomials with constant term 0. Then A is an ideal of $\mathbf{R}[x]$ and $A = \langle x \rangle$.

EXAMPLE 5 Let R be a commutative ring with unity and let $a_1, a_2, \ldots, a_n$ belong to R. Then $I = \langle a_1, a_2, \ldots, a_n \rangle = \{r_1a_1 + r_2a_2 + \cdots + r_na_n \mid r_i \in R\}$ is an ideal of R called the *ideal generated by* $a_1, a_2, \ldots, a_n$. The verification that I is an ideal is left as an easy exercise (Exercise 3).

EXAMPLE 6 Let $Z[x]$ denote the ring of all polynomials with integer coefficients and let I be the subset of $Z[x]$ of all polynomials with even constant terms. Then I is an ideal of $Z[x]$ and $I = \langle x, 2 \rangle$ (see Exercise 37).

EXAMPLE 7 Let R be the ring of all real-valued functions of a real variable. The subset S of all differentiable functions is a subring of R but not an ideal of R.

Factor Rings

Let R be a ring and let A be an ideal of R. Since R is a group under addition and A is a normal subgroup of R, we may form the factor group $R/A = \{r + A \mid r \in R\}$. The natural question at this point is: How may we form a ring of this group of cosets? The addition is already taken care of, and, by analogy with groups of cosets, we define the product of two cosets of $s + A$ and $t + A$ as $st + A$. The next theorem shows that this definition works as long as A is an ideal, and not just a subring, of R.

Theorem 14.2 Existence of Factor Rings

Let R be a ring and let A be a subring of R. The set of cosets $\{r + A \mid r \in R\}$ ***is a ring under the operations*** $(s + A) + (t + A) = s + t + A$ ***and*** $(s + A)(t + A) = st + A$ ***if and only if A is an ideal of R.***

PROOF We know that the set of cosets forms a group under addition. Once we know that multiplication is indeed a binary operation on the cosets, it is trivial to check that the multiplication is associative and that multiplication is distributive over addition. Hence, the proof boils down to showing that multiplication is well-defined if and only if A is an ideal of R. To do this, let us suppose that A is an ideal and let $s + A = s' + A$ and $t + A = t' + A$. Then we must show that $st + A = s't' + A$. Well, by definition, $s = s' + a$ and $t = t' + b$, where a and b belong to A. Then

$$st = (s' + a)(t' + b) = s't' + at' + s'b + ab,$$

and so

$$st + A = s't' + at' + s'b + ab + A = s't' + A,$$

since A absorbs $at' + s'b + ab$. Thus, multiplication is well-defined when A is an ideal.

On the other hand, suppose that A is a subring of R that is not an ideal of R. Then there exist elements $a \in A$ and $r \in R$ such that $ar \notin A$ or $ra \notin A$. For convenience, say $ar \notin A$. Consider the elements $a + A = 0 + A$ and $r + A$. Clearly, $(a + A)(r + A) = ar + A$ but $(0 + A) \cdot (r + A) = 0 \cdot r + A = A$. Since $ar + A \neq A$, the multiplication is not well-defined and the set of cosets is not a ring. ∎

Let's look at a few factor rings.

EXAMPLE 8 $Z/4Z = \{0 + 4Z, 1 + 4Z, 2 + 4Z, 3 + 4Z\}$. To see how to add and multiply, consider $2 + 4Z$ and $3 + 4Z$.

$$(2 + 4Z) + (3 + 4Z) = 5 + 4Z = 1 + 4 + 4Z = 1 + 4Z,$$
$$(2 + 4Z)(3 + 4Z) = 6 + 4Z = 2 + 4 + 4Z = 2 + 4Z.$$

One can readily see that the two operations are essentially modulo 4 arithmetic. ∎

EXAMPLE 9 $2Z/6Z = \{0 + 6Z, 2 + 6Z, 4 + 6Z\}$. Here the operations are essentially modulo 6 arithmetic. For example, $(4 + 6Z) + (4 + 6Z) = 2 + 6Z$ and $(4 + 6Z)(4 + 6Z) = 4 + 6Z$. ∎

Here is a noncommutative example of an ideal and factor ring.

EXAMPLE 10 Let $R = \left\{ \begin{bmatrix} a_1 & a_2 \\ a_3 & a_4 \end{bmatrix} \middle| a_i \in Z \right\}$ and let I be the subset of R consisting of matrices with even entries. It is easy to show that I is indeed an ideal of R (Exercise 21). Consider the factor ring R/I. The interesting question about this ring is: What is its size? We claim R/I has 16 elements; in fact, $R/I = \left\{ \begin{bmatrix} r_1 & r_2 \\ r_3 & r_4 \end{bmatrix} + I \mid r_i \in \{0, 1\} \right\}$. An example illustrates the typical situation. Which of the 16 elements is $\begin{bmatrix} 7 & 8 \\ 5 & -3 \end{bmatrix} + I$? Well, observe that $\begin{bmatrix} 7 & 8 \\ 5 & -3 \end{bmatrix} + I = \begin{bmatrix} 1 & 0 \\ 1 & 1 \end{bmatrix} + \begin{bmatrix} 6 & 8 \\ 4 & -4 \end{bmatrix} + I = \begin{bmatrix} 1 & 0 \\ 1 & 1 \end{bmatrix} + I$, since an ideal absorbs its own elements. The general case is left to the reader (Exercise 23). ∎

EXAMPLE 11 Consider the factor ring of the Gaussian integers $R = Z[i]/\langle 2 - i \rangle$. What does this ring look like? Of course, the elements of R have the form $a + bi + \langle 2 - i \rangle$, where a and b are integers, but the important question is: What do the *distinct* cosets look like? The fact that $2 - i + \langle 2 - i \rangle = 0 + \langle 2 - i \rangle$ means that *when dealing with coset representatives*, we may treat $2 - i$ as equivalent to 0, so that $2 = i$. For example, the coset $3 + 4i + \langle 2 - i \rangle = 3 + 8 + \langle 2 - i \rangle = 11 + \langle 2 - i \rangle$. Similarly, all the elements of R can be written in the form $a + \langle 2 - i \rangle$, where a is an integer. But we can further reduce the set of distinct coset representatives by observing that *when dealing with coset representatives*, $2 = i$ implies (by squaring both sides) that $4 = -1$ or $5 = 0$. Thus, the coset $3 + 4i + \langle 2 - i \rangle = 11 + \langle 2 - i \rangle = 1 + 5 + 5 + \langle 2 - i \rangle = 1 + \langle 2 - i \rangle$. In this way, we can show that every element of R is equal to one of the following cosets: $0 + \langle 2 - i \rangle$, $1 + \langle 2 - i \rangle$, $2 + \langle 2 - i \rangle$, $3 + \langle 2 - i \rangle$, $4 + \langle 2 - i \rangle$. Is any further reduction possible? To demonstrate that there is not, we will show that these five cosets are distinct. It suffices to show that $1 + \langle 2 - i \rangle$ has additive order 5. Since $5(1 + \langle 2 - i \rangle) = 5 + \langle 2 - i \rangle = 0 + \langle 2 - i \rangle$, $1 + \langle 2 - i \rangle$ has order 1 or 5. If the order is actually 1, then $1 + \langle 2 - i \rangle = 0 + \langle 2 - i \rangle$, so $1 \in \langle 2 - i \rangle$. Thus, $1 = (2 - i)(a + bi) = 2a + b + (-a + 2b)i$ for some integers a and b. But this equation implies that $1 = 2a + b$ and $0 = -a + 2b$, and solving these simultaneously yields $b = 1/5$, which is a contradiction. It should be clear that the ring R is essentially the same as the field Z_5. ∎

∎ **EXAMPLE 12** Let $\mathbf{R}[x]$ denote the ring of polynomials with real coefficients and let $\langle x^2 + 1\rangle$ denote the principal ideal generated by $x^2 + 1$; that is,

$$\langle x^2 + 1\rangle = \{f(x)(x^2 + 1) \mid f(x) \in \mathbf{R}[x]\}.$$

Then

$$\begin{aligned}\mathbf{R}[x]/\langle x^2 + 1\rangle &= \{g(x) + \langle x^2 + 1\rangle \mid g(x) \in \mathbf{R}[x]\}\\ &= \{ax + b + \langle x^2 + 1\rangle \mid a, b \in \mathbf{R}\}.\end{aligned}$$

To see this last equality, note that if $g(x)$ is any member of $\mathbf{R}[x]$, then we may write $g(x)$ in the form $q(x)(x^2 + 1) + r(x)$, where $q(x)$ is the quotient and $r(x)$ is the remainder upon dividing $g(x)$ by $x^2 + 1$. In particular, $r(x) = 0$ or the degree of $r(x)$ is less than 2, so that $r(x) = ax + b$ for some a and b in $\mathbf{R}$. Thus,

$$\begin{aligned}g(x) + \langle x^2 + 1\rangle &= q(x)(x^2 + 1) + r(x) + \langle x^2 + 1\rangle\\ &= r(x) + \langle x^2 + 1\rangle,\end{aligned}$$

since the ideal $\langle x^2 + 1\rangle$ absorbs the term $q(x)(x^2 + 1)$.

How is multiplication done? Since

$$x^2 + 1 + \langle x^2 + 1\rangle = 0 + \langle x^2 + 1\rangle,$$

one should think of $x^2 + 1$ as 0 or, equivalently, as $x^2 = -1$. So, for example,

$$\begin{aligned}&(x + 3 + \langle x^2 + 1\rangle) \cdot (2x + 5 + \langle x^2 + 1\rangle)\\ &\quad = 2x^2 + 11x + 15 + \langle x^2 + 1\rangle = 11x + 13 + \langle x^2 + 1\rangle.\end{aligned}$$

In view of the fact that the elements of this ring have the form $ax + b + \langle x^2 + 1\rangle$, where $x^2 + \langle x^2 + 1\rangle = -1 + \langle x^2 + 1\rangle$, it is perhaps not surprising that this ring turns out to be algebraically the same ring as the ring of complex numbers. This observation was first made by Cauchy in 1847. ∎

Examples 11 and 12 illustrate one of the most important applications of factor rings—the construction of rings with highly desirable properties. In particular, we shall show how one may use factor rings to construct integral domains and fields.

Prime Ideals and Maximal Ideals

Definition Prime Ideal, Maximal Ideal

A *prime ideal* A of a commutative ring R is a proper ideal of R such that $a, b \in R$ and $ab \in A$ imply $a \in A$ or $b \in A$. A *maximal* ideal of a commutative ring R is a *proper* ideal of R such that, whenever B is an ideal of R and $A \subseteq B \subseteq R$, then $B = A$ or $B = R$.

So, the only ideal that properly contains a maximal ideal is the entire ring. The motivation for the definition of a prime ideal comes from the integers.

EXAMPLE 13 Let n be an integer greater than 1. Then, in the ring of integers, the ideal nZ is prime if and only if n is prime (Exercise 9). ($\{0\}$ is also a prime ideal of Z.) ■

EXAMPLE 14 The lattice of ideals of Z_{36} (Figure 14.1) shows that only $\langle 2\rangle$ and $\langle 3\rangle$ are maximal ideals. ■

EXAMPLE 15 The ideal $\langle x^2 + 1\rangle$ is maximal in $\mathbf{R}[x]$. To see this, assume that A is an ideal of $\mathbf{R}[x]$ that properly contains $\langle x^2 + 1\rangle$. We will prove that $A = \mathbf{R}[x]$ by showing that A contains some nonzero real number c. [This is the constant polynomial $h(x) = c$ for all x.] Then $1 = (1/c)c \in A$ and therefore, by Exercise 15, $A = \mathbf{R}[x]$. To this end, let $f(x) \in A$, but $f(x) \notin \langle x^2 + 1\rangle$. Then

$$f(x) = q(x)(x^2 + 1) + r(x),$$

where $r(x) \neq 0$ and the degree of $r(x)$ is less than 2. It follows that $r(x) = ax + b$, where a and b are not both 0, and

$$ax + b = r(x) = f(x) - q(x)(x^2 + 1) \in A.$$

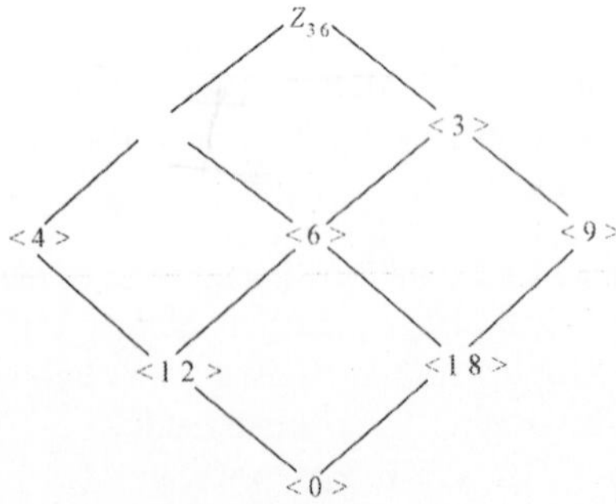

Figure 14.1

Thus,

$$a^2x^2 - b^2 = (ax + b)(ax - b) \in A \qquad \text{and} \qquad a^2(x^2 + 1) \in A.$$

So,

$$0 \neq a^2 + b^2 = (a^2x^2 + a^2) - (a^2x^2 - b^2) \in A.$$

■

EXAMPLE 16 The ideal $\langle x^2 + 1 \rangle$ is not prime in $Z_2[x]$, since it contains $(x + 1)^2 = x^2 + 2x + 1 = x^2 + 1$ but does not contain $x + 1$. ■

The next two theorems are useful for determining whether a particular ideal is prime or maximal.

Theorem 14.3 R/A Is an Integral Domain If and Only If A Is Prime

Let R be a commutative ring with unity and let A be an ideal of R. Then R/A is an integral domain if and only if A is prime.

PROOF Suppose that R/A is an integral domain and $ab \in A$. Then $(a + A)(b + A) = ab + A = A$, the zero element of the ring R/A. So, either $a + A = A$ or $b + A = A$; that is, either $a \in A$ or $b \in A$. Hence, A is prime.

To prove the other half of the theorem, we first observe that R/A is a commutative ring with unity for any proper ideal A. Thus, our task is simply to show that when A is prime, R/A has no zero-divisors. So, suppose that A is prime and $(a + A)(b + A) = 0 + A = A$. Then $ab \in A$ and, therefore, $a \in A$ or $b \in A$. Thus, one of $a + A$ or $b + A$ is the zero coset in R/A. ■

For maximal ideals, we can do even better.

Theorem 14.4 R/A Is a Field If and Only If A Is Maximal

Let R be a commutative ring with unity and let A be an ideal of R. Then R/A is a field if and only if A is maximal.

PROOF Suppose that R/A is a field and B is an ideal of R that properly contains A. Let $b \in B$ but $b \notin A$. Then $b + A$ is a nonzero element of R/A and, therefore, there exists an element $c + A$ such that $(b + A) \cdot (c + A) = 1 + A$, the multiplicative identity of R/A. Since $b \in B$, we have $bc \in B$. Because

$$1 + A = (b + A)(c + A) = bc + A,$$

we have $1 - bc \in A \subset B$. So, $1 = (1 - bc) + bc \in B$. By Exercise 15, $B = R$. This proves that A is maximal.

Now suppose that A is maximal and let $b \in R$ but $b \notin A$. It suffices to show that $b + A$ has a multiplicative inverse. (All other properties for a field follow trivially.) Consider $B = \{br + a \mid r \in R, a \in A\}$. This is an ideal of R that properly contains A (Exercise 25). Since A is maximal, we must have $B = R$. Thus, $1 \in B$, say, $1 = bc + a'$, where $a' \in A$. Then

$$1 + A = bc + a' + A = bc + A = (b + A)(c + A).$$

When a commutative ring has a unity, it follows from Theorems 14.3 and 14.4 that a maximal ideal is a prime ideal. The next example shows that a prime ideal need not be maximal.

EXAMPLE 17 The ideal $\langle x \rangle$ is a prime ideal in $Z[x]$ but not a maximal ideal in $Z[x]$. To verify this, we begin with the observation that $\langle x \rangle = \{f(x) \in Z[x] \mid f(0) = 0\}$ (see Exercise 29). Thus, if $g(x)h(x) \in \langle x \rangle$, then $g(0)h(0) = 0$. And since $g(0)$ and $h(0)$ are integers, we have $g(0) = 0$ or $h(0) = 0$.

To see that $\langle x \rangle$ is not maximal, we simply note that $\langle x \rangle \subset \langle x, 2 \rangle \subset Z[x]$ (see Exercise 37).

Exercises

Problems worthy of attack
prove their worth by hitting back.

PIET HEIN, "Problems," Grooks*

1. Verify that the set defined in Example 3 is an ideal.
2. Verify that the set A in Example 4 is an ideal and that $A = \langle x \rangle$.
3. Verify that the set I in Example 5 is an ideal and that if J is any ideal of R that contains $a_1, a_2, \ldots, a_n$, then $I \subseteq J$. (Hence, $\langle a_1, a_2, \ldots, a_n \rangle$ is the smallest ideal of R that contains $a_1, a_2, \ldots, a_n$.)
4. Find a subring of $Z \oplus Z$ that is not an ideal of $Z \oplus Z$.
5. Let $S = \{a + bi \mid a, b \in Z, b \text{ is even}\}$. Show that S is a subring of $Z[i]$, but not an ideal of $Z[i]$.
6. Find all maximal ideals in
   a. Z_8. **b.** Z_{10}. **c.** Z_{12}. **d.** Z_n.
7. Let a belong to a commutative ring R. Show that $aR = \{ar \mid r \in R\}$ is an ideal of R. If R is the ring of even integers, list the elements of $4R$.

8. Prove that the intersection of any set of ideals of a ring is an ideal.

9. If n is an integer greater than 1, show that $\langle n \rangle = nZ$ is a prime ideal of Z if and only if n is prime. (This exercise is referred to in this chapter.)

10. If A and B are ideals of a ring, show that the *sum* of A and B, $A + B = \{a + b \mid a \in A, b \in B\}$, is an ideal.

11. In the ring of integers, find a positive integer a such that

a. $\langle a \rangle = \langle 2 \rangle + \langle 3 \rangle$.

b. $\langle a \rangle = \langle 6 \rangle + \langle 8 \rangle$.

c. $\langle a \rangle = \langle m \rangle + \langle n \rangle$.

12. If A and B are ideals of a ring, show that the *product* of A and B, $AB = \{a_1b_1 + a_2b_2 + \cdots + a_nb_n \mid a_i \in A, b_i \in B, n \text{ a positive integer}\}$, is an ideal.

13. Find a positive integer a such that

a. $\langle a \rangle = \langle 3 \rangle\langle 4 \rangle$.

b. $\langle a \rangle = \langle 6 \rangle\langle 8 \rangle$.

c. $\langle a \rangle = \langle m \rangle\langle n \rangle$.

14. Let A and B be ideals of a ring. Prove that $AB \subseteq A \cap B$.

15. If A is an ideal of a ring R and 1 belongs to A, prove that $A = R$. (This exercise is referred to in this chapter.)

16. If A and B are ideals of a commutative ring R with unity and $A + B = R$, show that $A \cap B = AB$.

17. If an ideal I of a ring R contains a unit, show that $I = R$.

18. Suppose that in the ring Z, the ideal $\langle 35 \rangle$ is a proper ideal of J and J is a proper ideal of I. What are the possibilities for J? What are the possibilities for I?

19. Give an example of a ring that has exactly two maximal ideals.

20. Suppose that R is a commutative ring and $|R| = 30$. If I is an ideal of R and $|R| = 10$, prove that I is a maximal ideal.

21. Let R and I be as described in Example 10. Prove that I is an ideal of R.

22. Let $I = \langle 2 \rangle$. Prove that $I[x]$ is not a maximal ideal of $Z[x]$ even though I is a maximal ideal of Z.

23. Verify the claim made in Example 10 about the size of R/I.

24. Give an example of a commutative ring that has a maximal ideal that is not a prime ideal.

25. Show that the set B in the latter half of the proof of Theorem 14.4 is an ideal of R. (This exercise is referred to in this chapter.)

26. If R is a commutative ring with unity and A is a proper ideal of R, show that R/A is a commutative ring with unity.

27. Prove that the only ideals of a field F are $\{0\}$ and F itself.

28. Show that $\mathbf{R}[x]/\langle x^2 + 1\rangle$ is a field.

29. In $Z[x]$, the ring of polynomials with integer coefficients, let $I = \{f(x) \in Z[x] \mid f(0) = 0\}$. Prove that $I = \langle x\rangle$. (This exercise is referred to in this chapter and in Chapter 15.)

30. Show that $A = \{(3x, y) \mid x, y \in Z\}$ is a maximal ideal of $Z \oplus Z$. Generalize. What happens if $3x$ is replaced by $4x$? Generalize.

31. Let R be the ring of continuous functions from $\mathbf{R}$ to $\mathbf{R}$. Show that $A = \{f \in R \mid f(0) = 0\}$ is a maximal ideal of R.

32. Let $R = Z_8 \oplus Z_{30}$. Find all maximal ideals of R, and for each maximal ideal I, identify the size of the field R/I.

33. How many elements are in $Z[i]/\langle 3 + i\rangle$? Give reasons for your answer.

34. In $Z[x]$, the ring of polynomials with integer coefficients, let $I = \{f(x) \in Z[x] \mid f(0) = 0\}$. Prove that I is not a maximal ideal.

35. In $Z \oplus Z$, let $I = \{(a, 0) \mid a \in Z\}$. Show that I is a prime ideal but not a maximal ideal.

36. Let R be a ring and let I be an ideal of R. Prove that the factor ring R/I is commutative if and only if $rs - sr \in I$ for all r and s in R.

37. In $Z[x]$, let $I = \{f(x) \in Z[x] \mid f(0)$ is an even integer$\}$. Prove that $I = \langle x, 2\rangle$. Is I a prime ideal of $Z[x]$? Is I a maximal ideal? How many elements does $Z[x]/I$ have? (This exercise is referred to in this chapter.)

38. Prove that $I = \langle 2 + 2i\rangle$ is not a prime ideal of $Z[i]$. How many elements are in $Z[i]/I$? What is the characteristic of $Z[i]/I$?

39. In $Z_5[x]$, let $I = \langle x^2 + x + 2\rangle$. Find the multiplicative inverse of $2x + 3 + I$ in $Z_5[x]/I$.

40. Let R be a ring and let p be a fixed prime. Show that $I_p = \{r \in R \mid$ additive order of r is a power of $p\}$ is an ideal of R.

41. An integral domain D is called a *principal ideal domain* if every ideal of D has the form $\langle a\rangle = \{ad \mid d \in D\}$ for some a in D. Show that Z is a principal ideal domain. (This exercise is referred to in Chapter 18.)

42. Let $R = \left\{\begin{bmatrix} a & b \\ 0 & d \end{bmatrix} \middle| \, a, b, d \in Z\right\}$ and $S = \left\{\begin{bmatrix} r & s \\ 0 & t \end{bmatrix} \middle| \, r, s, t \in Z, s \text{ is even}\right\}$. If S is an ideal of R, what can you say about r and t?

43. If R and S are principal ideal domains, prove that $R \oplus S$ is a principal ideal ring. (See Exercise 41 for the definition.)

44. Let a and b belong to a commutative ring R. Prove that $\{x \in R \mid ax \in bR\}$ is an ideal.

45. Let R be a commutative ring and let A be any subset of R. Show that the *annihilator* of A, $\text{Ann}(A) = \{r \in R \mid ra = 0 \text{ for all } a \text{ in } A\}$, is an ideal.

46. Let R be a commutative ring and let A be any ideal of R. Show that the *nil radical* of A, $N(A) = \{r \in R \mid r^n \in A$ for some positive integer n (n depends on r)$\}$, is an ideal of R. [$N(\langle 0 \rangle)$ is called the *nil radical* of R.]

47. Let $R = Z_{27}$. Find
a. $N(\langle 0 \rangle)$. **b.** $N(\langle 3 \rangle)$. **c.** $N(\langle 9 \rangle)$.

48. Let $R = Z_{36}$. Find
a. $N(\langle 0 \rangle)$. **b.** $N(\langle 4 \rangle)$. **c.** $N(\langle 6 \rangle)$.

49. Let R be a commutative ring. Show that $R/N(\langle 0 \rangle)$ has no nonzero nilpotent elements.

50. Let A be an ideal of a commutative ring. Prove that $N(N(A)) = N(A)$.

51. Let $Z_2[x]$ be the ring of all polynomials with coefficients in Z_2 (that is, coefficients are 0 or 1, and addition and multiplication of coefficients are done modulo 2). Show that $Z_2[x]/\langle x^2 + x + 1 \rangle$ is a field.

52. List the elements of the field given in Exercise 51, and make an addition and multiplication table for the field.

53. Show that $Z_3[x]/\langle x^2 + x + 1 \rangle$ is not a field.

54. Let R be a commutative ring without unity, and let $a \in R$. Describe the smallest ideal I of R that contains a (that is, if J is any ideal that contains a, then $I \subseteq J$).

55. Let R be the ring of continuous functions from **R** to **R**. Let $A = \{f \in R \mid f(0)$ is an even integer$\}$. Show that A is a subring of R, but not an ideal of R.

56. Show that $Z[i]/\langle 1 - i \rangle$ is a field. How many elements does this field have?

57. If R is a principal ideal domain and I is an ideal of R, prove that every ideal of R/I is principal (see Exercise 41).

58. How many elements are in $Z_5[i]/\langle 1 + i \rangle$?

59. Let R be a commutative ring with unity that has the property that $a^2 = a$ for all a in R. Let I be a prime ideal in R. Show that $|R/I| = 2$.

60. Let R be a commutative ring with unity, and let I be a proper ideal with the property that every element of R that is not in I is a unit of R. Prove that I is the unique maximal ideal of R.

61. Let $I_0 = \{f(x) \in Z[x] \mid f(0) = 0\}$. For any positive integer n, show that there exists a sequence of strictly increasing ideals such that $I_0 \subset I_1 \subset I_2 \subset \cdots \subset I_n \subset Z[x]$.

62. Let $R = \{(a_1, a_2, a_3, \ldots)\}$, where each $a_i \in Z$. Let $I = \{(a_1, a_2, a_3, \ldots)\}$, where only a finite number of terms are nonzero. Prove that I is not a principal ideal of R.

63. Let R be a commutative ring with unity and let $a, b \in R$. Show that $\langle a, b \rangle$, the smallest ideal of R containing a and b, is $I = \{ra + sb \mid r, s \in R\}$. That is, show that I contains a and b and that any ideal that contains a and b also contains I.

Computer Exercises

Computer exercises for this chapter are available at the website:

http://www.d.umn.edu/~jgallian

Richard Dedekind

Richard Dedekind was not only a mathematician, but one of the wholly great in the history of mathematics, now and in the past, the last hero of a great epoch, the last pupil of Gauss, for four decades himself a classic, from whose works not only we, but our teachers and the teachers of our teachers, have drawn.

EDMUND LANDAU,
Commemorative Address to the Royal Society of Göttingen

This stamp was issued by East Germany in 1981 to commemorate the 150th anniversary of Dedekind's birth. Notice that it features the representation of an ideal as the product of powers of prime ideals.

RICHARD DEDEKIND was born on October 6, 1831, in Brunswick, Germany, the birthplace of Gauss. Dedekind was the youngest of four children of a law professor. His early interests were in chemistry and physics, but he obtained a doctor's degree in mathematics at the age of 21 under Gauss at the University of Göttingen. Dedekind continued his studies at Göttingen for a few years, and in 1854 he began to lecture there.

Dedekind spent the years 1858–1862 as a professor in Zürich. Then he accepted a position at an institute in Brunswick where he had once been a student. Although this school was less than university level, Dedekind remained there for the next 50 years. He died in Brunswick in 1916.

During his career, Dedekind made numerous fundamental contributions to mathematics. His treatment of irrational numbers, "Dedekind cuts," put analysis on a firm, logical foundation. His work on unique factorization led to the modern theory of algebraic numbers. He was a pioneer in the theory of rings and fields. The notion of ideals as well as the term itself are attributed to Dedekind. Mathematics historian Morris Kline has called him "the effective founder of abstract algebra."

To find more information about Dedekind, visit:

http://www-groups.dcs.st-and.ac.uk/~history/

Emmy Noether

The Granger Collection, New York

. . . she discovered methods which have proved of enormous importance in the development of the present-day younger generation of mathematicians.

ALBERT EINSTEIN, *The New York Time*

EMMY NOETHER was born on March 23, 1882, in Germany. When she entered the University of Erlangen, she was one of only two women among the 1000 students. Noether completed her doctorate in 1907.

In 1916, Noether went to Göttingen and, under the influence of David Hilbert and Felix Klein, became interested in general relativity. While there, she made a major contribution to physics with her theorem that whenever there is a symmetry in nature, there is also a conservation law, and vice versa. Hilbert tried unsuccessfully to obtain a faculty appointment at Göttingen for Noether, saying, "I do not see that the sex of the candidate is an argument against her admission as Privatdozent. After all, we are a university and not a bathing establishment."

It was not until she was 38 that Noether's true genius revealed itself. Over the next 13 years, she used an axiomatic method to develop a general theory of ideals and noncommutative algebras. With this abstract theory, Noether was able to weld together many important concepts. Her approach was even more important than the individual results. Hermann Weyl said of Noether, "She originated above all a new and epoch-making style of thinking in algebra."

With the rise of Hitler in 1933, Noether, a Jew, fled to the United States and took a position at Bryn Mawr College. She died suddenly on April 14, 1935, following an operation.

To find more information about Noether, visit:

http://www-groups.dcs.st-and.ac.uk/~history/

Supplementary Exercises for Chapters 12–14

If at first you do succeed—try to hide your astonishment.

HARRY F. BANKS

True/false questions for Chapters 12–14 are available on the Web at:

http://www.d.umn.edu/~jgallian/TF

1. Find all idempotents in Z_{10}, Z_{20}, and Z_{30}. (Recall that a is an idempotent if $a^2 = a$.)
2. If m and n are relatively prime integers greater than 1, prove that Z_{mn} has at least two idempotents besides 0 and 1.
3. Suppose that R is a ring in which $a^2 = 0$ implies $a = 0$. Show that R has no nonzero nilpotent elements. (Recall that b is nilpotent if $b^n = 0$ for some positive integer n.)
4. Let R be a commutative ring with more than one element. Prove that if for every nonzero element a of R we have $aR = R$, then R is a field.
5. Let A, B, and C be ideals of a ring R. If $AB \subseteq C$ and C is a prime ideal of R, show that $A \subseteq C$ or $B \subseteq C$. (Compare this with Euclid's Lemma in Chapter 0.)
6. Show, by example, that the intersection of two prime ideals need not be a prime ideal.
7. Let **R** denote the ring of real numbers. Determine all ideals of $\mathbf{R} \oplus \mathbf{R}$. What happens if **R** is replaced by any field F?
8. Determine all factor rings of Z.
9. Suppose that n is a square-free positive integer (that is, n is not divisible by the square of any prime). Prove that Z_n has no nonzero nilpotent elements.
10. Let R be a commutative ring with unity. Suppose that a is a unit and b is nilpotent. Show that $a + b$ is a unit. (*Hint:* See Exercise 29 in Chapter 12.)
11. Let A, B, and C be subrings of a ring R. If $A \subseteq B \cup C$, show that $A \subseteq B$ or $A \subseteq C$.
12. For any element a in a ring R, define $\langle a \rangle$ to be the smallest ideal of R that contains a. If R is a commutative ring with unity, show that $\langle a \rangle = aR = \{ar \mid r \in R\}$. Show, by example, that if R is commutative but does not have a unity, then $\langle a \rangle$ and aR may be different.
13. Let R be a ring with unity. Show that $\langle a \rangle = \{s_1at_1 + s_2at_2 + \cdots + s_nat_n \mid s_i, t_i \in R$ and n is a positive integer$\}$.
14. Show that $Z_n[x]$ has characteristic n.

15. Let A and B be ideals of a ring R. If $A \cap B = \{0\}$, show that $ab = 0$ when $a \in A$ and $b \in B$.
16. Show that the direct sum of two integral domains is not an integral domain.
17. Consider the ring $R = \{0, 2, 4, 6, 8, 10\}$ under addition and multiplication modulo 12. What is the characteristic of R?
18. What is the characteristic of $Z_m \oplus Z_n$? Generalize.
19. Let R be a commutative ring with unity. Suppose that the only ideals of R are $\{0\}$ and R. Show that R is a field.
20. Suppose that I is an ideal of J and that J is an ideal of R. Prove that if I has a unity, then I is an ideal of R. (Be careful not to assume that the unity of I is the unity of R. It need not be—see Exercise 2 in Chapter 12.)
21. Show that in the ring $Z[x]/\langle 2x + 1\rangle$, the element $x + \langle 2x + 1\rangle$ is a unit.
22. Let $a \in Z$. Show that $\langle a\rangle$ is not a maximal ideal in $Z[x]$.
23. Recall that an idempotent b in a ring is an element with the property that $b^2 = b$. Find a nontrivial idempotent (that is, not 0 and not 1) in $Q[x]/\langle x^4 + x^2\rangle$.
24. In a principal ideal domain, show that every nontrivial prime ideal is a maximal ideal.
25. Find an example of a commutative ring R with unity such that a, $b \in R$, $a \neq b$, $a^n = b^n$, and $a^m = b^m$, where n and m are positive integers that are relatively prime. (Compare with Exercise 39, part b, in Chapter 13.)
26. Let $Q(\sqrt[3]{2})$ denote the smallest subfield of $\mathbf{R}$ that contains Q and $\sqrt[3]{2}$. [That is, $Q(\sqrt[3]{2})$ is the subfield with the property that $Q(\sqrt[3]{2})$ contains Q and $\sqrt[3]{2}$ and if F is any subfield containing Q and $\sqrt[3]{2}$, then F contains $Q(\sqrt[3]{2})$.] Describe the elements of $Q(\sqrt[3]{2})$.
27. Let R be an integral domain with nonzero characteristic. If A is a proper ideal of R, show that R/A has the same characteristic as R.
28. Let F be a field of order p^n. Determine the group isomorphism class of F under the operation addition.
29. If R is a finite commutative ring with unity, prove that every prime ideal of R is a maximal ideal of R.
30. Let R be a noncommutative ring and let $C(R)$ be the center of R (see Exercise 19 in Chapter 12). Prove that the additive group of $R/C(R)$ is not cyclic.

31. Let

$$R = \left\{ \begin{bmatrix} a & b \\ c & d \end{bmatrix} \middle| \, a, b, c, d \in Z_2 \right\}$$

with ordinary matrix addition and multiplication modulo 2. Show that

$$A = \left\{ \begin{bmatrix} 1 & 0 \\ 0 & 0 \end{bmatrix} r \middle| \, r \in R \right\}$$

is not an ideal of R. (Hence, in Exercise 7 in Chapter 14, the commutativity assumption is necessary.)

32. If R is an integral domain and A is a proper ideal of R, must R/A be an integral domain?

33. Let $A = \{a + bi \mid a, b \in Z, a \bmod 2 = b \bmod 2\}$. How many elements does $Z[i]/A$ have? Show that A is a maximal ideal of $Z[i]$.

34. Suppose that R is a commutative ring with unity such that for each a in R there is a positive integer n greater than 1 (n depends on a) such that $a^n = a$. Prove that every prime ideal of R is a maximal ideal of R.

35. State a "finite subfield test"; that is, state conditions that guarantee that a finite subset of a field is a subfield.

36. Let F be a finite field with more than two elements. Prove that the sum of all of the elements of F is 0.

37. Show that if there are nonzero elements a and b in Z_n such that $a^2 + b^2 = 0$, then the ring $Z_n[i] = \{x + yi \mid x, y \in Z_n\}$ has zero-divisors. Use this fact to find a zero-divisor in $Z_{13}[i]$.

38. Suppose that R is a ring with no zero-divisors and that R contains a nonzero element b such that $b^2 = b$. Show that b is a unity for R.

39. Find the characteristic of $Z[i]/\langle 2 + i \rangle$.

40. Show that the characteristic of $Z[i]/\langle a + bi \rangle$ divides $a^2 + b^2$.

41. Show that $4x^2 + 6x + 3$ is a unit in $Z_8[x]$.

42. For any commutative ring R, $R[x, y]$ is the ring of polynomials in x and y with coefficients in R (that is, $R[x, y]$ consists of all finite sums of terms of the form ax^iy^j, where $a \in R$ and i and j are nonnegative integers). (For example, $x^4 - 3x^2y - y^3 \in Z[x, y]$.) Prove that $\langle x, y \rangle$ is a prime ideal in $Z[x, y]$ but not a maximal ideal in $Z[x, y]$.

43. Prove that $\langle x, y \rangle$ is a maximal ideal in $Z_5[x, y]$.

44. Prove that $\langle 2, x, y \rangle$ is a maximal ideal in $Z[x, y]$.

45. Let R and S be rings. Prove that (a, b) is nilpotent in $R \oplus S$ if and only if both a and b are nilpotent.

46. Let R and S be commutative rings. Prove that (a, b) is a zero-divisor in $R \oplus S$ if and only if a or b is a zero-divisor or exactly one of a or b is 0.
47. Determine all idempotents in Z_{p^k}, where p is a prime.
48. Let R be a commutative ring with unity 1. Show that a is an idempotent if and only if there exists an element b in R such that $ab = 0$ and $a + b = 1$.
49. Let $Z_n[\sqrt{2}] = \{a + b\sqrt{2} \mid a, b \in Z_n\}$. Define addition and multiplication as in $Z[\sqrt{2}]$, except that modulo n arithmetic is used to combine the coefficients. Show that $Z_3[\sqrt{2}]$ is a field but $Z_7[\sqrt{2}]$ is not.
50. Let p be a prime. Prove that every zero-divisor in Z_{p^n} is a nilpotent element.
51. If x is a nilpotent element in a commutative ring R, prove that rx is nilpotent for all r in R.
52. List the distinct elements in the ring $Z[x]/\langle 3, x^2 + 1\rangle$. Show that this ring is a field.

15 Ring Homomorphisms

> If there is one central idea which is common to all aspects of modern algebra it is the notion of homomorphism.
>
> I. N. HERSTEIN, *Topics in Algebra*

Definition and Examples

In our work with groups, we saw that one way to discover information about a group is to examine its interaction with other groups by way of homomorphisms. It should not be surprising to learn that this concept extends to rings with equally profitable results.

Just as a group homomorphism preserves the group operation, a ring homomorphism preserves the ring operations.

Definitions Ring Homomorphism, Ring Isomorphism

A *ring homomorphism* ϕ from a ring R to a ring S is a mapping from R to S that preserves the two ring operations; that is, for all a, b in R,

$$\phi(a + b) = \phi(a) + \phi(b) \qquad \text{and} \qquad \phi(ab) = \phi(a)\phi(b).$$

A ring homomorphism that is both one-to-one and onto is called a *ring isomorphism*.

As is the case for groups, in the preceding definition the operations on the left of the equal signs are those of R, whereas the operations on the right of the equal signs are those of S.

Again as with group theory, the roles of isomorphisms and homomorphisms are entirely distinct. An isomorphism is used to show that two rings are algebraically identical; a homomorphism is used to simplify a ring while retaining certain of its features.

A schematic representation of a ring homomorphism is given in Figure 15.1. The dashed arrows indicate the results of performing the ring operations.

The following examples illustrate ring homomorphisms. The reader should supply the missing details.

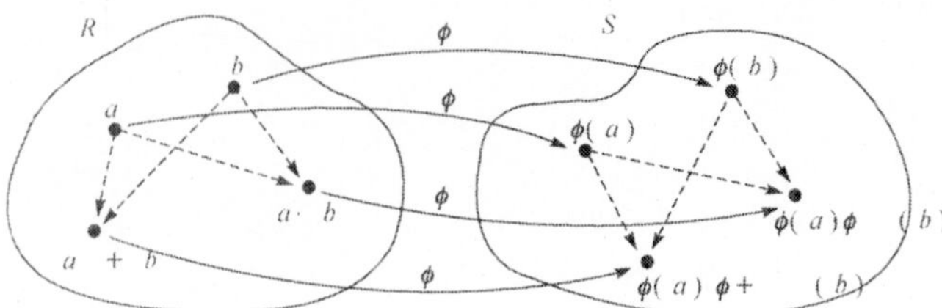

Figure 15.1

EXAMPLE 1 For any positive integer n, the mapping $k \to k \bmod n$ is a ring homomorphism from Z onto Z_n (see Exercise 9 in Chapter 0). This mapping is called the *natural homomorphism* from Z to Z_n. ∎

EXAMPLE 2 The mapping $a + bi \to a - bi$ is a ring isomorphism from the complex numbers onto the complex numbers (see Exercise 35 in Chapter 6). ∎

EXAMPLE 3 Let $\mathbf{R}[x]$ denote the ring of all polynomials with real coefficients. The mapping $f(x) \to f(1)$ is a ring homomorphism from $\mathbf{R}[x]$ onto $\mathbf{R}$. ∎

EXAMPLE 4 The correspondence $\phi\colon x \to 5x$ from Z_4 to Z_{10} is a ring homomorphism. Although showing that $\phi(x + y) = \phi(x) + \phi(y)$ appears to be accomplished by the simple statement that $5(x + y) = 5x + 5y$, we must bear in mind that the addition on the left is done modulo 4, whereas the addition on the right and the multiplication on both sides are done modulo 10. An analogous difficulty arises in showing that ϕ preserves multiplication. So, to verify that ϕ preserves both operations, we write $x + y = 4q_1 + r_1$ and $xy = 4q_2 + r_2$, where $0 \le r_1 < 4$ and $0 \le r_2 < 4$. Then $\phi(x + y) = \phi(r_1) = 5r_1 = 5(x + y - 4q_1) = 5x + 5y - 20q_1 = 5x + 5y = \phi(x) + \phi(y)$ in Z_{10}. Similarly, using the fact that $5 \cdot 5 = 5$ in Z_{10}, we have $\phi(xy) = \phi(r_2) = 5r_2 = 5(xy - 4q_2) = 5xy - 20q_2 = (5 \cdot 5)xy = 5x5y = \phi(x)\phi(y)$ in Z_{10}. ∎

EXAMPLE 5 We determine all ring homomorphisms from Z_{12} to Z_{30}. By Example 10 in Chapter 10, the only group homomorphisms from Z_{12} to Z_{30} are $x \to ax$, where $a = 0, 15, 10, 20, 5$, or 25. But, since $1 \cdot 1 = 1$ in Z_{12}, we must have $a \cdot a = a$ in Z_{30}. This requirement rules out 20 and 5 as possibilities for a. Finally, simple calculations show that each of the remaining four choices does yield a ring homomorphism. ∎

EXAMPLE 6 Let R be a commutative ring of characteristic 2. Then the mapping $a \to a^2$ is a ring homomorphism from R to R. ∎

EXAMPLE 7 Although $2Z$, the group of even integers under addition, is group-isomorphic to the group Z under addition, the ring $2Z$ is not ring-isomorphic to the ring Z. (Quick! What does Z have that $2Z$ doesn't?) ∎

Our next two examples are applications to number theory of the natural homomorphism given in Example 1.

EXAMPLE 8 Test for Divisibility by 9

An integer n with decimal representation $a_k a_{k-1} \cdots a_0$ is divisible by 9 if and only if $a_k + a_{k-1} + \cdots + a_0$ is divisible by 9. To verify this, observe that $n = a_k 10^k + a_{k-1} 10^{k-1} + \cdots + a_0$. Then, letting α denote the natural homomorphism from Z to Z_9 [in particular, $\alpha(10) = 1$], we note that n is divisible by 9 if and only if

$$\begin{aligned} 0 = \alpha(n) &= \alpha(a_k)(\alpha(10))^k + \alpha(a_{k-1})(\alpha(10))^{k-1} + \cdots + \alpha(a_0) \\ &= \alpha(a_k) + \alpha(a_{k-1}) + \cdots + \alpha(a_0) \\ &= \alpha(a_k + a_{k-1} + \cdots + a_0). \end{aligned}$$

But $\alpha(a_k + a_{k-1} + \cdots + a_0) = 0$ is equivalent to $a_k + a_{k-1} + \cdots + a_0$ being divisible by 9. ∎

EXAMPLE 9 Theorem of Gersonides

Among the most important unsolved problems in number theory is the so-called "abc conjecture." This conjecture is a natural generalization of a theorem first proved in the fourteenth century by the Rabbi Gersonides. Gersonides proved that the only pairs of positive integers that are powers of 2 and powers of 3 which differ by 1 are 1, 2; 2, 3; 3, 4; and 8, 9. That is, these four pairs are the only solutions to the equations $2^m = 3^n \pm 1$. To verify that this is so for $2^m = 3^n + 1$, observe that for all n we have $3^n \bmod 8 = 3$ or 1. Thus, $3^n + 1 \bmod 8 = 4$ or 2. On the other hand, for $m > 2$, we have $2^m \bmod 8 = 0$. To handle the case where $2^m = 3^n - 1$, we first note that for all n, $3^n \bmod 16 = 3, 9, 11$, or 1, depending on the value of $n \bmod 4$. Thus, $(3^n - 1) \bmod 16 = 2, 8, 10$, or 0. Since $2^m \bmod 16 = 0$ for $m \geq 4$, we have ruled out the cases where $n \bmod 4 = 1, 2$, or 3. Because $3^{4k} \bmod 5 = (3^4)^k \bmod 5 = 1^k \bmod 5 = 1$, we know that $(3^{4k} - 1) \bmod 5 = 0$. But the only values for $2^m \bmod 5$ are 2, 4, 3, and 1. This contradiction completes the proof. ∎

Properties of Ring Homomorphisms

Theorem 15.1 Properties of Ring Homomorphisms

Let ϕ be a ring homomorphism from a ring R to a ring S. Let A be a subring of R and let B be an ideal of S.

1. *For any $r \in R$ and any positive integer n, $\phi(nr) = n\phi(r)$ and $\phi(r^n) = (\phi(r))^n$.*
2. *$\phi(A) = \{\phi(a) \mid a \in A\}$ is a subring of S.*
3. *If A is an ideal and ϕ is onto S, then $\phi(A)$ is an ideal.*
4. *$\phi^{-1}(B) = \{r \in R \mid \phi(r) \in B\}$ is an ideal of R.*
5. *If R is commutative, then $\phi(R)$ is commutative.*
6. *If R has a unity 1, $S \neq \{0\}$, and ϕ is onto, then $\phi(1)$ is the unity of S.*
7. *ϕ is an isomorphism if and only if ϕ is onto and Ker $\phi = \{r \in R \mid \phi(r) = 0\} = \{0\}$.*
8. *If ϕ is an isomorphism from R onto S, then ϕ^{-1} is an isomorphism from S onto R.*

PROOF The proofs of these properties are similar to those given in Theorems 10.1 and 10.2 and are left as exercises (Exercise 1). ∎

The student should learn the various properties of Theorem 15.1 in words in addition to the symbols. Property 2 says that the homomorphic image of a subring is a subring. Property 4 says that the pullback of an ideal is an ideal, and so on.

The next three theorems parallel results we had for groups. The proofs are nearly identical to their group theory counterparts and are left as exercises (Exercises 2, 3, and 4).

Theorem 15.2 Kernels Are Ideals

Let ϕ be a ring homomorphism from a ring R to a ring S. Then Ker $\phi = \{r \in R \mid \phi(r) = 0\}$ is an ideal of R.

Theorem 15.3 First Isomorphism Theorem for Rings

Let ϕ be a ring homomorphism from R to S. Then the mapping from $R/\text{Ker } \phi$ to $\phi(R)$, given by $r + \text{Ker } \phi \to \phi(r)$, is an isomorphism. In symbols, $R/\text{Ker } \phi \approx \phi(R)$.

Theorem 15.4 Ideals Are Kernels

Every ideal of a ring R is the kernel of a ring homomorphism of R. In particular, an ideal A is the kernel of the mapping $r \to r + A$ from R to R/A.

The homomorphism from R to R/A given in Theorem 15.4 is called the *natural homomorphism* from R to R/A. Theorem 15.3 is often referred to as the Fundamental Theorem of Ring Homomorphisms.

In Example 17 in Chapter 14 we gave a direct proof that $\langle x \rangle$ is a prime ideal of $Z[x]$ but not a maximal ideal. In the following example we illustrate a better way to do this kind of problem.

EXAMPLE 10 Since the mapping ϕ from $Z[x]$ onto Z given by $\phi(f(x)) = f(0)$ is a ring homomorphism with Ker $\phi = \langle x \rangle$ (see Exercise 29 in Chapter 14), we have, by Theorem 15.3, $Z[x]/\langle x \rangle \approx Z$. And because Z is an integral domain but not a field, we know by Theorems 14.3 and 14.4 that the ideal $\langle x \rangle$ is prime but not maximal in $Z[x]$. ∎

Theorem 15.5 Homomorphism from Z to a Ring with Unity

Let R be a ring with unity 1. The mapping $\phi: Z \to R$ given by $n \to n \cdot 1$ is a ring homomorphism.

PROOF Since the multiplicative group property $a^{m+n} = a^m a^n$ translates to $(m + n)a = ma + na$ when the operation is addition, we have $\phi(m + n) = (m + n) \cdot 1 = m \cdot 1 + n \cdot 1$. So, ϕ preserves addition.

That ϕ also preserves multiplication follows from Exercise 15 in Chapter 12, which says that $(m \cdot a)(n \cdot b) = (mn) \cdot (ab)$ for all integers m and n. Thus, $\phi(mn) = (mn) \cdot 1 = (mn) \cdot ((1)(1)) = (m \cdot 1)(n \cdot 1) = \phi(m)\phi(n)$. So, ϕ preserves multiplication as well. ∎

Corollary 1 A Ring with Unity Contains Z_n or Z

If R is a ring with unity and the characteristic of R is $n > 0$, then R contains a subring isomorphic to Z_n. If the characteristic of R is 0, then R contains a subring isomorphic to Z.

PROOF Let 1 be the unity of R and let $S = \{k \cdot 1 \mid k \in Z\}$. Theorem 15.5 shows that the mapping ϕ from Z to S given by $\phi(k) = k \cdot 1$ is a homomorphism, and by the First Isomorphism Theorem for rings, we have $Z/\text{Ker } \phi \approx S$. But, clearly, Ker $\phi = \langle n \rangle$, where n is the additive order of 1 and, by Theorem 13.3, n is also the characteristic of R. So, when R has characteristic n, $S \approx Z/\langle n\rangle \approx Z_n$. When R has characteristic 0, $S \approx Z/\langle 0\rangle \approx Z$. ▮

Corollary 2 Z_m Is a Homomorphic Image of Z

For any positive integer m, the mapping of ϕ: $Z \to Z_m$ given by $x \to x \bmod m$ is a ring homomorphism.

PROOF This follows directly from the statement of Theorem 15.5, since in the ring Z_m, the integer $x \bmod m$ is $x \cdot 1$. (For example, in Z_3, if $x = 5$, we have $5 \cdot 1 = 1 + 1 + 1 + 1 + 1 = 2$.) ▮

Corollary 3 A Field Contains Z_p or Q (Steinitz, 1910)

If F is a field of characteristic p, then F contains a subfield isomorphic to Z_p. If F is a field of characteristic 0, then F contains a subfield isomorphic to the rational numbers.

PROOF By Corollary 1, F contains a subring isomorphic to Z_p if F has characteristic p, and F has a subring S isomorphic to Z if F has characteristic 0. In the latter case, let

$$T = \{ab^{-1} \mid a, b \in S, b \neq 0\}.$$

Then T is isomorphic to the rationals (Exercise 63). ▮

Since the intersection of all subfields of a field is itself a subfield (Exercise 11), every field has a smallest subfield (that is, a subfield that is contained in every subfield). This subfield is called the *prime subfield* of the field. It follows from Corollary 3 that the prime subfield of a field of characteristic p is isomorphic to Z_p, whereas the prime subfield of a field of characteristic 0 is isomorphic to Q. (See Exercise 67.)

The Field of Quotients

Although the integral domain Z is not a field, it is at least contained in a field—the field of rational numbers. And notice that the field of rational numbers is nothing more than quotients of integers. Can we mimic the construction of the rationals from the integers for other integral domains? Yes. The field constructed in Theorem 15.6 is called the *field of quotients of D*. Throughout the proof of Theorem 15.6, you should keep in mind that we are using the construction of the rationals from the integers as a model for our construction of the field of quotients of D.

Theorem 15.6 Field of Quotients

Let D be an integral domain. Then there exists a field F (called the field of quotients of D) that contains a subring isomorphic to D.

PROOF Let $S = \{(a, b) \mid a, b \in D, b \neq 0\}$. We define an equivalence relation on S by $(a, b) \equiv (c, d)$ if $ad = bc$ (compare with Example 17 in Chapter 0). Now, let F be the set of equivalence classes of S under the relation $\equiv$ and denote the equivalence class that contains (x, y) by x/y. We define addition and multiplication on F by

$$a/b + c/d = (ad + bc)/(bd) \qquad \text{and} \qquad a/b \cdot c/d = (ac)/(bd).$$

(Notice that here we need the fact that D is an integral domain to ensure that multiplication is closed; that is, $bd \neq 0$ whenever $b \neq 0$ and $d \neq 0$.)

Since there are many representations of any particular element of F (just as in the rationals, we have $1/2 = 3/6 = 4/8$), we must show that these two operations are well-defined. To do this, suppose that $a/b = a'/b'$ and $c/d = c'/d'$, so that $ab' = a'b$ and $cd' = c'd$. It then follows that

$$\begin{aligned}(ad + bc)b'd' &= adb'd' + bcb'd' = (ab')dd' + (cd')bb' \\ &= (a'b)dd' + (c'd)bb' = a'd'bd + b'c'bd \\ &= (a'd' + b'c')bd.\end{aligned}$$

Thus, by definition, we have

$$(ad + bc)/(bd) = (a'd' + b'c')/(b'd'),$$

and, therefore, addition is well-defined. We leave the verification that multiplication is well-defined as an exercise (Exercise 55). That F is a field is straightforward. Let 1 denote the unity of D. Then $0/1$ is the additive identity of F. The additive inverse of a/b is $-a/b$; the multiplicative inverse of a nonzero element a/b is b/a. The remaining field properties can be checked easily.

Finally, the mapping $\phi: D \to F$ given by $x \to x/1$ is a ring isomorphism from D to $\phi(D)$ (see Exercise 7). ■

■ **EXAMPLE 11** Let $D = Z[x]$. Then the field of quotients of D is $\{f(x)/g(x) \mid f(x), g(x) \in D$, where $g(x)$ is not the zero polynomial$\}$. ■

When F is a field, the field of quotients of $F[x]$ is traditionally denoted by $F(x)$.

EXAMPLE 12 Let p be a prime. Then $Z_p(x) = \{f(x)/g(x) \mid f(x), g(x) \in Z_p[x], g(x) \neq 0\}$ is an infinite field of characteristic p. ∎

Exercises

We can work it out.

JOHN LENNON AND PAUL McCARTNEY,
"We Can Work It Out," single*

1. Prove Theorem 15.1.
2. Prove Theorem 15.2.
3. Prove Theorem 15.3.
4. Prove Theorem 15.4.
5. Show that the correspondence $x \to 5x$ from Z_5 to Z_{10} does not preserve addition.
6. Show that the correspondence $x \to 3x$ from Z_4 to Z_{12} does not preserve multiplication.
7. Show that the mapping $\phi: D \to F$ in the proof of Theorem 15.6 is a ring homomorphism.
8. Prove that every ring homomorphism ϕ from Z_n to itself has the form $\phi(x) = ax$, where $a^2 = a$.
9. Suppose that ϕ is a ring homomorphism from Z_m to Z_n. Prove that if $\phi(1) = a$, then $a^2 = a$. Give an example to show that the converse is false.
10. **a.** Is the ring $2Z$ isomorphic to the ring $3Z$?
    b. Is the ring $2Z$ isomorphic to the ring $4Z$?
11. Prove that the intersection of any collection of subfields of a field F is a subfield of F. (This exercise is referred to in this chapter.)
12. Let $Z_3[i] = \{a + bi \mid a, b \in Z_3\}$ (see Example 9 in Chapter 13). Show that the field $Z_3[i]$ is ring-isomorphic to the field $Z_3[x]/\langle x^2 + 1\rangle$.
13. Let

$$S = \left\{ \begin{bmatrix} a & b \\ -b & a \end{bmatrix} \middle| a, b \in \mathbf{R} \right\}.$$

Show that $\phi: \mathbf{C} \to S$ given by

$$\phi(a + bi) = \begin{bmatrix} a & b \\ -b & a \end{bmatrix}$$

is a ring isomorphism.

14. Let $Z[\sqrt{2}] = \{a + b\sqrt{2} \mid a, b \in Z\}$ and

$$H = \left\{ \begin{bmatrix} a & 2b \\ b & a \end{bmatrix} \middle| a, b \in Z \right\}.$$

Show that $Z[\sqrt{2}]$ and H are isomorphic as rings.

15. Consider the mapping from $M_2(Z)$ into Z given by $\begin{bmatrix} a & b \\ c & d \end{bmatrix} \to a$. Prove or disprove that this is a ring homomorphism.

16. Let $R = \left\{ \begin{bmatrix} a & b \\ 0 & c \end{bmatrix} \middle| a, b, c \in Z \right\}$. Prove or disprove that the mapping $\begin{bmatrix} a & b \\ 0 & c \end{bmatrix} \to a$ is a ring homomorphism.

17. Is the mapping from Z_5 to Z_{30} given by $x \to 6x$ a ring homomorphism? Note that the image of the unity is the unity of the image but not the unity of Z_{30}.

18. Is the mapping from Z_{10} to Z_{10} given by $x \to 2x$ a ring homomorphism?

19. Describe the kernel of the homomorphism given in Example 3.

20. Recall that a ring element a is called an idempotent if $a^2 = a$. Prove that a ring homomorphism carries an idempotent to an idempotent.

21. Determine all ring homomorphisms from Z_6 to Z_6. Determine all ring homomorphisms from Z_{20} to Z_{30}.

22. Determine all ring isomorphisms from Z_n to itself.

23. Determine all ring homomorphisms from Z to Z.

24. Suppose ϕ is a ring homomorphism from $Z \oplus Z$ into $Z \oplus Z$. What are the possibilities for $\phi((1, 0))$?

25. Determine all ring homomorphisms from $Z \oplus Z$ into $Z \oplus Z$.

26. In Z, let $A = \langle 2 \rangle$ and $B = \langle 8 \rangle$. Show that the group A/B is isomorphic to the group Z_4 but that the ring A/B is not ring-isomorphic to the ring Z_4.

27. Let R be a ring with unity and let ϕ be a ring homomorphism from R onto S where S has more than one element. Prove that S has a unity.

28. Show that $(Z \oplus Z)/(\langle a \rangle \oplus \langle b \rangle)$ is ring-isomorphic to $Z_a \oplus Z_b$.

29. Determine all ring homomorphisms from $Z \oplus Z$ to Z.

30. Prove that the sum of the squares of three consecutive integers cannot be a square.

31. Let m be a positive integer and let n be an integer obtained from m by rearranging the digits of m in some way. (For example, 72345 is a rearrangement of 35274.) Show that $m - n$ is divisible by 9.

32. (Test for Divisibility by 11) Let n be an integer with decimal representation $a_k a_{k-1} \cdots a_1 a_0$. Prove that n is divisible by 11 if and only if $a_0 - a_1 + a_2 - \cdots (-1)^k a_k$ is divisible by 11.

33. Show that the number 7,176,825,942,116,027,211 is divisible by 9 but not divisible by 11.

34. Show that the number 9,897,654,527,609,805 is divisible by 99.

35. (Test for Divisibility by 3) Let n be an integer with decimal representation $a_k a_{k-1} \cdots a_1 a_0$. Prove that n is divisible by 3 if and only if $a_k + a_{k-1} + \cdots + a_1 + a_0$ is divisible by 3.

36. (Test for Divisibility by 4) Let n be an integer with decimal representation $a_k a_{k-1} \cdots a_1 a_0$. Prove that n is divisible by 4 if and only if $a_1 a_0$ is divisible by 4.

37. Show that no integer of the form 111,111,111, . . .,111 is prime.

38. Consider an integer n of the form a,111,111,111,111,111,111, 111,111,12b. Find values for a and b such that n is divisible by 99.

39. Suppose n is a positive integer written in the form $n = a_k 3^k + a_{k-1} 3^{k-1} + \cdots + a_1 3 + a_0$, where each of the a_i's is 0, 1, or 2 (the base 3 representative of n). Show that n is even if and only if $a_k + a_{k-1} + \cdots + a_1 + a_0$ is even.

40. Find an analog of the condition given in the previous exercise for characterizing divisibility by 4.

41. In your head, determine $(2 \cdot 10^{75} + 2)^{100}$ mod 3 and $(10^{100} + 1)^{99}$ mod 3.

42. Determine all ring homomorphisms from Q to Q.

43. Let R and S be commutative rings with unity. If ϕ is a homomorphism from R onto S and the characteristic of R is nonzero, prove that the characteristic of S divides the characteristic of R.

44. Let R be a commutative ring of prime characteristic p. Show that the *Frobenius* map $x \to x^p$ is a ring homomorphism from R to R.

45. Is there a ring homomorphism from the reals to some ring whose kernel is the integers?

46. Show that a homomorphism from a field onto a ring with more than one element must be an isomorphism.

47. Suppose that R and S are commutative rings with unities. Let ϕ be a ring homomorphism from R onto S and let A be an ideal of S.

   a. If A is prime in S, show that $\phi^{-1}(A) = \{x \in R \mid \phi(x) \in A\}$ is prime in R.

   b. If A is maximal in S, show that $\phi^{-1}(A)$ is maximal in R.

48. A *principal ideal ring* is a ring with the property that every ideal has the form $\langle a \rangle$. Show that the homomorphic image of a principal ideal ring is a principal ideal ring.

49. Let R and S be rings.
   a. Show that the mapping from $R \oplus S$ onto R given by $(a, b) \to a$ is a ring homomorphism.
   b. Show that the mapping from R to $R \oplus S$ given by $a \to (a, 0)$ is a one-to-one ring homomorphism.
   c. Show that $R \oplus S$ is ring-isomorphic to $S \oplus R$.

50. Show that if m and n are distinct positive integers, then mZ is not ring-isomorphic to nZ.

51. Prove or disprove that the field of real numbers is ring-isomorphic to the field of complex numbers.

52. Show that the only ring automorphism of the real numbers is the identity mapping.

53. Determine all ring homomorphisms from **R** to **R**.

54. Suppose that n divides m and that a is an idempotent of Z_n (that is, $a^2 = a$). Show that the mapping $x \to ax$ is a ring homomorphism from Z_m to Z_n. Show that the same correspondence need not yield a ring homomorphism if n does not divide m.

55. Show that the operation of multiplication defined in the proof of Theorem 15.6 is well-defined.

56. Let $Q[\sqrt{2}] = \{a + b\sqrt{2} \mid a, b \in Q\}$ and $Q[\sqrt{5}] = \{a + b\sqrt{5} \mid a, b \in Q\}$. Show that these two rings are not ring-isomorphic.

57. Let $Z[i] = \{a + bi \mid a, b \in Z\}$. Show that the field of quotients of $Z[i]$ is ring-isomorphic to $Q[i] = \{r + si \mid r, s \in Q\}$. (This exercise is referred to in Chapter 18.)

58. Let F be a field. Show that the field of quotients of F is ring-isomorphic to F.

59. Let D be an integral domain and let F be the field of quotients of D. Show that if E is any field that contains D, then E contains a subfield that is ring-isomorphic to F. (Thus, the field of quotients of an integral domain D is the smallest field containing D.)

60. Explain why a commutative ring with unity that is not an integral domain cannot be contained in a field. (Compare with Theorem 15.6.)

61. Show that the relation $\equiv$ defined in the proof of Theorem 15.6 is an equivalence relation.

62. Give an example of a ring without unity that is contained in a field.

63. Prove that the set T in the proof of Corollary 3 to Theorem 15.5 is ring-isomorphic to the field of rational numbers.

64. Suppose that ϕ: $R \to S$ is a ring homomorphism and that the image of ϕ is not $\{0\}$. If R has a unity and S is an integral domain, show that ϕ carries the unity of R to the unity of S. Give an example to show that the preceding statement need not be true if S is not an integral domain.

65. Let $f(x) \in \mathbf{R}[x]$. If $a + bi$ is a complex zero of $f(x)$ (here $i = \sqrt{-1}$), show that $a - bi$ is a zero of $f(x)$. (This exercise is referred to in Chapter 32.)

66. Let $R = \left\{ \begin{bmatrix} a & b \\ b & a \end{bmatrix} \middle| \, a, b \in Z \right\}$, and let ϕ be the mapping that takes $\begin{bmatrix} a & b \\ b & a \end{bmatrix}$ to $a - b$.

a. Show that ϕ is a homomorphism.
b. Determine the kernel of ϕ.
c. Show that $R/\text{Ker}\ \phi$ is isomorphic to Z.
d. Is Ker ϕ a prime ideal?
e. Is Ker ϕ a maximal ideal?

67. Show that the prime subfield of a field of characteristic p is ring-isomorphic to Z_p and that the prime subfield of a field of characteristic 0 is ring-isomorphic to Q. (This exercise is referred to in this chapter.)

68. Let n be a positive integer. Show that there is a ring isomorphism from Z_2 to a subring of Z_{2n} if and only if n is odd.

69. Show that Z_{mn} is ring-isomorphic to $Z_m \oplus Z_n$ when m and n are relatively prime.

70. Prove that every integer with decimal representation of the form *abcabc* (for example, 916916) is divisible by 11.

Suggested Readings

J. A. Gallian and J. Van Buskirk, "The Number of Homomorphisms from Z_m into Z_n," *American Mathematical Monthly* 91 (1984): 196–197.

In this article, formulas are given for the number of group homomorphisms from Z_m into Z_n and the number of ring homomorphisms from Z_m into Z_n.

Lillian Kinkade and Joyce Wagner, "When Polynomial Rings Are Principal Ideal Rings," *Journal of Undergraduate Mathematics* 23 (1991): 59–62.

In this article written by undergraduates, it is shown that $R[x]$ is a principal ideal ring if and only if $R \approx R_1 \oplus R_2 \oplus \cdots \oplus R_n$, where each R_i is a field.

Mohammad Saleh and Hasan Yousef, "The Number of Ring Homomorphisms from $Z_{m1} \oplus \cdots \oplus Z_{mr}$ into $Z_{k1} \oplus \cdots \oplus Z_{ks}$," *American Mathematical Monthly* 105 (1998): 259–260.

This article gives a formula for the number described in the title.

Suggested Website

http://www.d.umn.edu/~jgallian/puzzle

This site has a math puzzle that is based on the ideas presented in this chapter. The user selects an integer and then proceeds through a series of steps to produce a new integer. Finally, another integer is created by using all but one of the digits of the previous integer in any order. The software then determines the digit not used.

16 Polynomial Rings

Wit lies in recognizing the resemblance among things which differ and the difference between things which are alike.

MADAME DE STAËL

Notation and Terminology

One of the mathematical concepts that students are most familiar with and most comfortable with is that of a polynomial. In high school, students study polynomials with integer coefficients, rational coefficients, real coefficients, and perhaps even complex coefficients. In earlier chapters of this book, we introduced something that was probably new—polynomials with coefficients from Z_n. Notice that all of these sets of polynomials are rings, and, in each case, the set of coefficients is also a ring. In this chapter, we abstract all of these examples into one.

Definition Ring of Polynomials over *R*

Let R be a commutative ring. The set of formal symbols

$$R[x] = \{a_n x^n + a_{n-1}x^{n-1} + \cdots + a_1 x + a_0 \mid a_i \in R, \\ n \text{ is a nonnegative integer}\}$$

is called the *ring of polynomials over R in the indeterminate x.*

Two elements

$$a_n x^n + a_{n-1}x^{n-1} + \cdots + a_1 x + a_0$$

and

$$b_m x^m + b_{m-1}x^{m-1} + \cdots + b_1 x + b_0$$

of $R[x]$ are considered equal if and only if $a_i = b_i$ for all nonnegative integers i. (Define $a_i = 0$ when $i > n$ and $b_i = 0$ when $i > m$.)

In this definition, the symbols $x, x^2, \ldots, x^n$ do not represent "unknown" elements or variables from the ring R. Rather, their purpose is to serve as convenient placeholders that separate the ring elements $a_n, a_{n-1}, \ldots, a_0$. We could have avoided the x's by defining a polynomial as an infinite sequence $a_0, a_1, a_2, \ldots, a_n, 0, 0, 0, \ldots$, but our method takes advantage of the student's experience in manipulating polynomials where x does represent a variable. The disadvantage of our method is that one must be careful not to confuse a polynomial with the function determined by a polynomial. For example, in $Z_3[x]$, the polynomials $f(x) = x$ and $g(x) = x^3$ determine the same function from Z_3 to Z_3, since $f(a) = g(a)$ for all a in Z_3.† But $f(x)$ and $g(x)$ are different elements of $Z_3[x]$. Also, in the ring $Z_n[x]$, be careful to reduce only the coefficients and not the exponents modulo n. For example, in $Z_3[x]$, $5x = 2x$, but $x^5 \neq x^2$.

To make $R[x]$ into a ring, we define addition and multiplication in the usual way.

Definition Addition and Multiplication in $R[x]$

Let R be a commutative ring and let

$$f(x) = a_nx^n + a_{n-1}x^{n-1} + \cdots + a_1x + a_0$$

and

$$g(x) = b_mx^m + b_{m-1}x^{m-1} + \cdots + b_1x + b_0$$

belong to $R[x]$. Then

$$f(x) + g(x) = (a_s + b_s)x^s + (a_{s-1} + b_{s-1})x^{s-1} + \cdots + (a_1 + b_1)x + a_0 + b_0,$$

where s is the maximum of m and n, $a_i = 0$ for $i > n$, and $b_i = 0$ for $i > m$. Also,

$$f(x)g(x) = c_{m+n}x^{m+n} + c_{m+n-1}x^{m+n-1} + \cdots + c_1x + c_0,$$

where

$$c_k = a_kb_0 + a_{k-1}b_1 + \cdots + a_1b_{k-1} + a_0b_k$$

for $k = 0, \ldots, m + n$.

Although the definition of multiplication might appear complicated, it is just a formalization of the familiar process of using the distributive

†In general, given $f(x)$ in $R[x]$ and a in R, $f(a)$ means substitute a for x in the formula for $f(x)$. This substitution is a homomorphism from $R[x]$ to R.

property and collecting like terms. So, just multiply polynomials over a commutative ring R in the same way that polynomials are always multiplied. Here is an example.

Consider $f(x) = 2x^3 + x^2 + 2x + 2$ and $g(x) = 2x^2 + 2x + 1$ in $Z_3[x]$. Then, in our preceding notation, $a_5 = 0$, $a_4 = 0$, $a_3 = 2$, $a_2 = 1$, $a_1 = 2$, $a_0 = 2$, and $b_5 = 0$, $b_4 = 0$, $b_3 = 0$, $b_2 = 2$, $b_1 = 2$, $b_0 = 1$. Now, using the definitions and remembering that addition and multiplication of the coefficients are done modulo 3, we have

$$\begin{aligned} f(x) + g(x) &= (2 + 0)x^3 + (1 + 2)x^2 + (2 + 2)x + (2 + 1) \\ &= 2x^3 + 0x^2 + 1x + 0 \\ &= 2x^3 + x \end{aligned}$$

and

$$\begin{aligned} f(x) \cdot g(x) &= (0 \cdot 1 + 0 \cdot 2 + 2 \cdot 2 + 1 \cdot 0 + 2 \cdot 0 + 2 \cdot 0)x^5 \\ &\quad + (0 \cdot 1 + 2 \cdot 2 + 1 \cdot 2 + 2 \cdot 0 + 2 \cdot 0)x^4 \\ &\quad + (2 \cdot 1 + 1 \cdot 2 + 2 \cdot 2 + 2 \cdot 0)x^3 \\ &\quad + (1 \cdot 1 + 2 \cdot 2 + 2 \cdot 2)x^2 + (2 \cdot 1 + 2 \cdot 2)x + 2 \cdot 1 \\ &= x^5 + 0x^4 + 2x^3 + 0x^2 + 0x + 2 \\ &= x^5 + 2x^3 + 2. \end{aligned}$$

Our definitions for addition and multiplication of polynomials were formulated so that they are commutative and associative, and so that multiplication is distributive over addition. We leave the verification that $R[x]$ is a ring to the reader.

It is time to introduce some terminology for polynomials. If

$$f(x) = a_n x^n + a_{n-1}x^{n-1} + \cdots + a_1 x + a_0,$$

where $a_n \neq 0$, we say that $f(x)$ has *degree n*; the term a_n is called the *leading coefficient* of $f(x)$, and if the leading coefficient is the multiplicative identity element of R, we say that $f(x)$ is a *monic* polynomial. The polynomial $f(x) = 0$ has no degree. Polynomials of the form $f(x) = a_0$ are called *constant*. We often write $\deg f(x) = n$ to indicate that $f(x)$ has degree n. As with polynomials with real coefficients, we may insert or delete terms of the form $0x^k$; $1x^k$ is the same as x^k; and $+(-a_k)x^k$ is the same as $-a_k x^k$.

Very often properties of R carry over to $R[x]$. Our first theorem is a case in point.

Theorem 16.1 *D* an Integral Domain Implies *D*[*x*] an Integral Domain

If D is an integral domain, then D[x] is an integral domain.

PROOF Since we already know that $D[x]$ is a ring, all we need to show is that $D[x]$ is commutative with a unity and has no zero-divisors. Clearly, $D[x]$ is commutative whenever D is. If 1 is the unity element of D, it is obvious that $f(x) = 1$ is the unity element of $D[x]$. Finally, suppose that

$$f(x) = a_n x^n + a_{n-1}x^{n-1} + \cdots + a_0$$

and

$$g(x) = b_m x^m + b_{m-1}x^{m-1} + \cdots + b_0,$$

where $a_n \neq 0$ and $b_m \neq 0$. Then, by definition, $f(x)g(x)$ has leading coefficient $a_n b_m$ and, since D is an integral domain, $a_n b_m \neq 0$. ∎

The Division Algorithm and Consequences

One of the properties of integers that we have used repeatedly is the division algorithm: If a and b are integers and $b \neq 0$, then there exist unique integers q and r such that $a = bq + r$, where $0 \leq r < |b|$. The next theorem is the analogous statement for polynomials over a field.

Theorem 16.2 Division Algorithm for *F*[*x*]

Let F be a field and let $f(x), g(x) \in F[x]$ with $g(x) \neq 0$. Then there exist unique polynomials $q(x)$ and $r(x)$ in $F[x]$ such that $f(x) = g(x)q(x) + r(x)$ and either $r(x) = 0$ or $\deg r(x) < \deg g(x)$.

PROOF We begin by showing the existence of $q(x)$ and $r(x)$. If $f(x) = 0$ or $\deg f(x) < \deg g(x)$, we simply set $q(x) = 0$ and $r(x) = f(x)$. So, we may assume that $n = \deg f(x) \geq \deg g(x) = m$ and let $f(x) = a_n x^n + \cdots + a_0$ and $g(x) = b_m x^m + \cdots + b_0$. The idea behind this proof is to begin just as if you were going to "long divide" $g(x)$ into $f(x)$, then use the Second Principle of Mathematical Induction on $\deg f(x)$ to finish up. Thus, resorting to long division, we let $f_1(x) =$

$f(x) - a_n b_m^{-1} x^{n-m} g(x)$.[†] Then, $f_1(x) = 0$ or $\deg f_1(x) < \deg f(x)$; so, by our induction hypothesis, there exist $q_1(x)$ and $r_1(x)$ in $F[x]$ such that $f_1(x) = g(x)q_1(x) + r_1(x)$, where $r_1(x) = 0$ or $\deg r_1(x) < \deg g(x)$. [Technically, we should get the induction started by proving the case in which $\deg f(x) = 0$, but this is trivial.] Thus,

$$\begin{aligned} f(x) &= a_n b_m^{-1} x^{n-m} g(x) + f_1(x) \\ &= a_n b_m^{-1} x^{n-m} g(x) + q_1(x)g(x) + r_1(x) \\ &= [a_n b_m^{-1} x^{n-m} + q_1(x)]g(x) + r_1(x). \end{aligned}$$

So, the polynomials $q(x) = a_n b_m^{-1} x^{n-m} + q_1(x)$ and $r(x) = r_1(x)$ have the desired properties.

To prove uniqueness, suppose that $f(x) = g(x)q(x) + r(x)$ and $f(x) = g(x)\bar{q}(x) + \bar{r}(x)$, where $r(x) = 0$ or $\deg r(x) < \deg g(x)$ and $\bar{r}(x) = 0$ or $\deg \bar{r}(x) < \deg g(x)$. Then, subtracting these two equations, we obtain

$$0 = g(x)[q(x) - \bar{q}(x)] + [r(x) - \bar{r}(x)]$$

or

$$\bar{r}(x) - r(x) = g(x)[q(x) - \bar{q}(x)].$$

Thus, $\bar{r}(x) - r(x)$ is 0, or the degree of $\bar{r}(x) - r(x)$ is at least that of $g(x)$. Since the latter is clearly impossible, we have $\bar{r}(x) = r(x)$ and $q(x) = \bar{q}(x)$ as well. ■

The polynomials $q(x)$ and $r(x)$ in the division algorithm are called the *quotient* and *remainder* in the division of $f(x)$ by $g(x)$. When the ring of coefficients of a polynomial ring is a field, we can use the long division process to determine the quotient and remainder.

[†]For example,

$$\begin{array}{r} (3/2)x^2 \\ 2x^2 + 2 \overline{)\,3x^4 \qquad\quad + x + 1} \\ \underline{3x^4 + 3x^2 } \\ -3x^2 + x + 1 \end{array}$$

So,

$$-3x^2 + x + 1 = 3x^4 + x + 1 - (3/2)x^2(2x^2 + 2)$$

In general,

$$\begin{array}{r} a_n b_m^{-1} x^{n-m} \\ b_m x^m + \cdots \overline{)\, a_n x^n + \cdots} \\ \underline{a_n x^n + \cdots} \\ f_1(x) \end{array}$$

So,

$$f_1(x) = (a_n x^n + \cdots) - a_n b_m^{-1} x^{n-m}(b_m x^m + \cdots)$$

EXAMPLE 1 To find the quotient and remainder upon dividing $f(x) = 3x^4 + x^3 + 2x^2 + 1$ by $g(x) = x^2 + 4x + 2$, where $f(x)$ and $g(x)$ belong to $Z_5[x]$, we may proceed by long division, provided we keep in mind that addition and multiplication are done modulo 5. Thus,

$$\begin{array}{r|l} & 3x^2 + 4x \\ \hline x^2 + 4x + 2 & 3x^4 + x^3 + 2x^2 \qquad + 1 \\ & 3x^4 + 2x^3 + x^2 \\ \hline & \qquad 4x^3 + x^2 \qquad + 1 \\ & \qquad 4x^3 + x^2 + 3x \\ \hline & \qquad\qquad 2x + 1 \end{array}$$

So, $3x^2 + 4x$ is the quotient and $2x + 1$ is the remainder. Therefore,

$$3x^4 + x^3 + 2x^2 + 1 = (x^2 + 4x + 2)(3x^2 + 4x) + 2x + 1.$$ ∎

Let D be an integral domain. If $f(x)$ and $g(x) \in D[x]$, we say that $g(x)$ *divides* $f(x)$ in $D[x]$ [and write $g(x) \mid f(x)$] if there exists an $h(x) \in D[x]$ such that $f(x) = g(x)h(x)$. In this case, we also call $g(x)$ a *factor* of $f(x)$. An element a is a *zero* (or a *root*) of a polynomial $f(x)$ if $f(a) = 0$. [Recall that $f(a)$ means substitute a for x in the expression for $f(x)$.] When F is a field, $a \in F$, and $f(x) \in F[x]$, we say that a is a *zero of multiplicity* k ($k \geq 1$) if $(x - a)^k$ is a factor of $f(x)$ but $(x - a)^{k+1}$ is not a factor of $f(x)$. With these definitions, we may now give several important corollaries of the division algorithm. No doubt you have seen these for the special case where F is the field of real numbers.

Corollary 1 Remainder Theorem

Let F be a field, $a \in F$, and $f(x) \in F[x]$. Then $f(a)$ is the remainder in the division of $f(x)$ by $x - a$.

PROOF The proof of Corollary 1 is left as an exercise (Exercise 5). ∎

Corollary 2 Factor Theorem

Let F be a field, $a \in F$, and $f(x) \in F[x]$. Then a is a zero of $f(x)$ if and only if $x - a$ is a factor of $f(x)$.

PROOF The proof of Corollary 2 is left as an exercise (Exercise 9). ∎

Corollary 3 Polynomials of Degree n Have at Most n Zeros

A polynomial of degree n over a field has at most n zeros, counting multiplicity.

PROOF We proceed by induction on n. Clearly, a polynomial of degree 0 over a field has no zeros. Now suppose that $f(x)$ is a polynomial of degree n over a field and a is a zero of $f(x)$ of multiplicity k. Then, $f(x) = (x - a)^k q(x)$ and $q(a) \neq 0$; and, since $n = \deg f(x) = \deg (x - a)^k q(x) = k + \deg q(x)$, we have $k \leq n$ (see Exercise 19). If $f(x)$ has no zeros other than a, we are done. On the other hand, if $b \neq a$ and b is a zero of $f(x)$, then $0 = f(b) = (b - a)^k q(b)$, so that b is also a zero of $q(x)$ with the same multiplicity as it has for $f(x)$ (see Exercise 21). By the Second Principle of Mathematical Induction, we know that $q(x)$ has at most $\deg q(x) = n - k$ zeros, counting multiplicity. Thus, $f(x)$ has at most $k + n - k = n$ zeros, counting multiplicity. ∎

We remark that Corollary 3 is not true for arbitrary polynomial rings. For example, the polynomial $x^2 + 3x + 2$ has four zeros in Z_6. (See Exercise 3.) Lagrange was the first to prove Corollary 3 for polynomials in $Z_p[x]$.

EXAMPLE 2 The Complex Zeros of $x^n - 1$

We find all complex zeros of $x^n - 1$. Let $\omega = \cos(360°/n) + i\sin(360°/n)$. It follows from DeMoivre's Theorem (see Example 10 in Chapter 0) that $\omega^n = 1$ and $\omega^k \neq 1$ for $1 \leq k < n$. Thus, each of $1, \omega, \omega^2, \ldots, \omega^{n-1}$ is a zero of $x^n - 1$ and, by Corollary 3, there are no others. ∎

The complex number ω in Example 2 is called a *primitive nth root of unity.*

We conclude this chapter with an important theoretical application of the division algorithm, but first an important definition.

Definition Principal Ideal Domain (PID)

A *principal ideal domain* is an integral domain R in which every ideal has the form $\langle a \rangle = \{ra \mid r \in R\}$ for some a in R.

Theorem 16.3 $F[x]$ Is a PID

Let F be a field. Then $F[x]$ is a principal ideal domain.

PROOF By Theorem 16.1, we know that $F[x]$ is an integral domain. Now, let I be an ideal in $F[x]$. If $I = \{0\}$, then $I = \langle 0 \rangle$. If $I \neq \{0\}$, then among all the elements of I, let $g(x)$ be one of minimum degree. We will show that $I = \langle g(x) \rangle$. Since $g(x) \in I$, we have $\langle g(x) \rangle \subseteq I$. Now let $f(x) \in I$. Then, by the division algorithm, we may write $f(x) = g(x)q(x) + r(x)$, where $r(x) = 0$ or $\deg r(x) < \deg g(x)$. Since $r(x) = f(x) - g(x)q(x) \in I$, the minimality of $\deg g(x)$ implies that the latter condition cannot hold. So, $r(x) = 0$ and, therefore, $f(x) \in \langle g(x) \rangle$. This shows that $I \subseteq \langle g(x) \rangle$. ▮

The proof of Theorem 16.3 also establishes the following.

Theorem 16.4 Criterion for $I = \langle g(x) \rangle$

Let F be a field, I a nonzero ideal in $F[x]$, and $g(x)$ an element of $F[x]$. Then, $I = \langle g(x) \rangle$ if and only if $g(x)$ is a nonzero polynomial of minimum degree in I.

As an application of the First Isomorphism Theorem for Rings (Theorem 15.3) and Theorem 16.4, we verify the remark we made in Example 12 in Chapter 14 that the ring $\mathbf{R}[x]/\langle x^2 + 1 \rangle$ is isomorphic to the ring of complex numbers.

EXAMPLE 3 Consider the homomorphism ϕ from $\mathbf{R}[x]$ onto $\mathbf{C}$ given by $f(x) \to f(i)$ (that is, evaluate a polynomial in $\mathbf{R}[x]$ at i). Then $x^2 + 1 \in \operatorname{Ker} \phi$ and is clearly a polynomial of minimum degree in $\operatorname{Ker} \phi$. Thus, $\operatorname{Ker} \phi = \langle x^2 + 1 \rangle$ and $\mathbf{R}[x]/\langle x^2 + 1 \rangle$ is isomorphic to $\mathbf{C}$. ▮

Exercises

If I feel unhappy, I do mathematics to become happy. If I am happy, I do mathematics to keep happy.

PAUL TURÁN

1. Let $f(x) = 4x^3 + 2x^2 + x + 3$ and $g(x) = 3x^4 + 3x^3 + 3x^2 + x + 4$, where $f(x), g(x) \in Z_5[x]$. Compute $f(x) + g(x)$ and $f(x) \cdot g(x)$.
2. In $Z_3[x]$, show that the distinct polynomials $x^4 + x$ and $x^2 + x$ determine the same function from Z_3 to Z_3.
3. Show that $x^2 + 3x + 2$ has four zeros in Z_6. (This exercise is referred to in this chapter.)

4. If R is a commutative ring, show that the characteristic of $R[x]$ is the same as the characteristic of R.
5. Prove Corollary 1 of Theorem 16.2.
6. List all the polynomials of degree 2 in $Z_2[x]$. Which of these are equal as functions from Z_2 to Z_2?
7. Find two distinct cubic polynomials over Z_2 that determine the same function from Z_2 to Z_2.
8. For any positive integer n, how many polynomials are there of degree n over Z_2? How many distinct polynomial functions from Z_2 to Z_2 are there?
9. Prove Corollary 2 of Theorem 16.2.
10. Let R be a commutative ring. Show that $R[x]$ has a subring isomorphic to R.
11. If $\phi: R \to S$ is a ring homomorphism, define $\overline{\phi}: R[x] \to S[x]$ by $(a_nx^n + \cdots + a_0) \to \phi(a_n)x^n + \cdots + \phi(a_0)$. Show that $\overline{\phi}$ is a ring homomorphism. (This exercise is referred to in Chapter 33.)
12. If the rings R and S are isomorphic, show that $R[x]$ and $S[x]$ are isomorphic.
13. Let $f(x) = 5x^4 + 3x^3 + 1$ and $g(x) = 3x^2 + 2x + 1$ in $Z_7[x]$. Determine the quotient and remainder upon dividing $f(x)$ by $g(x)$.
14. Let $f(x)$ and $g(x)$ be cubic polynomials with integer coefficients such that $f(a) = g(a)$ for four integer values of a. Prove that $f(x) = g(x)$. Generalize.
15. Show that the polynomial $2x + 1$ in $Z_4[x]$ has a multiplicative inverse in $Z_4[x]$.
16. Are there any nonconstant polynomials in $Z[x]$ that have multiplicative inverses? Explain your answer.
17. Let p be a prime. Are there any nonconstant polynomials in $Z_p[x]$ that have multiplicative inverses? Explain your answer.
18. Show that Corollary 3 of Theorem 16.2 is false for any commutative ring that has a zero divisor.
19. (Degree Rule) Let D be an integral domain and $f(x), g(x) \in D[x]$. Prove that $\deg(f(x) \cdot g(x)) = \deg f(x) + \deg g(x)$. Show, by example, that for commutative ring R it is possible that $\deg f(x)g(x) < \deg f(x) + \deg g(x)$, where $f(x)$ and $g(x)$ are nonzero elements in $R[x]$. (This exercise is referred to in this chapter, Chapter 17, and Chapter 18.)
20. Prove that the ideal $\langle x \rangle$ in $Q[x]$ is maximal.

21. Let $f(x)$ belong to $F[x]$, where F is a field. Let a be a zero of $f(x)$ of multiplicity n, and write $f(x) = (x - a)^n q(x)$. If $b \neq a$ is a zero of $q(x)$, show that b has the same multiplicity as a zero of $q(x)$ as it does for $f(x)$. (This exercise is referred to in this chapter.)

22. Prove that for any positive integer n, a field F can have at most a finite number of elements of multiplicative order at most n.

23. Let F be an infinite field and let $f(x) \in F[x]$. If $f(a) = 0$ for infinitely many elements a of F, show that $f(x) = 0$.

24. Let F be an infinite field and let $f(x), g(x) \in F[x]$. If $f(a) = g(a)$ for infinitely many elements a of F, show that $f(x) = g(x)$.

25. Let F be a field and let $p(x) \in F[x]$. If $f(x), g(x) \in F[x]$ and $\deg f(x) < \deg p(x)$ and $\deg g(x) < \deg p(x)$, show that $f(x) + \langle p(x)\rangle = g(x) + \langle p(x)\rangle$ implies $f(x) = g(x)$. (This exercise is referred to in Chapter 20.)

26. Prove that $Z[x]$ is not a principal ideal domain. (Compare this with Theorem 16.3.)

27. Find a polynomial with integer coefficients that has 1/2 and $-1/3$ as zeros.

28. Let $f(x) \in \mathbf{R}[x]$. Suppose that $f(a) = 0$ but $f'(a) \neq 0$, where $f'(x)$ is the derivative of $f(x)$. Show that a is a zero of $f(x)$ of multiplicity 1.

29. Show that Corollary 2 of Theorem 16.2 is true over any commutative ring with unity.

30. Show that Corollary 3 of Theorem 16.2 is true for polynomials over integral domains.

31. Let F be a field and let

$$I = \{a_nx^n + a_{n-1}x^{n-1} + \cdots + a_0 \mid a_n, a_{n-1}, \ldots, a_0 \in F \text{ and } a_n + a_{n-1} + \cdots + a_0 = 0\}.$$

Show that I is an ideal of $F[x]$ and find a generator for I.

32. Let F be a field and let $f(x) = a_nx^n + a_{n-1}x^{n-1} + \cdots + a_0 \in F[x]$. Prove that $x - 1$ is a factor of $f(x)$ if and only if $a_n + a_{n-1} + \cdots + a_0 = 0$.

33. Let m be a fixed positive integer. For any integer a, let $\bar{a}$ denote $a \bmod m$. Show that the mapping of ϕ: $Z[x] \to Z_m[x]$ given by

$$\phi(a_nx^n + a_{n-1}x^{n-1} + \cdots + a_0) = \bar{a}_nx^n + \bar{a}_{n-1}x^{n-1} + \cdots + \bar{a}_0$$

is a ring homomorphism. (This exercise is referred to in Chapters 17 and 33.)

34. Find infinitely many polynomials $f(x)$ in $Z_3[x]$ such that $f(a) = 0$ for all a in Z_3.

35. For every prime p, show that

$$x^{p-1} - 1 = (x - 1)(x - 2) \cdots [x - (p - 1)]$$

in $Z_p[x]$.

36. Let ϕ be the ring homomorphism from $Z[x]$ to Z given by $\phi(f(x)) = f(1)$. Find a polynomial $g(x)$ in $Z[x]$ such that Ker $\phi = \langle g(x) \rangle$. Is there more than one possibility for $g(x)$? To what familiar ring is $Z[x]/\text{Ker } \phi$ isomorphic? Do this exercise with Z replaced by Q.

37. Give an example of a field that properly contains the field of complex numbers **C**.

38. (Wilson's Theorem) For every integer $n > 1$, prove that $(n - 1)! \bmod n = n - 1$ if and only if n is prime.

39. For every prime p, show that $(p - 2)! \bmod p = 1$.

40. Find the remainder upon dividing 98! by 101.

41. Prove that $(50!)^2 \bmod 101 = -1 \bmod 101$.

42. If I is an ideal of a ring R, prove that $I[x]$ is an ideal of $R[x]$.

43. Give an example of a commutative ring R with unity and a maximal ideal I of R such that $I[x]$ is not a maximal ideal of $R[x]$.

44. Let R be a commutative ring with unity. If I is a prime ideal of R, prove that $I[x]$ is a prime ideal of $R[x]$.

45. Let F be a field, and let $f(x)$ and $g(x)$ belong to $F[x]$. If there is no polynomial of positive degree in $F[x]$ that divides both $f(x)$ and $g(x)$ [in this case, $f(x)$ and $g(x)$ are said to be *relatively prime*], prove that there exist polynomials $h(x)$ and $k(x)$ in $F[x]$ with the property that $f(x)h(x) + g(x)k(x) = 1$. (This exercise is referred to in Chapter 20.)

46. Prove that $Q[x]/\langle x^2 - 2 \rangle$ is ring-isomorphic to $Q[\sqrt{2}] = \{a + b\sqrt{2} \mid a, b \in Q\}$.

47. Let $f(x) \in \mathbf{R}[x]$. If $f(a) = 0$ and $f'(a) = 0$ [$f'(a)$ is the derivative of $f(x)$ at a], show that $(x - a)^2$ divides $f(x)$.

48. Let F be a field and let $I = \{f(x) \in F[x] \mid f(a) = 0 \text{ for all } a \text{ in } F\}$. Prove that I is an ideal in $F[x]$. Prove that I is infinite when F is finite and $I = \{0\}$ when F is infinite. When F is finite, find a monic polynomial $g(x)$ such that $I = \langle g(x) \rangle$.

49. Let $g(x)$ and $h(x)$ belong to $Z[x]$ and let $h(x)$ be monic. If $h(x)$ divides $g(x)$ in $Q[x]$, show that $h(x)$ divides $g(x)$ in $Z[x]$. (This exercise is referred to in Chapter 33.)

50. Let R be a ring and x be an indeterminate. Prove that the rings $R[x]$ and $R[x^2]$ are ring-isomorphic.

51. Let $f(x)$ be a nonconstant element of $Z[x]$. Prove that $f(x)$ takes on infinitely many values in Z.

52. Let $f(x)$ be a nonconstant element in $Z[x]$. Prove that $\langle f(x)\rangle$ is not maximal in $Z[x]$.
53. Let $Z_n[x]$ be the ring of polynomials in x with coefficients from Z_n and ordinary addition and multiplication. If n can be written in the form t^2m, show that $tmx + 1$ is a unit in $Z_n[x]$.
54. Let $f(x)$ belong to $Z_p[x]$. Prove that if $f(b) = 0$, then $f(b^p) = 0$.
55. Suppose $f(x)$ is a polynomial with odd integer coefficients and even degree. Prove that $f(x)$ has no rational zeros.
56. For any field F, recall that $F(x)$ denotes the field of quotients of the ring $F[x]$. Prove that there is no element in $F(x)$ whose square is x.
57. Let F be a field. Show that there exist $a, b \in F$ with the property that $x^2 + x + 1$ divides $x^{43} + ax + b$.
58. Let $f(x) = a_mx^m + a_{m-1}x^{m-1} + \cdots + a_0$ and $g(x) = b_n + b_{n-1}x^{n-1} + \cdots + b_0$ belong to $Q[x]$ and suppose that $f(x)g(x)$ belongs to $Z[x]$. Prove that a_ib_j is an integer for every i and j.
59. Let $f(x)$ belong to $Z[x]$. If $a \bmod m = b \bmod m$, prove that $f(a) \bmod m = f(b) \bmod m$. Prove that if both $f(0)$ and $f(1)$ are odd, then f has no zero in Z.
60. Find the remainder when x^{51} is divided by $x + 4$ in $Z_7[x]$.
61. Show that 1 is the only solution of $x^{25} - 1 = 0$ in Z_{37}.

Saunders Mac Lane

The 1986 Steele Prize for cumulative influence is awarded to Saunders Mac Lane for his many contributions to algebra and algebraic topology, and in particular for his pioneering work in homological and categorical algebra.

Citation for the Steele Prize

SAUNDERS MAC LANE ranks among the most influential mathematicians in the 20th century. He was born on August 4, 1909, in Norwich, Connecticut. In 1933, at the height of the Depression, he was newly married; despite having degrees from Yale, the University of Chicago, and the University of Göttingen, he had no prospects for a position at a college or university. After applying for employment as a master at a private preparatory school for boys, Mac Lane received a two-year instructorship at Harvard in 1934. He then spent a year at Cornell and a year at the University of Chicago before returning to Harvard in 1938. In 1947, he went back to Chicago permanently.

Much of Mac Lane's work focuses on the interconnections among algebra, topology, and geometry. His book *Survey of Modern Algebra,* coauthored with Garrett Birkhoff, influenced generations of mathematicians and is now a classic. Mac Lane served as president of the Mathematical Association of America and the American Mathematical Society. He was elected to the National Academy of Sciences, received the National Medal of Science and the American Mathematical Society's Steele Prize for Lifetime Achievement, and supervised 41 Ph.D. theses. Mac Lane died April 14, 2005, at age 95.

To find more information about Mac Lane, visit:

http://www-groups.dcs.st-and.ac.uk/~history/

17 Factorization of Polynomials

The value of a principle is the number of things it will explain.

RALPH WALDO EMERSON

Reducibility Tests

In high school, students spend much time factoring polynomials and finding their zeros. In this chapter, we consider the same problems in a more abstract setting.

To discuss factorization of polynomials, we must first introduce the polynomial analog of a prime integer.

> Definition **Irreducible Polynomial, Reducible Polynomial**
> **Let D be an integral domain. A polynomial $f(x)$ from $D[x]$ that is neither the zero polynomial nor a unit in $D[x]$ is said to be *irreducible over D* if, whenever $f(x)$ is expressed as a product $f(x) = g(x)h(x)$, with $g(x)$ and $h(x)$ from $D[x]$, then $g(x)$ or $h(x)$ is a unit in $D[x]$. A nonzero, nonunit element of $D[x]$ that is not irreducible over D is called *reducible over D*.**

In the case that an integral domain is a field F, it is equivalent and more convenient to define a nonconstant $f(x) \in F[x]$ to be irreducible if $f(x)$ cannot be expressed as a product of two polynomials of lower degree.

EXAMPLE 1 The polynomial $f(x) = 2x^2 + 4$ is irreducible over Q but reducible over Z, since $2x^2 + 4 = 2(x^2 + 2)$ and neither 2 nor $x^2 + 2$ is a unit in $Z[x]$. ∎

EXAMPLE 2 The polynomial $f(x) = 2x^2 + 4$ is irreducible over **R** but reducible over **C**. ∎

EXAMPLE 3 The polynomial $x^2 - 2$ is irreducible over Q but reducible over **R**. ∎

EXAMPLE 4 The polynomial $x^2 + 1$ is irreducible over Z_3 but reducible over Z_5. ∎

In general, it is a difficult problem to decide whether or not a particular polynomial is reducible over an integral domain, but there are special cases when it is easy. Our first theorem is a case in point. It applies to the four preceding examples.

Theorem 17.1 Reducibility Test for Degrees 2 and 3

Let F be a field. If $f(x) \in F[x]$ and $\deg f(x)$ is 2 or 3, then $f(x)$ is reducible over F if and only if $f(x)$ has a zero in F.

PROOF Suppose that $f(x) = g(x)h(x)$, where both $g(x)$ and $h(x)$ belong to $F[x]$ and have degrees less than that of $f(x)$. Since $\deg f(x) = \deg g(x) + \deg h(x)$ (Exercise 19 in Chapter 16) and $\deg f(x)$ is 2 or 3, at least one of $g(x)$ and $h(x)$ has degree 1. Say $g(x) = ax + b$. Then, clearly, $-a^{-1}b$ is a zero of $g(x)$ and therefore a zero of $f(x)$ as well.

Conversely, suppose that $f(a) = 0$, where $a \in F$. Then, by the Factor Theorem, we know that $x - a$ is a factor of $f(x)$ and, therefore, $f(x)$ is reducible over F. ∎

Theorem 17.1 is particularly easy to use when the field is Z_p, because in this case we can check for reducibility of $f(x)$ by simply testing to see if $f(a) = 0$ for $a = 0, 1, \ldots, p - 1$. For example, since 2 is a zero of $x^2 + 1$ over Z_5, $x^2 + 1$ is reducible over Z_5. On the other hand, because neither 0, 1, nor 2 is a zero of $x^2 + 1$ over Z_3, $x^2 + 1$ is irreducible over Z_3.

Note that polynomials of degree larger than 3 may be reducible over a field even though they do not have zeros in the field. For example, in $Q[x]$, the polynomial $x^4 + 2x^2 + 1$ is equal to $(x^2 + 1)^2$, but has no zeros in Q.

Our next three tests deal with polynomials with integer coefficients. To simplify the proof of the first of these, we introduce some terminology and isolate a portion of the argument in the form of a lemma.

Definition Content of a Polynomial, Primitive Polynomial

The *content* of a nonzero polynomial $a_nx^n + a_{n-1}x^{n-1} + \cdots + a_0$, where the a's are integers, is the greatest common divisor of the integers $a_n, a_{n-1}, \ldots, a_0$. A *primitive polynomial* is an element of $Z[x]$ with content 1.

Gauss's Lemma

The product of two primitive polynomials is primitive.

PROOF Let $f(x)$ and $g(x)$ be primitive polynomials, and suppose that $f(x)g(x)$ is not primitive. Let p be a prime divisor of the content of $f(x)g(x)$, and let $\overline{f}(x)$, $\overline{g}(x)$, and $\overline{f(x)g(x)}$ be the polynomials obtained from $f(x)$, $g(x)$, and $f(x)g(x)$ by reducing the coefficients modulo p. Then, $\overline{f}(x)$ and $\overline{g}(x)$ belong to the integral domain $Z_p[x]$ and $\overline{f}(x)\overline{g}(x) = \overline{f(x)g(x)} = 0$, the zero element of $Z_p[x]$ (see Exercise 33 in Chapter 16). Thus, $\overline{f}(x) = 0$ or $\overline{g}(x) = 0$. This means that either p divides every coefficient of $f(x)$ or p divides every coefficient of $g(x)$. Hence, either $f(x)$ is not primitive or $g(x)$ is not primitive. This contradiction completes the proof. ■

Remember that the question of reducibility depends on which ring of coefficients one permits. Thus, $x^2 - 2$ is irreducible over Z but reducible over $Q[\sqrt{2}]$. In Chapter 20, we will prove that every polynomial of degree greater than 1 with coefficients from an integral domain is reducible over some field. Theorem 17.2 shows that in the case of polynomials irreducible over Z, this field must be larger than the field of rational numbers.

Theorem 17.2 Reducibility over Q Implies Reducibility over Z

Let $f(x) \in Z[x]$. If $f(x)$ is reducible over Q, then it is reducible over Z.

PROOF Suppose that $f(x) = g(x)h(x)$, where $g(x)$ and $h(x) \in Q[x]$. Clearly, we may assume that $f(x)$ is primitive because we can divide both $f(x)$ and $g(x)$ by the content of $f(x)$. Let a be the least common multiple of the denominators of the coefficients of $g(x)$, and b the least common multiple of the denominators of the coefficients of $h(x)$. Then $abf(x) = ag(x) \cdot bh(x)$, where $ag(x)$ and $bh(x) \in Z[x]$. Let c_1 be the content of $ag(x)$ and let c_2 be the content of $bh(x)$. Then $ag(x) = c_1g_1(x)$ and $bh(x) = c_2h_1(x)$, where both $g_1(x)$ and $h_1(x)$ are primitive, and $abf(x) = c_1c_2g_1(x)h_1(x)$. Since $f(x)$ is primitive, the content of $abf(x)$ is ab. Also, since the product of two primitive polynomials is primitive, it follows that the content of $c_1c_2g_1(x)h_1(x)$ is c_1c_2. Thus, $ab = c_1c_2$ and $f(x) = g_1(x)h_1(x)$, where $g_1(x)$ and $h_1(x) \in Z[x]$ and $\deg g_1(x) = \deg g(x)$ and $\deg h_1(x) = \deg h(x)$. ■

EXAMPLE 5 We illustrate the proof of Theorem 17.2 by tracing through it for the polynomial $f(x) = 6x^2 + x - 2 = (3x - 3/2)(2x + 4/3) = g(x)h(x)$. In this case we have $a = 2$, $b = 3$, $c_1 = 3$, $c_2 = 2$, $g_1(x) = 2x - 1$, and $h_1(x) = 3x + 2$, so that $2 \cdot 3(6x^2 + x - 2) = 3 \cdot 2(2x - 1)(3x + 2)$ or $6x^2 + x - 2 = (2x - 1)(3x + 2)$. ∎

Irreducibility Tests

Theorem 17.1 reduces the question of irreducibility of a polynomial of degree 2 or 3 to one of finding a zero. The next theorem often allows us to simplify the problem even further.

Theorem 17.3 Mod *p* Irreducibility Test

Let p be a prime and suppose that $f(x) \in Z[x]$ with $\deg f(x) \geq 1$. Let $\bar{f}(x)$ be the polynomial in $Z_p[x]$ obtained from $f(x)$ by reducing all the coefficients of $f(x)$ modulo p. If $\bar{f}(x)$ is irreducible over Z_p and $\deg \bar{f}(x) = \deg f(x)$, then $f(x)$ is irreducible over Q.

PROOF It follows from the proof of Theorem 17.2 that if $f(x)$ is reducible over Q, then $f(x) = g(x)h(x)$ with $g(x), h(x) \in Z[x]$, and both $g(x)$ and $h(x)$ have degree less than that of $f(x)$. Let $\bar{f}(x)$, $\bar{g}(x)$, and $\bar{h}(x)$ be the polynomials obtained from $f(x)$, $g(x)$, and $h(x)$ by reducing all the coefficients modulo p. Since $\deg f(x) = \deg \bar{f}(x)$, we have $\deg \bar{g}(x) \leq \deg g(x) < \deg \bar{f}(x)$ and $\deg \bar{h}(x) \leq \deg h(x) < \deg \bar{f}(x)$. But, $\bar{f}(x) = \bar{g}(x)\bar{h}(x)$, and this contradicts our assumption that $\bar{f}(x)$ is irreducible over Z_p. ∎

EXAMPLE 6 Let $f(x) = 21x^3 - 3x^2 + 2x + 9$. Then, over Z_2, we have $\bar{f}(x) = x^3 + x^2 + 1$ and, since $\bar{f}(0) = 1$ and $\bar{f}(1) = 1$, we see that $\bar{f}(x)$ is irreducible over Z_2. Thus, $f(x)$ is irreducible over Q. Notice that, over Z_3, $\bar{f}(x) = 2x$ is irreducible, but we may *not* apply Theorem 17.3 to conclude that $f(x)$ is irreducible over Q. ∎

Be careful not to use the converse of Theorem 17.3. If $f(x) \in Z[x]$ and $\bar{f}(x)$ is reducible over Z_p for some p, $f(x)$ may still be irreducible over Q. For example, consider $f(x) = 21x^3 - 3x^2 + 2x + 8$. Then, over Z_2, $\bar{f}(x) = x^3 + x^2 = x^2(x + 1)$. But over Z_5, $\bar{f}(x)$ has no zeros and therefore is irreducible over Z_5. So, $f(x)$ is irreducible over Q. Note that this example shows that the Mod p Irreducibility Test may fail for some p and work for others. To conclude that a particular $f(x)$ in $Z[x]$ is irreducible over Q, all we need to do is find a single p for which the corresponding polynomial $\overline{f}(x)$ in Z_p is irreducible. However, this is not always possible, since $f(x) = x^4 + 1$ is irreducible over Q but reducible over Z_p for *every* prime p. (See Exercise 29.)

The Mod p Irreducibility Test can also be helpful in checking for irreducibility of polynomials of degree greater than 3 and polynomials with rational coefficients.

EXAMPLE 7 Let $f(x) = (3/7)x^4 - (2/7)x^2 + (9/35)x + 3/5$. We will show that $f(x)$ is irreducible over Q. First, let $h(x) = 35f(x) = 15x^4 - 10x^2 + 9x + 21$. Then $f(x)$ is irreducible over Q if $h(x)$ is irreducible over Z. Next, applying the Mod 2 Irreducibility Test to $h(x)$, we get $\overline{h}(x) = x^4 + x + 1$. Clearly, $\overline{h}(x)$ has no zeros in Z_2. Furthermore, $\overline{h}(x)$ has no quadratic factor in $Z_2[x]$ either. [For if so, the factor would have to be either $x^2 + x + 1$ or $x^2 + 1$. Long division shows that $x^2 + x + 1$ is not a factor, and $x^2 + 1$ cannot be a factor because it has a zero, whereas $\overline{h}(x)$ does not.] Thus, $\overline{h}(x)$ is irreducible over $Z_2[x]$. This guarantees that $h(x)$ is irreducible over Q. ∎

EXAMPLE 8 Let $f(x) = x^5 + 2x + 4$. Obviously, neither Theorem 17.1 nor the Mod 2 Irreducibility Test helps here. Let's try mod 3. Substitution of 0, 1, and 2 into $\overline{f}(x)$ does not yield 0, so there are no linear factors. But $\overline{f}(x)$ may have a quadratic factor. If so, we may assume it has the form $x^2 + ax + b$ (see Exercise 5). This gives nine possibilities to check. We can immediately rule out each of the nine that has a zero over Z_3, since $\overline{f}(x)$ does not have one. This leaves only $x^2 + 1$, $x^2 + x + 2$, and $x^2 + 2x + 2$ to check. These are eliminated by long division. So, since $\overline{f}(x)$ is irreducible over Z_3, $f(x)$ is irreducible over Q. (Why is it unnecessary to check for cubic or fourth-degree factors?) ∎

Another important irreducibility test is the following one, credited to Ferdinand Eisenstein (1823–1852), a student of Gauss. The corollary was first proved by Gauss by a different method.

Theorem 17.4 Eisenstein's Criterion (1850)

Let

$$f(x) = a_n x^n + a_{n-1}x^{n-1} + \cdots + a_0 \in Z[x].$$

If there is a prime p such that $p \nmid a_n$, $p \mid a_{n-1}, \ldots, p \mid a_0$ ***and*** $p^2 \nmid a_0$, ***then*** $f(x)$ ***is irreducible over*** Q.

PROOF If $f(x)$ is reducible over Q, we know by Theorem 17.2 that there exist elements $g(x)$ and $h(x)$ in $Z[x]$ such that $f(x) = g(x)h(x)$, $1 \leq \deg g(x)$, and $1 \leq \deg h(x) < n$. Say $g(x) = b_rx^r + \cdots + b_0$ and $h(x) = c_sx^s + \cdots + c_0$. Then, since $p \mid a_0$, $p^2 \nmid a_0$, and $a_0 = b_0c_0$, it follows that p divides one of b_0 and c_0 but not the other. Let us say $p \mid b_0$ and $p \nmid c_0$. Also, since $p \nmid a_n = b_rc_s$, we know that $p \nmid b_r$. So, there is a least integer t such that $p \nmid b_t$. Now, consider $a_t = b_tc_0 + b_{t-1}c_1 + \cdots + b_0c_t$. By assumption, p divides a_t and, by choice of t, every summand on the right after the first one is divisible by p. Clearly, this forces p to divide b_tc_0 as well. This is impossible, however, since p is prime and p divides neither b_t nor c_0. ∎

Corollary Irreducibility of *p*th Cyclotomic Polynomial

For any prime p, the pth cyclotomic polynomial

$$\Phi_p(x) = \frac{x^p - 1}{x - 1} = x^{p-1} + x^{p-2} + \cdots + x + 1$$

is irreducible over Q.

PROOF Let

$$f(x) = \Phi_p(x+1) = \frac{(x+1)^p - 1}{(x+1) - 1} = x^{p-1} + \binom{p}{1}x^{p-2} + \binom{p}{2}x^{p-3} + \cdots + \binom{p}{1}.$$

Then, since every coefficient except that of x^{p-1} is divisible by p and the constant term is not divisible by p^2, by Eisenstein's Criterion, $f(x)$ is irreducible over Q. So, if $\Phi_p(x) = g(x)h(x)$ were a nontrivial factorization of $\Phi_p(x)$ over Q, then $f(x) = \Phi_p(x+1) = g(x+1) \cdot h(x+1)$ would be a nontrivial factorization of $f(x)$ over Q. Since this is impossible, we conclude that $\Phi_p(x)$ is irreducible over Q. ∎

EXAMPLE 9 The polynomial $3x^5 + 15x^4 - 20x^3 + 10x + 20$ is irreducible over Q because $5 \nmid 3$ and $25 \nmid 20$ but 5 does divide 15, -20, 10, and 20. ∎

The principal reason for our interest in irreducible polynomials stems from the fact that there is an intimate connection among them, maximal ideals, and fields. This connection is revealed in the next theorem and its first corollary.

Theorem 17.5 $\langle p(x)\rangle$ Is Maximal If and Only If $p(x)$ Is Irreducible

Let F be a field and let $p(x) \in F[x]$. Then $\langle p(x)\rangle$ is a maximal ideal in $F[x]$ if and only if $p(x)$ is irreducible over F.

PROOF Suppose first that $\langle p(x)\rangle$ is a maximal ideal in $F[x]$. Clearly, $p(x)$ is neither the zero polynomial nor a unit in $F[x]$, because neither $\{0\}$ nor $F[x]$ is a maximal ideal in $F[x]$. If $p(x) = g(x)h(x)$ is a factorization of $p(x)$ over F, then $\langle p(x)\rangle \subseteq \langle g(x)\rangle \subseteq F[x]$. Thus, $\langle p(x)\rangle = \langle g(x)\rangle$ or $F[x] = \langle g(x)\rangle$. In the first case, we must have $\deg p(x) = \deg g(x)$. In the second case, it follows that $\deg g(x) = 0$ and, consequently, $\deg h(x) = \deg p(x)$. Thus, $p(x)$ cannot be written as a product of two polynomials in $F[x]$ of lower degree.

Now, suppose that $p(x)$ is irreducible over F. Let I be any ideal of $F[x]$ such that $\langle p(x)\rangle \subseteq I \subseteq F[x]$. Because $F[x]$ is a principal ideal domain, we know that $I = \langle g(x)\rangle$ for some $g(x)$ in $F[x]$. So, $p(x) \in \langle g(x)\rangle$ and, therefore, $p(x) = g(x)h(x)$, where $h(x) \in F[x]$. Since $p(x)$ is irreducible over F, it follows that either $g(x)$ is a constant or $h(x)$ is a constant. In the first case, we have $I = F[x]$; in the second case, we have $\langle p(x)\rangle = \langle g(x)\rangle = I$. So, $\langle p(x)\rangle$ is maximal in $F[x]$. ∎

Corollary 1 $F[x]/\langle p(x)\rangle$ Is a Field

Let F be a field and $p(x)$ be an irreducible polynomial over F. Then $F[x]/\langle p(x)\rangle$ is a field.

PROOF This follows directly from Theorems 17.5 and 14.4. ∎

The next corollary is a polynomial analog of Euclid's Lemma for primes (see Chapter 0).

Corollary 2 $p(x) \mid a(x)b(x)$ Implies $p(x) \mid a(x)$ or $p(x) \mid b(x)$

Let F be a field and let $p(x), a(x), b(x) \in F[x]$. If $p(x)$ is irreducible over F and $p(x) \mid a(x)b(x)$, then $p(x) \mid a(x)$ or $p(x) \mid b(x)$.

PROOF Since $p(x)$ is irreducible, $F[x]/\langle p(x)\rangle$ is a field and, therefore, an integral domain. From Theorem 14.3, we know that $\langle p(x)\rangle$ is a prime ideal, and since $p(x)$ divides $a(x)b(x)$, we have $a(x)b(x) \in \langle p(x)\rangle$. Thus, $a(x) \in \langle p(x)\rangle$ or $b(x) \in \langle p(x)\rangle$. This means that $p(x) \mid a(x)$ or $p(x) \mid b(x)$. ∎

The next two examples put the theory to work.

EXAMPLE 10 We construct a field with eight elements. By Theorem 17.1 and Corollary 1 of Theorem 17.5, it suffices to find a cubic polynomial over Z_2 that has no zero in Z_2. By inspection, $x^3 + x + 1$ fills the bill. Thus, $Z_2[x]/\langle x^3 + x + 1\rangle = \{ax^2 + bx + c + \langle x^3 + x + 1\rangle \mid a, b, c \in Z_2\}$ is a field with eight elements. For practice, let us do a few calculations in this field. Since the sum of two polynomials of the form $ax^2 + bx + c$ is another one of the same form, addition is easy. For example,

$$(x^2 + x + 1 + \langle x^3 + x + 1\rangle) + (x^2 + 1 + \langle x^3 + x + 1\rangle) = x + \langle x^3 + x + 1\rangle.$$

On the other hand, multiplication of two coset representatives need not yield one of the original eight coset representatives:

$$(x^2 + x + 1 + \langle x^3 + x + 1\rangle) \cdot (x^2 + 1 + \langle x^3 + x + 1\rangle) = x^4 + x^3 + x + 1 + \langle x^3 + x + 1\rangle = x^4 + \langle x^3 + x + 1\rangle$$

(since the ideal absorbs the last three terms). How do we express this in the form $ax^2 + bx + c + \langle x^3 + x + 1\rangle$? One way is to long divide x^4 by $x^3 + x + 1$ to obtain the remainder of $x^2 + x$ (just as one reduces $12 + \langle 5\rangle$ to $2 + \langle 5\rangle$ by dividing 12 by 5 to obtain the remainder 2). Another way is to observe that $x^3 + x + 1 + \langle x^3 + x + 1\rangle = 0 + \langle x^3 + x + 1\rangle$ implies $x^3 + \langle x^3 + x + 1\rangle = x + 1 + \langle x^3 + x + 1\rangle$. Thus, we may multiply both sides by x to obtain

$$x^4 + \langle x^3 + x + 1\rangle = x^2 + x + \langle x^3 + x + 1\rangle.$$

Similarly,

$$\begin{aligned}(x^2 + x + \langle x^3 + x + 1\rangle) \cdot (x + \langle x^3 + x + 1\rangle) &= x^3 + x^2 + \langle x^3 + x + 1\rangle \\ &= x^2 + x + 1 + \langle x^3 + x + 1\rangle.\end{aligned}$$

A partial multiplication table for this field is given in Table 17.1. To simplify the notation, we indicate a coset by its representative only.

Table 17.1 A Partial Multiplication Table for Example 10

	1	x	$x+1$	x^2	x^2+1	x^2+x	x^2+x+1
1	1	x	$x+1$	x^2	x^2+1	x^2+x	x^2+x+1
x	x	x^2	x^2+x	$x+1$	1	x^2+x+1	x^2+1
$x+1$	$x+1$	x^2+x	x^2+1	x^2+x+1	x^2	1	x
x^2	x^2	$x+1$	x^2+x+1	x^2+x	x	x^2+1	1
x^2+1	x^2+1	1	x^2	x	x^2+x+1	$x+1$	x^2+x

(Complete the table yourself. Keep in mind that x^3 can be replaced by $x + 1$ and x^4 by $x^2 + x$.) ▮

▮ **EXAMPLE 11** Since $x^2 + 1$ has no zero in Z_3, it is irreducible over Z_3. Thus, $Z_3[x]/\langle x^2 + 1\rangle$ is a field. Analogous to Example 12 in Chapter 14, $Z_3[x]/\langle x^2 + 1\rangle = \{ax + b + \langle x^2 + 1\rangle \mid a, b \in Z_3\}$. Thus, this field has nine elements. A multiplication table for this field can be obtained from Table 13.1 by replacing i by x. (Why does this work?) ▮

Unique Factorization in $Z[x]$

As a further application of the ideas presented in this chapter, we next prove that $Z[x]$ has an important factorization property. In Chapter 18, we will study this property in greater depth. The first proof of Theorem 17.6 was given by Gauss. In reading this theorem and its proof, keep in mind that the units in $Z[x]$ are precisely $f(x) = 1$ and $f(x) = -1$ (see Exercise 25 in Chapter 12), the irreducible polynomials of degree 0 over Z are precisely those of the form $f(x) = p$ and $f(x) = -p$ where p is a prime, and every nonconstant polynomial from $Z[x]$ that is irreducible over Z is primitive (see Exercise 3).

▮ Theorem 17.6 Unique Factorization in $Z[x]$

Every polynomial in $Z[x]$ that is not the zero polynomial or a unit in $Z[x]$ can be written in the form $b_1b_2 \cdots b_sp_1(x)p_2(x) \cdots p_m(x)$, where the b_i's are irreducible polynomials of degree 0 and the $p_i(x)$'s are irreducible polynomials of positive degree. Furthermore, if

$$b_1b_2 \cdots b_sp_1(x)p_2(x) \cdots p_m(x) = c_1c_2 \cdots c_tq_1(x)q_2(x) \cdots q_n(x),$$

where the b_i's and c_i's are irreducible polynomials of degree 0 and the $p_i(x)$'s and $q_i(x)$'s are irreducible polynomials of positive degree, then $s = t$, $m = n$, and, after renumbering the c's and $q(x)$'s, we have $b_i = \pm c_i$ for $i = 1, \ldots, s$ and $p_i(x) = \pm q_i(x)$ for $i = 1, \ldots, m$.

PROOF Let $f(x)$ be a nonzero, nonunit polynomial from $Z[x]$. If $\deg f(x) = 0$, then $f(x)$ is constant and the result follows from the Fundamental Theorem of Arithmetic. If $\deg f(x) > 0$, let b denote the content of $f(x)$, and let $b_1b_2 \cdots b_s$ be the factorization of b as a product of primes. Then, $f(x) = b_1b_2 \cdots b_sf_1(x)$, where $f_1(x)$ belongs to $Z[x]$, is primitive and $\deg f_1(x) = \deg f(x)$. Thus, to prove the existence portion of the theorem, it suffices to show that a primitive polynomial $f(x)$ of positive degree can be written as a product of irreducible polynomials of positive degree. We proceed by induction on $\deg f(x)$. If $\deg f(x) = 1$, then $f(x)$ is already irreducible and we are done. Now suppose that every primitive polynomial of degree less than $\deg f(x)$ can be written as a product of irreducibles of positive degree. If $f(x)$ is irreducible, there is nothing to prove. Otherwise, $f(x) = g(x)h(x)$, where both $g(x)$ and $h(x)$ are primitive and have degree less than that of $f(x)$. Thus, by induction, both $g(x)$ and $h(x)$ can be written as a product of irreducibles of positive degree. Clearly, then, $f(x)$ is also such a product.

To prove the uniqueness portion of the theorem, suppose that $f(x) = b_1b_2 \cdots b_sp_1(x)p_2(x) \cdots p_m(x) = c_1c_2 \cdots c_tq_1(x)q_2(x) \cdots q_n(x)$, where the b_i's and c_i's are irreducible polynomials of degree 0 and the $p_i(x)$'s and $q_i(x)$'s are irreducible polynomials of positive degree. Let $b = b_1b_2 \cdots b_s$ and $c = c_1c_2 \cdots c_t$. Since the $p(x)$'s and $q(x)$'s are primitive, it follows from Gauss's Lemma that $p_1(x)p_2(x) \cdots p_m(x)$ and $q_1(x)q_2(x) \cdots q_n(x)$ are primitive. Hence, both b and c must equal plus or minus the content of $f(x)$ and, therefore, are equal in absolute value. It then follows from the Fundamental Theorem of Arithmetic that $s = t$ and, after renumbering, $b_i = \pm c_i$ for $i = 1, 2, \ldots, s$. Thus, by canceling the constant terms in the two factorizations for $f(x)$, we have $p_1(x)p_2(x) \cdots p_m(x) = \pm q_1(x)\, q_2(x) \cdots q_n(x)$. Now, viewing the $p(x)$'s and $q(x)$'s as elements of $Q[x]$ and noting that $p_1(x)$ divides $q_1(x) \cdots q_n(x)$, it follows from Corollary 2 of Theorem 17.5 and induction (see Exercise 28) that $p_1(x) \mid q_i(x)$ for some i. By renumbering, we may assume $i = 1$. Then, since $q_1(x)$ is irreducible, we have $q_1(x) = (r/s)p_1(x)$, where $r, s \in Z$. However, because both $q_1(x)$ and $p_1(x)$ are primitive, we must have $r/s = \pm 1$. So, $q_1(x) = \pm p_1(x)$. Also, after canceling, we have $p_2(x) \cdots p_m(x) = \pm q_2(x) \cdots q_n(x)$. Now, we may repeat the argument above with $p_2(x)$ in place of $p_1(x)$. If $m < n$, after m such steps we would have 1 on the left and a nonconstant polynomial on the right. Clearly, this is impossible. On the other hand, if $m > n$, after n steps we would have ± 1 on the right and a nonconstant polynomial on the left—another impossibility. So, $m = n$ and $p_i(x) = \pm q_i(x)$ after suitable renumbering of the $q(x)$'s. ∎

Weird Dice: An Application of Unique Factorization

EXAMPLE 12 Consider an ordinary pair of dice whose faces are labeled 1 through 6. The probability of rolling a sum of 2 is 1/36, the probability of rolling a sum of 3 is 2/36, and so on. In a 1978 issue of

Scientific American [1], Martin Gardner remarked that if one were to label the six faces of one cube with the integers 1, 2, 2, 3, 3, 4 and the six faces of another cube with the integers 1, 3, 4, 5, 6, 8, then the probability of obtaining any particular sum with these dice (called *Sicherman dice*) would be the same as the probability of rolling that sum with ordinary dice (that is, 1/36 for a 2, 2/36 for a 3, and so on). See Figure 17.1. In this example, we show how the Sicherman labels can be derived, and that they are the only possible such labels besides 1 through 6. To do so, we utilize the fact that $Z[x]$ has the unique factorization property.

	1	2	3	4	5	6
1	2	3	4	5	6	7
2	3	4	5	6	7	8
3	4	5	6	7	8	9
4	5	6	7	8	9	10
5	6	7	8	9	10	11
6	7	8	9	10	11	12

	1	2	2	3	3	4
1	2	3	3	4	4	5
3	4	5	5	6	6	7
4	5	6	6	7	7	8
5	6	7	7	8	8	9
6	7	8	8	9	9	10
8	9	10	10	11	11	12

Figure 17.1

To begin, let us ask ourselves how we may obtain a sum of 6, say, with an ordinary pair of dice. Well, there are five possibilities for the two faces: (5, 1), (4, 2), (3, 3), (2, 4), and (1, 5). Next we consider the product of the two polynomials created by using the ordinary dice labels as exponents:

$$(x^6 + x^5 + x^4 + x^3 + x^2 + x)(x^6 + x^5 + x^4 + x^3 + x^2 + x).$$

Observe that we pick up the term x^6 in this product in precisely the following ways: $x^5 \cdot x^1, x^4 \cdot x^2, x^3 \cdot x^3, x^2 \cdot x^4, x^1 \cdot x^5$. Notice the correspondence between pairs of labels whose sums are 6 and pairs of terms whose products are x^6. This correspondence is one-to-one, and it is valid for all sums and all dice—including the Sicherman dice and any other dice that yield the desired probabilities. So, let $a_1, a_2, a_3, a_4, a_5, a_6$ and $b_1, b_2, b_3, b_4, b_5, b_6$ be any two lists of positive integer labels for the faces of a pair of cubes with the property that the probability of rolling any particular sum with these dice (let us call them *weird dice*) is the same as the probability of rolling that sum with ordinary dice labeled 1 through 6. Using our observation about products of polynomials, this means that

$$\begin{aligned}(x^6 + x^5 + x^4 + x^3 + x^2 + x)&(x^6 + x^5 + x^4 + x^3 + x^2 + x)\\ = (x^{a_1} + x^{a_2} + x^{a_3} + x^{a_4} + x^{a_5} + x^{a_6}) &\cdot\\ (x^{b_1} + x^{b_2} + x^{b_3} + x^{b_4} + x^{b_5} + x^{b_6}).& \qquad (1)\end{aligned}$$

Now all we have to do is solve this equation for the a's and b's. Here is where unique factorization in $Z[x]$ comes in. The polynomial $x^6 + x^5 + x^4 + x^3 + x^2 + x$ factors uniquely into irreducibles as

$$x(x + 1)(x^2 + x + 1)(x^2 - x + 1)$$

so that the left-hand side of Equation (1) has the irreducible factorization

$$x^2(x + 1)^2(x^2 + x + 1)^2(x^2 - x + 1)^2.$$

So, by Theorem 17.6, this means that these factors are the only possible irreducible factors of $P(x) = x^{a_1} + x^{a_2} + x^{a_3} + x^{a_4} + x^{a_5} + x^{a_6}$. Thus, $P(x)$ has the form

$$x^q(x + 1)^r(x^2 + x + 1)^t(x^2 - x + 1)^u,$$

where $0 \leq q, r, t, u \leq 2$.

To restrict further the possibilities for these four parameters, we evaluate $P(1)$ in two ways. $P(1) = 1^{a_1} + 1^{a_2} + \cdots + 1^{a_6} = 6$ and $P(1) = 1^q2^r3^t1^u$. Clearly, this means that $r = 1$ and $t = 1$. What about q? Evaluating $P(0)$ in two ways shows that $q \neq 0$. On the other hand, if $q = 2$, the smallest possible sum one could roll with the corresponding labels for dice would be 3. Since this violates our assumption, we have now reduced our list of possibilities for q, r, t, and u to $q = 1$, $r = 1$, $t = 1$, and $u = 0, 1, 2$. Let's consider each of these possibilities in turn.

When $u = 0$, $P(x) = x^4 + x^3 + x^3 + x^2 + x^2 + x$, so the die labels are 4, 3, 3, 2, 2, 1—a Sicherman die.

When $u = 1$, $P(x) = x^6 + x^5 + x^4 + x^3 + x^2 + x$, so the die labels are 6, 5, 4, 3, 2, 1—an ordinary die.

When $u = 2$, $P(x) = x^8 + x^6 + x^5 + x^4 + x^3 + x$, so the die labels are 8, 6, 5, 4, 3, 1—the other Sicherman die.

This proves that the Sicherman dice do give the same probabilities as ordinary dice *and* that they are the *only* other pair of dice that have this property. ∎

Exercises

No matter how good you are at something, there's always about a million people better than you.

HOMER SIMPSON

1. Verify the assertion made in Example 2.
2. Suppose that D is an integral domain and F is a field containing D. If $f(x) \in D[x]$ and $f(x)$ is irreducible over F but reducible over D, what can you say about the factorization of $f(x)$ over D?

3. Show that a nonconstant polynomial from $Z[x]$ that is irreducible over Z is primitive. (This exercise is referred to in this chapter.)

4. Suppose that $f(x) = x^n + a_{n-1}x^{n-1} + \cdots + a_0 \in Z[x]$. If r is rational and $x - r$ divides $f(x)$, show that r is an integer.

5. Let F be a field and let a be a nonzero element of F.

a. If $af(x)$ is irreducible over F, prove that $f(x)$ is irreducible over F.

b. If $f(ax)$ is irreducible over F, prove that $f(x)$ is irreducible over F.

c. If $f(x + a)$ is irreducible over F, prove that $f(x)$ is irreducible over F.

d. Use part c to prove that $8x^3 - 6x + 1$ is irreducible over Q.

(This exercise is referred to in this chapter.)

6. Let F be a field and $f(x) \in F[x]$. Show that, as far as deciding upon the irreducibility of $f(x)$ over F is concerned, we may assume that $f(x)$ is monic. (This assumption is useful when one uses a computer to check for irreducibility.)

7. Explain how the Mod p Irreducibility Test (Theorem 17.3) can be used to test members of $Q[x]$ for irreducibility.

8. Suppose that $f(x) \in Z_p[x]$ and $f(x)$ is irreducible over Z_p, where p is a prime. If $\deg f(x) = n$, prove that $Z_p[x]/\langle f(x)\rangle$ is a field with p^n elements.

9. Construct a field of order 25.

10. Construct a field of order 27.

11. Show that $x^3 + x^2 + x + 1$ is reducible over Q. Does this fact contradict the corollary to Theorem 17.4?

12. Determine which of the polynomials below is (are) irreducible over Q.

a. $x^5 + 9x^4 + 12x^2 + 6$

b. $x^4 + x + 1$

c. $x^4 + 3x^2 + 3$

d. $x^5 + 5x^2 + 1$

e. $(5/2)x^5 + (9/2)x^4 + 15x^3 + (3/7)x^2 + 6x + 3/14$

13. Show that $x^4 + 1$ is irreducible over Q but reducible over **R**. (This exercise is referred to in this chapter.)

14. Show that $x^2 + x + 4$ is irreducible over Z_{11}.

15. Let $f(x) = x^3 + 6 \in Z_7[x]$. Write $f(x)$ as a product of irreducible polynomials over Z_7.

16. Let $f(x) = x^3 + x^2 + x + 1 \in Z_2[x]$. Write $f(x)$ as a product of irreducible polynomials over Z_2.

17. Let p be a prime.
   a. Show that the number of reducible polynomials over Z_p of the form $x^2 + ax + b$ is $p(p + 1)/2$.
   b. Determine the number of reducible quadratic polynomials over Z_p.

18. Let p be a prime.
   a. Determine the number of irreducible polynomials over Z_p of the form $x^2 + ax + b$.
   b. Determine the number of irreducible quadratic polynomials over Z_p.

19. Show that for every prime p there exists a field of order p^2.

20. Prove that, for every positive integer n, there are infinitely many polynomials of degree n in $Z[x]$ that are irreducible over Q.

21. Show that the field given in Example 11 in this chapter is isomorphic to the field given in Example 9 in Chapter 13.

22. Let $f(x) \in Z_p[x]$. Prove that if $f(x)$ has no factor of the form $x^2 + ax + b$, then it has no quadratic factor over Z_p.

23. Find all monic irreducible polynomials of degree 2 over Z_3.

24. Given that π is not the zero of a nonzero polynomial with rational coefficients, prove that π^2 cannot be written in the form $a\pi + b$, where a and b are rational.

25. Find all the zeros and their multiplicities of $x^5 + 4x^4 + 4x^3 - x^2 - 4x + 1$ over Z_5.

26. Find all zeros of $f(x) = 3x^2 + x + 4$ over Z_7 by substitution. Find all zeros of $f(x)$ by using the quadratic formula $(-b \pm \sqrt{b^2 - 4ac}) \cdot (2a)^{-1}$ (all calculations are done in Z_7). Do your answers agree? Should they? Find all zeros of $g(x) = 2x^2 + x + 3$ over Z_5 by substitution. Try the quadratic formula on $g(x)$. Do your answers agree? State necessary and sufficient conditions for the quadratic formula to yield the zeros of a quadratic from $Z_p[x]$, where p is a prime greater than 2.

27. (Rational Root Theorem) Let

$$f(x) = a_nx^n + a_{n-1}x^{n-1} + \cdots + a_0 \in Z[x]$$

and $a_n \neq 0$. Prove that if r and s are relatively prime integers and $f(r/s) = 0$, then $r \mid a_0$ and $s \mid a_n$.

28. Let F be a field and let $p(x), a_1(x), a_2(x), \ldots, a_k(x) \in F[x]$, where $p(x)$ is irreducible over F. If $p(x) \mid a_1(x)a_2(x) \cdots a_k(x)$, show that $p(x)$ divides some $a_i(x)$. (This exercise is referred to in the proof of Theorem 17.6.)

29. Show that $x^4 + 1$ is reducible over Z_p for every prime p. (This exercise is referred to in this chapter.)

30. If p is a prime, prove that $x^{p-1} - x^{p-2} + x^{p-3} - \cdots - x + 1$ is irreducible over Q.

31. Let F be a field and let $p(x)$ be irreducible over F. If E is a field that contains F and there is an element a in E such that $p(a) = 0$, show that the mapping ϕ: $F[x] \to E$ given by $f(x) \to f(a)$ is a ring homomorphism with kernel $\langle p(x) \rangle$. (This exercise is referred to in Chapter 20.)

32. Prove that the ideal $\langle x^2 + 1 \rangle$ is prime in $Z[x]$ but not maximal in $Z[x]$.

33. Let F be a field and let $p(x)$ be irreducible over F. Show that $\{a + \langle p(x) \rangle \mid a \in F\}$ is a subfield of $F[x]/\langle p(x) \rangle$ isomorphic to F. (This exercise is referred to in Chapter 20.)

34. Let F be a field and let $f(x)$ be a polynomial in $F[x]$ that is reducible over F. Prove that $\langle f(x) \rangle$ is not a prime ideal in $F[x]$.

35. Example 1 in this chapter shows the converse of Theorem 17.2 is not true. That is, a polynomial $f(x)$ in $Z[x]$ can be reducible over Z but irreducible over Q. State a condition on $f(x)$ that makes the converse true.

36. Suppose there is a real number r with the property that $r + 1/r$ is an odd integer. Prove that r is irrational.

37. In the game of Monopoly, would the probabilities of landing on various properties be different if the game were played with Sicherman dice instead of ordinary dice? Why?

38. Carry out the analysis given in Example 12 for a pair of tetrahedrons instead of a pair of cubes. (Define ordinary tetrahedral dice as the ones labeled 1 through 4.)

39. Suppose in Example 12 that we begin with n $(n > 2)$ ordinary dice each labeled 1 through 6, instead of just two. Show that the only possible labels that produce the same probabilities as n ordinary dice are the labels 1 through 6 and the Sicherman labels.

40. Show that one two-sided die labeled with 1 and 4 and another 18-sided die labeled with 1, 2, 2, 3, 3, 3, 4, 4, 4, 5, 5, 5, 6, 6, 6, 7, 7, 8 yield the same probabilities as an ordinary pair of cubes labeled 1 through 6. Carry out an analysis similar to that given in Example 12 to derive these labels.

Computer Exercises

Computer exercises for this chapter are available at the website:

http://www.d.umn.edu/~jgallian

Reference

1. Martin Gardner, "Mathematical Games," *Scientific American* 238/2 (1978): 19–32.

Suggested Readings

Duane Broline, "Renumbering the Faces of Dice," *Mathematics Magazine* 52 (1979): 312–315.

In this article, the author extends the analysis we carried out in Example 12 to dice in the shape of Platonic solids.

J. A. Gallian and D. J. Rusin, "Cyclotomic Polynomials and Nonstandard Dice," *Discrete Mathematics* 27 (1979): 245–259.

Here Example 12 is generalized to the case of n dice each with m labels for all n and m greater than 1.

Randall Swift and Brian Fowler, "Relabeling Dice," *The College Mathematics Journal* 30 (1999): 204–208.

The authors use the method presented in this chapter to derive positive integer labels for a pair of dice that are not six-sided but give the same probabilities for the sum of the faces as a pair of cubes labeled 1 through 6.

Serge Lang

Bogdan Oporowski

> Lang's *Algebra* changed the way graduate algebra is taught.... It has affected all subsequent graduate-level algebra books.
>
> *Citation for the Steele Prize*

SERGE LANG was a prolific mathematician, inspiring teacher, and political activist. He was born near Paris on May 19, 1927. His family moved to Los Angeles when he was a teenager. Lang received a B.A. in physics from Caltech in 1946 and a Ph.D. in mathematics from Princeton in 1951 under Emil Artin (see the biography in Chapter 19). His first permanent position was at Columbia University in 1955, but in 1971 Lang resigned his position at Columbia as a protest against Columbia's handling of Vietnam antiwar protesters. He joined Yale University in 1972 and remained there until his retirement.

Lang made significant contributions to number theory, algebraic geometry, differential geometry, and analysis. He wrote more than 120 research articles and 60 books. His most famous and influential book was his graduate-level *Algebra*. Lang was a prize-winning teacher known for his extraordinary devotion to students. Lang often got into heated discussions about mathematics, the arts, and politics. In one incident, he threatened to hit a fellow mathematician with a bronze bust for not conceding it was self-evident that the Beatles were greater musicians than Beethoven.

Among Lang's honors were the Steele Prize for Mathematical Exposition from the American Mathematical Society, the Cole Prize in Algebra (see Chapter 25), and election to the National Academy of Sciences. Lang died on September 25, 2005, at the age of 78.

For more information about Lang, visit:

http://wikipedia.org/wiki/ Serge_Lang

18 Divisibility in Integral Domains

Give me a fruitful error anytime, full of seeds, bursting with its own corrections. You can keep your sterile truth for yourself.

VILFREDO PARETO

Irreducibles, Primes

In the preceding two chapters, we focused on factoring polynomials over the integers or a field. Several of those results—unique factorization in $Z[x]$ and the division algorithm for $F[x]$, for instance—are natural counterparts to theorems about the integers. In this chapter and the next, we examine factoring in a more abstract setting.

> **Definition Associates, Irreducibles, Primes**
> **Elements a and b of an integral domain D are called *associates* if $a = ub$, where u is a unit of D. A nonzero element a of an integral domain D is called an *irreducible* if a is not a unit and, whenever $b, c \in D$ with $a = bc$, then b or c is a unit. A nonzero element a of an integral domain D is called a *prime* if a is not a unit and $a \mid bc$ implies $a \mid b$ or $a \mid c$.**

Roughly speaking, an irreducible is an element that can be factored only in a trivial way. Notice that an element a is a prime if and only if $\langle a \rangle$ is a prime ideal.

Relating the definitions above to the integers may seem a bit confusing, since in Chapter 0 we defined a positive integer to be a prime if it satisfies our definition of an irreducible, and we proved that a prime integer satisfies the definition of a prime in an integral domain (Euclid's Lemma). The source of the confusion is that in the case of the integers, the concepts of irreducibles and primes are equivalent, but in general, as we will soon see, they are not.

The distinction between primes and irreducibles is best illustrated by integral domains of the form $Z[\sqrt{d}] = \{a + b\sqrt{d} \mid a, b \in Z\}$, where d is not 1 and is not divisible by the square of a prime. (These rings are of fundamental importance in number theory.) To analyze these rings, we need a convenient method of determining their units, irreducibles, and primes. To do this, we define a function N, called the *norm*, from $Z[\sqrt{d}]$ into the nonnegative integers by $N(a + b\sqrt{d}) = |a^2 - db^2|$. We leave it to the reader (Exercise 1) to verify the following four properties: $N(x) = 0$ if and only if $x = 0$; $N(xy) = N(x)N(y)$ for all x and y; x is a unit if and only if $N(x) = 1$; and, if $N(x)$ is prime, then x is irreducible in $Z[\sqrt{d}]$.

EXAMPLE 1 We exhibit an irreducible in $Z[\sqrt{-3}]$ that is not prime. Here, $N(a + b\sqrt{-3}) = a^2 + 3b^2$. Consider $1 + \sqrt{-3}$. Suppose that we can factor this as xy, where neither x nor y is a unit. Then $N(xy) = N(x)N(y) = N(1 + \sqrt{-3}) = 4$, and it follows that $N(x) = 2$. But there are no integers a and b that satisfy $a^2 + 3b^2 = 2$. Thus, x or y is a unit and $1 + \sqrt{-3}$ is an irreducible. To verify that it is not prime, we observe that $(1 + \sqrt{-3})(1 - \sqrt{-3}) = 4 = 2 \cdot 2$, so that $1 + \sqrt{-3}$ divides $2 \cdot 2$. On the other hand, for integers a and b to exist so that $2 = (1 + \sqrt{-3})(a + b\sqrt{-3}) = (a - 3b) + (a + b)\sqrt{-3}$, we must have $a - 3b = 2$ and $a + b = 0$, which is impossible. ∎

Showing that an element of a ring of the form $Z[\sqrt{d}]$ is irreducible is more difficult when $d > 1$. The next example illustrates one method of doing this. The example also shows that the converse of the fourth property above for the norm is not true. That is, it shows that x may be irreducible even if $N(x)$ is not prime.

EXAMPLE 2 The element 7 is irreducible in the ring $Z[\sqrt{5}]$. To verify this assertion, suppose that $7 = xy$, where neither x nor y is a unit. Then $49 = N(7) = N(x)N(y)$, and since x is not a unit, we cannot have $N(x) = 1$. This leaves only the case $N(x) = 7$. Let $x = a + b\sqrt{5}$. Then there are integers a and b satisfying $|a^2 - 5b^2| = 7$. This means that $a^2 - 5b^2 = \pm 7$. Viewing this equation modulo 5 and trying all possible cases for a reveals that the only solution is $a = 0$. But this means that a is divisible by 5, and this implies that $|a^2 - 5b^2| = 7$ is divisible by 5, which is false. ∎

Example 1 raises the question of whether or not there is an integral domain containing a prime that is not an irreducible. The answer: no.

Theorem 18.1 Prime Implies Irreducible

In an integral domain, every prime is an irreducible.

PROOF Suppose that a is a prime in an integral domain and $a = bc$. We must show that b or c is a unit. By the definition of prime, we know that $a \mid b$ or $a \mid c$. Say $at = b$. Then $1b = b = at = (bc)t = b(ct)$ and, by cancellation, $1 = ct$. Thus, c is a unit. ∎

Recall that a principal ideal domain is an integral domain in which every ideal has the form $\langle a \rangle$. The next theorem reveals a circumstance in which primes and irreducibles are equivalent.

Theorem 18.2 PID Implies Irreducible Equals Prime

In a principal ideal domain, an element is an irreducible if and only if it is a prime.

PROOF Theorem 18.1 shows that primes are irreducibles. To prove the converse, let a be an irreducible element of a principal ideal domain D and suppose that $a \mid bc$. We must show that $a \mid b$ or $a \mid c$. Consider the ideal $I = \{ax + by \mid x, y \in D\}$ and let $\langle d \rangle = I$. Since $a \in I$, we can write $a = dr$, and because a is irreducible, d is a unit or r is a unit. If d is a unit, then $I = D$ and we may write $1 = ax + by$. Then $c = acx + bcy$, and since a divides both terms on the right, a also divides c.

On the other hand, if r is a unit, then $\langle a \rangle = \langle d \rangle = I$, and, because $b \in I$, there is an element t in D such that $at = b$. Thus, a divides b. ∎

It is an easy consequence of the respective division algorithms for Z and $F[x]$, where F is a field, that Z and $F[x]$ are principal ideal domains (see Exercise 41 in Chapter 14 and Theorem 16.3). Our next example shows, however, that one of the most familiar rings is not a principal ideal domain.

EXAMPLE 3 We show that $Z[x]$ is not a principal ideal domain. Consider the ideal $I = \langle 2, x \rangle$. We claim that I is not of the form $\langle h(x) \rangle$. If this were so, there would be $f(x)$ and $g(x)$ in $Z[x]$ such that $2 = h(x)f(x)$ and $x = h(x)g(x)$, since both 2 and x belong to I. By the degree rule (Exercise 19 in Chapter 16), $0 = \deg 2 = \deg h(x) + \deg f(x)$, so that $h(x)$ is a constant polynomial. To determine which constant, we observe that $2 = h(1)f(1)$. Thus, $h(1) = \pm 1$ or ± 2. Since 1 is not in I, we must have $h(x) = \pm 2$. But then $x = \pm 2g(x)$, which is nonsense. ∎

We have previously proved that the integral domains Z and $Z[x]$ have important factorization properties: Every integer greater than 1 can be uniquely factored as a product of irreducibles (that is, primes), and every nonzero, nonunit polynomial can be uniquely factored as a product of irreducible polynomials. It is natural to ask whether all integral domains have this property. The question of unique factorization in integral domains first arose with the efforts to solve a famous problem in number theory that goes by the name Fermat's Last Theorem.

Historical Discussion of Fermat's Last Theorem

There are infinitely many nonzero integers x, y, z that satisfy the equation $x^2 + y^2 = z^2$. But what about the equation $x^3 + y^3 = z^3$ or, more generally, $x^n + y^n = z^n$, where n is an integer greater than 2 and x, y, z are nonzero integers? Well, no one has ever found a single solution of this equation, and for more than three centuries many have tried to prove that there is none. The tremendous effort put forth by the likes of Euler, Legendre, Abel, Gauss, Dirichlet, Cauchy, Kummer, Kronecker, and Hilbert to prove that there are no solutions to this equation has greatly influenced the development of ring theory.

About a thousand years ago, Arab mathematicians gave an incorrect proof that there were no solutions when $n = 3$. The problem lay dormant until 1637, when the French mathematician Pierre de Fermat (1601–1665) wrote in the margin of a book, ". . . it is impossible to separate a cube into two cubes, a fourth power into two fourth powers, or, generally, any power above the second into two powers of the same degree: I have discovered a truly marvelous demonstration [of this general theorem] which this margin is too narrow to contain."

Because Fermat gave no proof, many mathematicians tried to prove the result. The case where $n = 3$ was done by Euler in 1770, although his proof was incomplete. The case where $n = 4$ is elementary and was done by Fermat himself. The case where $n = 5$ was done in 1825 by Dirichlet, who had just turned 20, and by Legendre, who was past 70. Since the validity of the case for a particular integer implies the validity for all multiples of that integer, the next case of interest was $n = 7$. This case resisted the efforts of the best mathematicians until it was done by Gabriel Lamé in 1839. In 1847, Lamé stirred excitement by announcing that he had completely solved the problem. His approach was to factor the expression $x^p + y^p$, where p is an odd prime, into

$$(x + y)(x + \alpha y) \cdots (x + \alpha^{p-1}y),$$

where α is the complex number $\cos(2\pi/p) + i \sin(2\pi/p)$. Thus, his factorization took place in the ring $Z[\alpha] = \{a_0 + a_1\alpha + \cdots + a_{p-1}\alpha^{p-1} \mid a_i \in Z\}$. But Lamé made the mistake of assuming that, in such a ring, factorization into the product of irreducibles is unique. In fact, three years earlier, Ernst Eduard Kummer had proved that this is not always the case. Undaunted by the failure of unique factorization, Kummer began developing a theory to "save" factorization by creating a new type of number. Within a few weeks of Lamé's announcement, Kummer had shown that Fermat's Last Theorem is true for all primes of a special type. This proved that the theorem was true for all exponents less than 100, prime or not, except for 37, 59, 67, and 74. Kummer's work has led to the theory of ideals as we know it today.

Over the centuries, many proposed proofs have not held up under scrutiny. The famous number theorist Edmund Landau received so many of these that he had a form printed with "On page ____, lines ____ to ____, you will find there is a mistake." Martin Gardner, "Mathematical Games" columnist of *Scientific American,* had postcards printed to decline requests from readers asking him to examine their proofs.

Recent discoveries tying Fermat's Last Theorem closely to modern mathematical theories gave hope that these theories might eventually lead to a proof. In March 1988, newspapers and scientific publications worldwide carried news of a proof by Yoichi Miyaoka (see Figure 18.1). Within weeks, however, Miyaoka's proof was shown to be invalid. In June 1993, excitement spread through the mathematics community with the announcement that Andrew Wiles of Princeton University had proved Fermat's Last Theorem (see Figure 18.2). The Princeton mathematics department chairperson was quoted as saying, "When we heard it, people started walking on air." But once again a proof did not hold up under scrutiny. This story does have a happy ending. The mathematical community has agreed on the validity of the revised proof given by Wiles and Richard Taylor in September of 1994.

In view of the fact that so many eminent mathematicians were unable to prove Fermat's Last Theorem, despite the availability of the vastly powerful theories, it seems highly improbable that Fermat had a correct proof. Most likely, he made the error that his successors made of assuming that the properties of integers, such as unique factorization, carry over to integral domains in general.

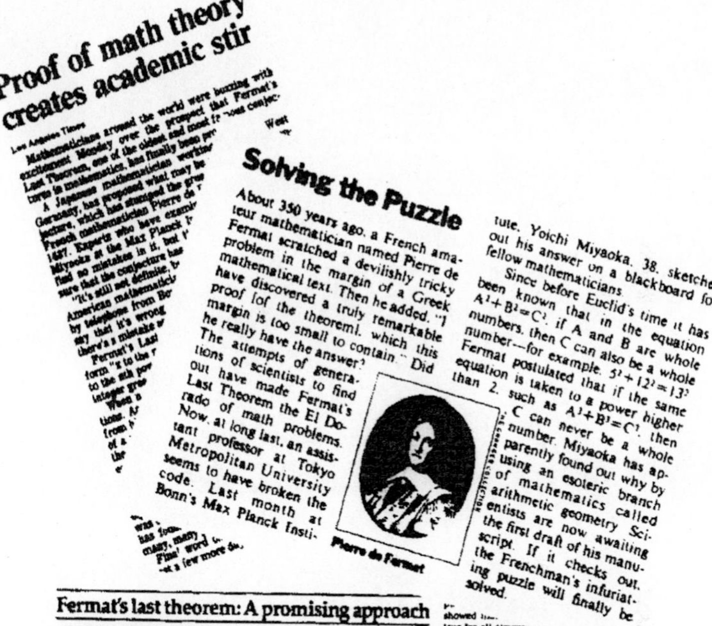

Proof of math theory creates academic stir

Los Angeles Times

Solving the Puzzle

About 350 years ago, a French amateur mathematician named Pierre de Fermat scratched a devilishly tricky problem in the margin of a Greek mathematical text. Then he added, "I have discovered a truly remarkable proof [of the theorem], which this margin is too small to contain." Did he really have the answer? The attempts of generations of scientists to find out have made Fermat's Last Theorem the El Dorado of math problems. Now, at long last, an assistant professor at Tokyo Metropolitan University seems to have broken the code. Last month at Bonn's Max Planck Institute, Yoichi Miyaoka, 38, sketched out his answer on a blackboard for fellow mathematicians.

Since before Euclid's time it has been known that in the equation $A^2+B^2=C^2$, if A and B are whole numbers, then C can also be a whole number—for example, $5^2+12^2=13^2$. Fermat postulated that if the same equation is taken to a power higher than 2, such as $A^3+B^3=C^3$, then C can never be a whole number. Miyaoka has apparently found out why by using an esoteric branch of mathematics called arithmetic geometry. Scientists are now awaiting the first draft of his manuscript. If it checks out, the Frenchman's infuriating puzzle will finally be solved.

Pierre de Fermat

Fermat's last theorem: A promising approach

The end of a centuries-long search for a proof of Fermat's last theorem, one of the most famous unsolved problems in mathematics, may at last be in sight. A Japanese mathematician, Yoichi Miyaoka of the Tokyo Metropolitan University, has proposed a proof for a key link in a chain of reasoning that establishes the theorem's truth. If Miyaoka's proof survives the mathematical community's intense scrutiny, then Fermat's conjecture (as it ought to be called until a proof is firmly established) can truly be called a theorem.

Miyaoka's method builds on work done by several Russian mathematicians and links important ideas in three mathematical fields: number theory, algebra and geometry. Though highly technical, his argument fills fewer than a dozen manuscript pages – short for such a significant mathematical proof. Miyaoka recently presented a sketch of his ideas at a seminar at the Max Planck Institute for Mathematics in Bonn, West Germany.

"It looks very nice," mathematician Don B. Zagier of the Max Planck Institute told SCIENCE NEWS. "There are many nice ideas, but it's very subtle, and there could easily be a mistake. It'll certainly take days, if not weeks, until the proof's completely checked."

Fermat's conjecture is related to a statement by the ancient Greek mathematician Diophantus, who observed that there are positive integers, x, y and z, that satisfy the equation $x^2 + y^2 = z^2$. For example, if $x = 3$ and $y = 4$, then $z = 5$. In fact, this equation has an infinite number of such solutions.

In the 17th century, French amateur mathematician Pierre de Fermat, while reading a book by Diophantus, scribbled a note in a margin proposing that there are no positive-integer solutions to the equation $x^n + y^n = z^n$, when n is greater than 2. In other words, when $n = 3$, no set of positive integers satisfies the equation $x^3 + y^3 = z^3$, and so on. Then, in a tantalizing sentence that was to haunt mathematicians for centuries to come, Fermat added that although he had a wonderful proof for the theorem, he didn't have enough room to write it out.

Later mathematicians found proofs for a number of special cases, and a com-[illegible] showed [illegible] true for all exponen[illegible]

But despite the efforts of [illegible] mathematicians, a proof for the general case remained elusive (SN: 6/20/87, p.397).

In 1983, Gerd Faltings, now at Princeton (N.J.) University, opened up a new direction in the search for a proof. As one consequence of his proof of the Mordell conjecture (SN: 7/23/83, p.58), he showed that if there are any solutions to Fermat's equations, then there are only a finite number of them for each value of n. However, that was still far from the assertion that there are no such solutions.

Some of the key ideas for Faltings' proof came from the work of Russian mathematician S. Arakelov, who was looking for connections between prime numbers, curves and geometrical surfaces. Both Arakelov and Faltings found that analogs of certain classical theorems already well established for geometrical surfaces could apply to curves and provide information about statements, such as Fermat's last theorem, that involve only integers.

About a year ago, A.N. Parshin of the Steklov Institute in Moscow, following

Times Magazine, Science News #133, LA Times

Figure 18.1

AP Photo/Charles Rex Arbogast

Figure 18.2 Andrew Wiles

Unique Factorization Domains

We now have the necessary terminology to formalize the idea of unique factorization.

> **Definition Unique Factorization Domain (UFD)**
> **An integral domain D is a *unique factorization domain* if**
>
> 1. **every nonzero element of D that is not a unit can be written as a product of irreducibles of D; and**
> 2. **the factorization into irreducibles is unique up to associates and the order in which the factors appear.**

Another way to formulate part 2 of this definition is the following: If $p_1^{n_1}p_2^{n_2} \cdots p_r^{n_r}$ and $q_1^{m_1}q_2^{m_2} \cdots q_s^{m_s}$ are two factorizations of some element as a product of irreducibles, where no two of the p_i's are associates and no two of the q_j's are associates, then $r = s$, each p_i is an associate of one and only one q_j, and $n_i = m_j$.

Of course, the Fundamental Theorem of Arithmetic tells us that the ring of integers is a unique factorization domain, and Theorem 17.6 says that $Z[x]$ is a unique factorization domain. In fact, as we shall soon see, most of the integral domains we have encountered are unique factorization domains.

Before proving our next theorem, we need the ascending chain condition for ideals.

Lemma Ascending Chain Condition for a PID

In a principal ideal domain, any strictly increasing chain of ideals $I_1 \subset I_2 \subset \cdots$ must be finite in length.

PROOF Let $I_1 \subset I_2 \subset \cdots$ be a chain of strictly increasing ideals in an integral domain D, and let I be the union of all the ideals in this chain. We leave it as an exercise (Exercise 3) to verify that I is an ideal of D.

Then, since D is a principal ideal domain, there is an element a in D such that $I = \langle a \rangle$. Because $a \in I$ and $I = \cup I_k$, a belongs to some member of the chain, say $a \in I_n$. Clearly, then, for any member I_i of the chain, we have $I_i \subseteq I = \langle a \rangle \subseteq I_n$, so that I_n must be the last member of the chain. ∎

Theorem 18.3 PID Implies UFD

Every principal ideal domain is a unique factorization domain.

PROOF Let D be a principal ideal domain and let a_0 be any nonzero nonunit in D. We will show that a_0 is a product of irreducibles (the product might consist of only one factor). We begin by showing that a_0 has at least one irreducible factor. If a_0 is irreducible, we are done. Thus, we may assume that $a_0 = b_1 a_1$, where neither b_1 nor a_1 is a unit and a_1 is nonzero. If a_1 is not irreducible, then we can write $a_1 = b_2 a_2$, where neither b_2 nor a_2 is a unit and a_2 is nonzero. Continuing in this fashion, we obtain a sequence $b_1, b_2, \ldots$ of elements that are not units in D and a sequence $a_0, a_1, a_2, \ldots$ of nonzero elements of D with $a_n = b_{n+1}\, a_{n+1}$ for each n. Hence, $\langle a_0 \rangle \subset \langle a_1 \rangle \subset \cdots$ is a strictly increasing chain of ideals (see Exercise 5), which, by the preceding lemma, must be finite, say, $\langle a_0 \rangle \subset \langle a_1 \rangle \subset \cdots \subset \langle a_r \rangle$. In particular, a_r is an irreducible factor of a_0. This argument shows that every nonzero nonunit in D has at least one irreducible factor.

Now write $a_0 = p_1 c_1$, where p_1 is irreducible and c_1 is not a unit. If c_1 is not irreducible, then we can write $c_1 = p_2 c_2$, where p_2 is irreducible and c_2 is not a unit. Continuing in this fashion, we obtain, as before, a strictly increasing sequence $\langle a_0 \rangle \subset \langle c_1 \rangle \subset \langle c_2 \rangle \subset \cdots$, which must end in a finite number of steps. Let us say that the sequence ends with $\langle c_s \rangle$. Then c_s is irreducible and $a_0 = p_1p_2 \cdots p_sc_s$, where each p_i is also irreducible. This completes the proof that every nonzero nonunit of a principal ideal domain is a product of irreducibles.

It remains to be shown that the factorization is unique up to associates and the order in which the factors appear. To do this, suppose that some element a of D can be written

$$a = p_1p_2 \cdots p_r = q_1q_2 \cdots q_s,$$

where the p's and q's are irreducible and repetition is permitted. We use induction on r. If $r = 1$, then a is irreducible and, clearly, $s = 1$ and $p_1 = q_1$. So we may assume that any element that can be expressed as a product of fewer than r irreducible factors can be so expressed in only one way (up to order and associates). Since D is a principal ideal domain, by Theorem 18.2, each irreducible p_i in the product $p_1p_2 \cdots p_r$ is prime. Then because p_1 divides $q_1q_2 \cdots q_s$, p_1 must divide some q_i (see Exercise 33), say $p_1 \mid q_1$. Then, $q_1 = up_1$, where u is a unit of D. Since

$$up_1p_2 \cdots p_r = uq_1q_2 \cdots q_s = q_1(uq_2) \cdots q_s$$

and

$$up_1 = q_1,$$

we have, by cancellation,

$$p_2 \cdots p_r = (uq_2) \cdots q_s.$$

The induction hypothesis now tells us that these two factorizations are identical up to associates and the order in which the factors appear. Hence, the same is true about the two factorizations of a. ∎

In the existence portion of the proof of Theorem 18.3, the only way we used the fact that the integral domain D is a principal ideal domain was to say that D has the property that there is no infinite, strictly increasing chain of ideals in D. An integral domain with this property is called a *Noetherian domain*, in honor of Emmy Noether, who inaugurated the use of chain conditions in algebra. Noetherian domains are of the utmost importance in algebraic geometry. One reason for this is that, for many important rings R, the polynomial ring $R[x]$ is a Noetherian domain but not a principal ideal domain. One such example is $Z[x]$. In particular, $Z[x]$ shows that a UFD need not be a PID (see Example 3).

As an immediate corollary of Theorem 18.3, we have the following fact.

Corollary $F[x]$ Is a UFD

Let F be a field. Then F[x] is a unique factorization domain.

PROOF By Theorem 16.3, $F[x]$ is a principal ideal domain. So, $F[x]$ is a unique factorization domain, as well. ∎

As an application of the preceding corollary, we give an elegant proof, due to Richard Singer, of Eisenstein's Criterion (Theorem 17.4).

EXAMPLE 4 Let

$$f(x) = a_n x^n + a_{n-1}x^{n-1} + \cdots + a_0 \in Z[x],$$

and suppose that p is prime such that

$$p \nmid a_n, p \mid a_{n-1}, \ldots, p \mid a_0 \qquad \text{and} \qquad p^2 \nmid a_0.$$

We will prove that $f(x)$ is irreducible over Q. If $f(x)$ is reducible over Q, we know by Theorem 17.2 that there exist elements $g(x)$ and $h(x)$ in $Z[x]$ such that $f(x) = g(x)h(x)$, $1 \le \deg g(x) < n$, and $1 \le \deg h(x) < n$. Let $\overline{f}(x)$, $\overline{g}(x)$, and $\overline{h}(x)$ be the polynomials in $Z_p[x]$ obtained from $f(x)$, $g(x)$, and $h(x)$ by reducing all coefficients modulo p. Then, since p divides all the coefficients of $f(x)$ except a_n, we have $\overline{a}_n x^n = \overline{f}(x) = \overline{g}(x) \cdot \overline{h}(x)$. Since Z_p is a field, $Z_p[x]$ is a unique factorization domain. Thus, $x \mid \overline{g}(x)$ and $x \mid \overline{h}(x)$. So, $\overline{g}(0) = \overline{h}(0) = 0$ and, therefore, $p \mid g(0)$ and $p \mid h(0)$. But then $p^2 \mid g(0)h(0) = f(0) = a_0$, which is a contradiction. ∎

Euclidean Domains

Another important kind of integral domain is a Euclidean domain.

Definition Euclidean Domain (ED)

An integral domain D is called a *Euclidean domain* if there is a function d (called the *measure*) from the nonzero elements of D to the nonnegative integers such that

1. **$d(a) \le d(ab)$ for all nonzero a, b in D; and**
2. **if $a, b \in D$, $b \neq 0$, then there exist elements q and r in D such that $a = bq + r$, where $r = 0$ or $d(r) < d(b)$.**

EXAMPLE 5 The ring Z is a Euclidean domain with $d(a) = |a|$ (the absolute value of a). ∎

EXAMPLE 6 Let F be a field. Then $F[x]$ is a Euclidean domain with $d(f(x)) = \deg f(x)$ (see Theorem 16.2). ∎

Examples 5 and 6 illustrate just one of many similarities between the rings Z and $F[x]$. Additional similarities are summarized in Table 18.1.

Table 18.1 Similarities Between Z and $F[x]$

Z		$F[x]$
Euclidean domain: $d(a) = \|a\|$	↔	Euclidean domain: $d(f(x)) = \deg f(x)$
Units: a is a unit if and only if $\|a\| = 1$		Units: $f(x)$ is a unit if and only if $\deg f(x) = 0$
Division algorithm: For $a, b \in Z$, $b \neq 0$, there exist $q, r \in Z$ such that $a = bq + r$, $0 \leq r < \|b\|$	↔	Division algorithm: For $f(x)$, $g(x) \in F[x]$, $g(x) \neq 0$, there exist $q(x)$, $r(x) \in F[x]$ such that $f(x) = g(x)q(x) + r(x)$, $0 \leq \deg r(x) < \deg g(x)$ or $r(x) = 0$
PID: Every nonzero ideal $I = \langle a \rangle$, where $a \neq 0$ and $\|a\|$ is minimum	↔	PID: Every nonzero ideal $I = \langle f(x) \rangle$, where $\deg f(x)$ is minimum
Prime: No nontrivial factors	↔	Irreducible: No nontrivial factors
UFD: Every element is a "unique" product of primes	↔	UFD: Every element is a "unique" product of irreducibles

EXAMPLE 7 The ring of Gaussian integers

$$Z[i] = \{a + bi \mid a, b \in Z\}$$

is a Euclidean domain with $d(a + bi) = a^2 + b^2$. Unlike the previous two examples, in this example the function d does not obviously satisfy the necessary conditions. That $d(x) \leq d(xy)$ for $x, y \in Z[i]$ follows directly from the fact that $d(xy) = d(x)d(y)$ (Exercise 7). To verify that condition 2 holds, observe that if $x, y \in Z[i]$ and $y \neq 0$, then $xy^{-1} \in Q[i]$, the field of quotients of $Z[i]$ (Exercise 57 in Chapter 15). Say $xy^{-1} = s + ti$, where $s, t \in Q$. Now let m be the integer nearest s, and let n be the integer nearest t. (These integers may not be uniquely determined, but that does not matter.) Thus, $|m - s| \leq 1/2$ and $|n - t| \leq 1/2$. Then

$$\begin{aligned} xy^{-1} = s + ti &= (m - m + s) + (n - n + t)i \\ &= (m + ni) + [(s - m) + (t - n)i]. \end{aligned}$$

So,

$$x = (m + ni)y + [(s - m) + (t - n)i]y.$$

We claim that the division condition of the definition of a Euclidean domain is satisfied with $q = m + ni$ and

$$r = [(s - m) + (t - n)i]y.$$

Clearly, q belongs to $Z[i]$, and since $r = x - qy$, so does r. Finally,

$$\begin{aligned} d(r) &= d([(s - m) + (t - n)i])d(y) \\ &= [(s - m)^2 + (t - n)^2]d(y) \\ &\leq \left(\frac{1}{4} + \frac{1}{4}\right)d(y) < d(y). \end{aligned}$$

■

Theorem 18.4 ED Implies PID

Every Euclidean domain is a principal ideal domain.

PROOF Let D be a Euclidean domain and I a nonzero ideal of D. Among all the nonzero elements of I, let a be such that $d(a)$ is a minimum. Then $I = \langle a \rangle$. For, if $b \in I$, there are elements q and r such that $b = aq + r$, where $r = 0$ or $d(r) < d(a)$. But $r = b - aq \in I$, so $d(r)$ cannot be less than $d(a)$. Thus, $r = 0$ and $b \in \langle a \rangle$. Finally, the zero ideal is $\langle 0 \rangle$. ■

Although it is not easy to verify, we remark that there are principal ideal domains that are not Euclidean domains. The first such example was given by T. Motzkin in 1949. A more accessible account of Motzkin's result can be found in [2].

As an immediate consequence of Theorems 18.3 and 18.4, we have the following important result.

Corollary ED Implies UFD

Every Euclidean domain is a unique factorization domain.

We may summarize our theorems and remarks as follows:

$$\text{ED} \Rightarrow \text{PID} \Rightarrow \text{UFD};$$
$$\text{UFD} \not\Rightarrow \text{PID} \not\Rightarrow \text{ED}.$$

(You can remember these implications by listing the types alphabetically.)

In Chapter 17, we proved that $Z[x]$ is a unique factorization domain. Since Z is a unique factorization domain, the next theorem is a broad generalization of this fact. The proof is similar to that of the special case, and we therefore omit it.

Theorem 18.5 D a UFD Implies $D[x]$ a UFD

If D is a unique factorization domain, then $D[x]$ is a unique factorization domain.

We conclude this chapter with an example of an integral domain that is not a unique factorization domain.

EXAMPLE 8 The ring $Z[\sqrt{-5}] = \{a + b\sqrt{-5} \mid a, b \in Z\}$ is an integral domain but not a unique factorization domain. It is straightforward that $Z[\sqrt{-5}]$ is an integral domain (see Exercise 11 in Chapter 13). To verify that unique factorization does not hold, we mimic the method used in Example 1 with $N(a + b\sqrt{-5}) = a^2 + 5b^2$. Since $N(xy) = N(x)N(y)$ and $N(x) = 1$ if and only if x is a unit (see Exercise 1), it follows that the only units of $Z[\sqrt{-5}]$ are ± 1.

Now consider the following factorizations:

$$46 = 2 \cdot 23,$$
$$46 = (1 + 3\sqrt{-5})(1 - 3\sqrt{-5}).$$

We claim that each of these four factors is irreducible over $Z[\sqrt{-5}]$. Suppose that, say, $2 = xy$, where $x, y \in Z[\sqrt{-5}]$ and neither is a unit. Then $4 = N(2) = N(x)N(y)$ and, therefore, $N(x) = N(y) = 2$, which is impossible. Likewise, if $23 = xy$ were a nontrivial factorization, then $N(x) = 23$. Thus, there would be integers a and b such that $a^2 + 5b^2 = 23$. Clearly, no such integers exist. The same argument applies to $1 \pm 3\sqrt{-5}$. ∎

In light of Examples 7 and 8, one can't help but wonder for which $d < 0$ is $Z[\sqrt{d}]$ a unique factorization domain. The answer is only when $d = -1$ or -2 (see [1], p. 297). The case where $d = -1$ was first proved, naturally enough, by Gauss.

Exercises

I tell them that if they will occupy themselves with the study of mathematics they will find in it the best remedy against lust of the flesh.

THOMAS MANN, *The Magic Mountain*

1. For the ring $Z[\sqrt{d}] = \{a + b\sqrt{d} \mid a, b \in Z\}$, where $d \neq 1$ and d is not divisible by the square of a prime, prove that the norm $N(a + b\sqrt{d}) = |a^2 - db^2|$ satisfies the four assertions made preceding Example 1. (This exercise is referred to in this chapter.)
2. In an integral domain, show that a and b are associates if and only if $\langle a \rangle = \langle b \rangle$.
3. Show that the union of a chain $I_1 \subset I_2 \subset \cdots$ of ideals of a ring R is an ideal of R. (This exercise is referred to in this chapter.)
4. In an integral domain, show that the product of an irreducible and a unit is an irreducible.
5. Suppose that a and b belong to an integral domain, $b \neq 0$, and a is not a unit. Show that $\langle ab \rangle$ is a proper subset of $\langle b \rangle$. (This exercise is referred to in this chapter.)
6. Let D be an integral domain. Define $a \sim b$ if a and b are associates. Show that this defines an equivalence relation on D.
7. In the notation of Example 7, show that $d(xy) = d(x)d(y)$.
8. Let D be a Euclidean domain with measure d. Prove that u is a unit in D if and only if $d(u) = d(1)$.
9. Let D be a Euclidean domain with measure d. Show that if a and b are associates in D, then $d(a) = d(b)$.
10. Let D be a principal ideal domain and let $p \in D$. Prove that $\langle p \rangle$ is a maximal ideal in D if and only if p is irreducible.
11. Trace through the argument given in Example 7 to find q and r in $Z[i]$ such that $3 - 4i = (2 + 5i)q + r$ and $d(r) < d(2 + 5i)$.
12. Let D be a principal ideal domain. Show that every proper ideal of D is contained in a maximal ideal of D.
13. In $Z[\sqrt{-5}]$, show that 21 does not factor uniquely as a product of irreducibles.
14. Show that $1 - i$ is an irreducible in $Z[i]$.
15. Show that $Z[\sqrt{-6}]$ is not a unique factorization domain. (*Hint:* Factor 10 in two ways.) Why does this show that $Z[\sqrt{-6}]$ is not a principal ideal domain?
16. Give an example of a unique factorization domain with a subdomain that does not have a unique factorization.

17. In $Z[i]$, show that 3 is irreducible but 2 and 5 are not.
18. Prove that 7 is irreducible in $Z[\sqrt{6}]$, even though $N(7)$ is not prime.
19. Prove that if p is a prime in Z that can be written in the form $a^2 + b^2$, then $a + bi$ is irreducible in $Z[i]$. Find three primes that have this property and the corresponding irreducibles.
20. Prove that $Z[\sqrt{-3}]$ is not a principal ideal domain.
21. In $Z[\sqrt{-5}]$, prove that $1 + 3\sqrt{-5}$ is irreducible but not prime.
22. In $Z[\sqrt{5}]$, prove that both 2 and $1 + \sqrt{5}$ are irreducible but not prime.
23. Prove that $Z[\sqrt{5}]$ is not a unique factorization domain.
24. Let F be a field. Show that in $F[x]$ a prime ideal is a maximal ideal.
25. Let d be an integer less than -1 that is not divisible by the square of a prime. Prove that the only units of $Z[\sqrt{d}]$ are $+1$ and -1.
26. In $Z[\sqrt{2}] = \{a + b\sqrt{2} \mid a, b \in Z\}$, show that every element of the form $(3 + 2\sqrt{2})^n$ is a unit, where n is a positive integer.
27. If a and b belong to $Z[\sqrt{d}]$, where d is not divisible by the square of a prime and ab is a unit, prove that a and b are units.
28. For a commutative ring with unity we may define associates, irreducibles, and primes exactly as we did for integral domains. With these definitions, show that both 2 and 3 are prime in Z_{12} but 2 is irreducible and 3 is not.
29. Let n be a positive integer and p a prime that divides n. Prove that p is prime in Z_n. (See Exercise 28).
30. Let p be a prime divisor of a positive integer n. Prove that p is irreducible in Z_n if and only if p^2 divides n. (See Exercise 28).
31. Prove or disprove that if D is a principal ideal domain, then $D[x]$ is a principal ideal domain.
32. Determine the units in $Z[i]$.
33. Let p be a prime in an integral domain. If $p \mid a_1a_2 \cdots a_n$, prove that p divides some a_i. (This exercise is referred to in this chapter.)
34. Show that $3x^2 + 4x + 3 \in Z_5[x]$ factors as $(3x + 2)(x + 4)$ and $(4x + 1)(2x + 3)$. Explain why this does not contradict the corollary of Theorem 18.3.
35. Let D be a principal ideal domain and p an irreducible element of D. Prove that $D/\langle p\rangle$ is a field.
36. Show that an integral domain with the property that every strictly decreasing chain of ideals $I_1 \supset I_2 \supset \cdots$ must be finite in length is a field.
37. An ideal A of a commutative ring R with unity is said to be *finitely generated* if there exist elements $a_1, a_2, \ldots, a_n$ of A such that

$A = \langle a_1, a_2, \ldots, a_n \rangle$. An integral domain R is said to satisfy the *ascending chain condition* if every strictly increasing chain of ideals $I_1 \subset I_2 \subset \cdots$ must be finite in length. Show that an integral domain R satisfies the ascending chain condition if and only if every ideal of R is finitely generated.

38. Prove or disprove that a subdomain of a Euclidean domain is a Euclidean domain.

39. Show that for any nontrivial ideal I of $Z[i]$, $Z[i]/I$ is finite.

40. Find the inverse of $1 + \sqrt{2}$ in $Z[\sqrt{2}]$. What is the multiplicative order of $1 + \sqrt{2}$?

41. In $Z[\sqrt{-7}]$, show that $N(6 + 2\sqrt{-7}) = N(1 + 3\sqrt{-7})$ but $6 + 2\sqrt{-7}$ and $1 + 3\sqrt{-7}$ are not associates.

42. Let $R = Z \oplus Z \oplus \cdots$ (the collection of all sequences of integers under componentwise addition and multiplication). Show that R has ideals $I_1, I_2, I_3, \ldots$ with the property that $I_1 \subset I_2 \subset I_3 \subset \cdots$. (Thus R does not have the ascending chain condition.)

43. Prove that in a unique factorization domain, an element is irreducible if and only if it is prime.

44. Let F be a field and let R be the integral domain in $F[x]$ generated by x^2 and x^3. (That is, R is contained in every integral domain in $F[x]$ that contains x^2 and x^3.) Show that R is not a unique factorization domain.

45. Prove that for every field F, there are infinitely many irreducible elements in $F[x]$.

46. Find a mistake in the statement shown in Figure 18.2.

Computer Exercise

Software for a computer exercise is available at the website:

http://www.d.umn.edu/~jgallian

References

1. H. M. Stark, *An Introduction to Number Theory*, Chicago: Markham, 1970.
2. J. C. Wilson, "A Principal Ideal Ring That Is Not a Euclidean Ring," *Mathematics Magazine* 46 (1973): 34–38.

Suggested Readings

Oscar Campoli, "A Principal Ideal Domain That Is Not a Euclidean Domain," *The American Mathematical Monthly* 95 (1988): 868–871.

The author shows that $\{a + b\theta \mid a, b \in Z, \theta = (1 + \sqrt{-19})/2\}$ is a PID that is not an ED.

Gina Kolata, "At Last, Shout of 'Eureka!' in Age-Old Math Mystery," *The New York Times,* June 24, 1993.

This front-page article reports on Andrew Wiles's announced proof of Fermat's Last Theorem.

C. Krauthhammer, "The Joy of Math, or Fermat's Revenge," *Time,* April 18, 1988: 92.

The demise of Miyaoka's proof of Fermat's Last Theorem is charmingly lamented.

Sahib Singh, "Non-Euclidean Domains: An Example," *Mathematics Magazine* 49 (1976): 243.

This article gives a short proof that $Z[\sqrt{-n}] = \{a + b\sqrt{-n} \mid a, b \in Z\}$ is an integral domain that is not Euclidean when $n > 2$ and $-n \bmod 4 = 2$ or $-n \bmod 4 = 3$.

Simon Singh and Kenneth Ribet, "Fermat's Last Stand," *Scientific American* 277 (1997): 68–73.

This article gives an accessible description of Andrew Wiles's proof of Fermat's Last Theorem.

Suggested Video

The Proof, Nova, **http://www.pbs.org/wgbh/nova/proof**

This documentary film shown on PBS's *Nova* program in 1997 chronicles the seven-year effort of Andrew Wiles to prove Fermat's Last Theorem. It can be viewed in five segments at **http://www.youtube.com**.

Suggested Websites

http://en.wikipedia.org/wiki/Fermat's_Last_Theorem

This website provides a concise history of the efforts to prove Fermat's Last Theorem. It includes photographs, references, and links.

Sophie Germain

Stock Montage

One of the very few women to overcome the prejudice and discrimination that tended to exclude women from the pursuit of higher mathematics in her time was Sophie Germain.

SOPHIE GERMAIN was born in Paris on April 1, 1776. She educated herself by reading the works of Newton and Euler in Latin and the lecture notes of Lagrange. In 1804, Germain wrote to Gauss about her work in number theory but used the pseudonym Monsieur LeBlanc because she feared that Gauss would not take seriously the efforts of a woman. Gauss gave Germain's results high praise and a few years later, upon learning her true identity, wrote to her:

> But how to describe to you my admiration and astonishment at seeing my esteemed correspondent Mr. LeBlanc metamorphose himself into this illustrious personage who gives such a brilliant example of what I would find it difficult to believe. A taste for the abstract sciences in general and above all the mysteries of numbers is excessively rare: it is not a subject which strikes everyone; the enchanting charms of this sublime science reveal themselves only to those who have the courage to go deeply into it. But when a person of the sex which, according to our customs and prejudices, must encounter infinitely more difficulties than men to familiarize herself with these thorny researches, succeeds nevertheless in surmounting these obstacles and penetrating the most obscure parts of them, then without doubt she must have the noblest courage, quite extraordinary talents, and a superior genius.*

Germain is best known for her work on Fermat's Last Theorem. She died on June 27, 1831, in Paris.

For more information about Germain, visit:

http://www-groups.dcs.st-and.ac.uk/~history

*Quote from *Math's Hidden Woman*, Nova Online, http://www.pbs.org/wgbh/nova/proof/germain.html (accessed Nov 5, 2008).

Andrew Wiles

For spectacular contributions to number theory and related fields, for major advances on fundamental conjectures, and for settling Fermat's Last Theorem.

Citation for the Wolf Prize

Postage stamp issued by the Czech Republic in honor of Fermat's Last Theorem.

In 1993, Andrew Wiles of Princeton electrified the mathematics community by announcing that he had proved Fermat's Last Theorem after seven years of effort. His proof, which ran 200 pages, relied heavily on ring theory and group theory. Because of Wiles's solid reputation and because his approach was based on deep results that had already shed much light on the problem, many experts in the field believed that Wiles had succeeded where so many others had failed. Wiles's achievement was reported in newspapers and magazines around the world. *The New York Times* ran a front-page story on it, and one TV network announced it on the evening news. Wiles even made *People* magazine's list of the 25 most intriguing people of 1993! In San Francisco a group of mathematicians rented a 1200-seat movie theater and sold tickets for $5.00 each for public lectures on the proof. Scalpers received as much as $25.00 a ticket for the sold-out event.

The bubble soon burst when experts had an opportunity to scrutinize Wiles's manuscript. By December, Wiles released a statement saying he was working to resolve a gap in the proof. In September of 1994, a paper by Wiles and Richard Taylor, a former student of his, circumvented the gap in the original proof. Since then, many experts have checked the proof and have found no errors. One mathematician was quoted as saying, "The exuberance is back." In 1997, Wiles's proof was the subject of a PBS *Nova* program.

Wiles was born in 1953 in Cambridge, England. He obtained his bachelor's degree at Oxford and his doctoral degree at Cambridge University in 1980. He was a professor at Harvard before moving to Princeton in 1982. In 2011, he became a research professor at Oxford. He has received many prestigious awards.

To find more information about Wiles, visit:

http://www-groups.dcs.st-and.ac.uk/~history/

Supplementary Exercises for Chapters 15–18

The intelligence is proved not by ease of learning, but by understanding what we learn.

JOSEPH WHITNEY

True/false questions for Chapters 15–18 are available on the Web at:

http://www.d.umn.edu/~jgallian/TF

1. Suppose that F is a field and there is a ring homomorphism from Z onto F. Show that F is isomorphic to Z_p for some prime p.
2. Let $Q[\sqrt{2}] = \{r + s\sqrt{2} \mid r, s \in Q\}$. Determine all ring automorphisms of $Q[\sqrt{2}]$.
3. (Second Isomorphism Theorem for Rings) Let A be a subring of R and let B be an ideal of R. Show that $A \cap B$ is an ideal of A and that $A/(A \cap B)$ is isomorphic to $(A + B)/B$. (Recall that $A + B = \{a + b \mid a \in A, b \in B\}$.)
4. (Third Isomorphism Theorem for Rings) Let A and B be ideals of a ring R with $B \subseteq A$. Show that A/B is an ideal of R/B and $(R/B)/(A/B)$ is isomorphic to R/A.
5. Let $f(x)$ and $g(x)$ be irreducible polynomials over a field F. If $f(x)$ and $g(x)$ are not associates, prove that $F[x]/\langle f(x)g(x)\rangle$ is isomorphic to $F[x]/\langle f(x)\rangle \oplus F[x]/\langle g(x)\rangle$.
6. (Chinese Remainder Theorem for Rings) If R is a commutative ring and I and J are two proper ideals with $I + J = R$, prove that $R/(I \cap J)$ is isomorphic to $R/I \oplus R/J$. Explain why Exercise 5 is a special case of this theorem.
7. Prove that the set of all polynomials whose coefficients are all even is a prime ideal in $Z[x]$.
8. Let $R = Z[\sqrt{-5}]$ and let $I = \{a + b\sqrt{-5} \mid a, b \in Z, a - b$ is even$\}$. Show that I is a maximal ideal of R.
9. Let R be a ring with unity and let a be a unit in R. Show that the mapping from R into itself given by $x \to axa^{-1}$ is a ring automorphism.
10. Let $a + b\sqrt{-5}$ belong to $Z[\sqrt{-5}]$ with $b \neq 0$. Show that 2 does not belong to $\langle a + b\sqrt{-5}\rangle$.
11. Show that $Z[i]/\langle 2 + i\rangle$ is a field. How many elements does it have?
12. Is the homomorphic image of a principal ideal domain a principal ideal domain?
13. For any $f(x) \in Z_p[x]$, show that $f(x^p) = (f(x))^p$.

14. Let p be a prime. Show that there is exactly one ring homomorphism from Z_m to Z_{p^k} if p^k does not divide m, and exactly two ring homomorphisms from Z_m to Z_{p^k} if p^k does divide m.

15. Recall that a is an idempotent if $a^2 = a$. Show that if $1 + k$ is an idempotent in Z_n, then $n - k$ is an idempotent in Z_n.

16. Show that Z_n (where $n > 1$) always has an even number of idempotents. (The number is 2^d, where d is the number of distinct prime divisors of n.)

17. Show that the equation $x^2 + y^2 = 2003$ has no solutions in the integers.

18. Prove that if both k and $k + 1$ are idempotents in Z_n and $k \neq 0$, then $n = 2k$.

19. Prove that $x^4 + 15x^3 + 7$ is irreducible over Q.

20. For any integers m and n, prove that the polynomial $x^3 + (5m + 1)x + 5n + 1$ is irreducible over Z.

21. Prove that $\langle\sqrt{2}\rangle$ is a maximal ideal in $Z[\sqrt{2}]$. How many elements are in the ring $Z[\sqrt{2}]/\langle\sqrt{2}\rangle$?

22. Prove that $Z[\sqrt{-2}]$ and $Z[\sqrt{2}]$ are unique factorization domains. (*Hint:* Mimic Example 7 in Chapter 18.)

23. Is $\langle 3\rangle$ a maximal ideal in $Z[i]$?

24. Express both 13 and $5 + i$ as products of irreducibles from $Z[i]$.

25. Let $R = \{a/b \mid a, b \in Z, 3 \nmid b\}$. Prove that R is an integral domain. Find its field of quotients.

26. Give an example of a ring that contains a subring isomorphic to Z and a subring isomorphic to Z_3.

27. Show that $Z[i]/\langle 3\rangle$ is not ring-isomorphic to $Z_3 \oplus Z_3$.

28. For any $n > 1$, prove that $R = \left\{\begin{bmatrix} a & 0 \\ 0 & b \end{bmatrix} \middle|\, a, b \in Z_n\right\}$ is ring-isomorphic to $Z_n \oplus Z_n$.

29. Suppose that R is a commutative ring and I is an ideal of R. Prove that $R[x]/I[x]$ is isomorphic to $(R/I)[x]$.

30. Find an ideal I of $Z_8[x]$ such that the factor ring $Z_8[x]/I$ is a field.

31. Find an ideal I of $Z_8[x]$ such that the factor ring $Z_8[x]/I$ is an integral domain but not a field.

32. Find an ideal I of $Z[x]$ such that $Z[x]/I$ is ring-isomorphic to Z_3.

PART 4

Fields

For online student resources, visit this textbook's website at www.CengageBrain.com

19 Vector Spaces

Still round the corner there may wait
A new road or a secret gate.

J. R. R. TOLKIEN, *The Fellowship of the Ring*

Definition and Examples

Abstract algebra has three basic components: groups, rings, and fields. Thus far we have covered groups and rings in some detail, and we have touched on the notion of a field. To explore fields more deeply, we need some rudiments of vector space theory that are covered in a linear algebra course. In this chapter, we provide a concise review of this material.

Definition Vector Space

A set V is said to be a *vector space* over a field F if V is an Abelian group under addition (denoted by $+$) and, if for each $a \in F$ and $v \in V$, there is an element av in V such that the following conditions hold for all a, b in F and all u, v in V.

1. $a(v + u) = av + au$
2. $(a + b)v = av + bv$
3. $a(bv) = (ab)v$
4. $1v = v$

The members of a vector space are called *vectors*. The members of the field are called *scalars*. The operation that combines a scalar a and a vector v to form the vector av is called *scalar multiplication*. In general, we will denote vectors by letters from the end of the alphabet, such as u, v, w, and scalars by letters from the beginning of the alphabet, such as a, b, c.

EXAMPLE 1 The set $\mathbf{R}^n = \{(a_1, a_2, \ldots, a_n) \mid a_i \in \mathbf{R}\}$ is a vector space over $\mathbf{R}$. Here the operations are the obvious ones:

$$(a_1, a_2, \ldots, a_n) + (b_1, b_2, \ldots, b_n) = (a_1 + b_1, a_2 + b_2, \ldots, a_n + b_n)$$

and

$$b(a_1, a_2, \ldots, a_n) = (ba_1, ba_2, \ldots, ba_n).$$

EXAMPLE 2 The set $M_2(Q)$ of 2×2 matrices with entries from Q is a vector space over Q. The operations are

$$\begin{bmatrix} a_1 & a_2 \\ a_3 & a_4 \end{bmatrix} + \begin{bmatrix} b_1 & b_2 \\ b_3 & b_4 \end{bmatrix} = \begin{bmatrix} a_1 + b_1 & a_2 + b_2 \\ a_3 + b_3 & a_4 + b_4 \end{bmatrix}$$

and

$$b\begin{bmatrix} a_1 & a_2 \\ a_3 & a_4 \end{bmatrix} = \begin{bmatrix} ba_1 & ba_2 \\ ba_3 & ba_4 \end{bmatrix}.$$

EXAMPLE 3 The set $Z_p[x]$ of polynomials with coefficients from Z_p is a vector space over Z_p, where p is a prime.

EXAMPLE 4 The set of complex numbers $\mathbf{C} = \{a + bi \mid a, b \in \mathbf{R}\}$ is a vector space over $\mathbf{R}$. The vector addition and scalar multiplication are the usual addition and multiplication of complex numbers.

The next example is a generalization of Example 4. Although it appears rather trivial, it is of the utmost importance in the theory of fields.

EXAMPLE 5 Let E be a field and let F be a subfield of E. Then E is a vector space over F. The vector addition and scalar multiplication are the operations of E.

Subspaces

Of course, there is a natural analog of subgroup and subring.

Definition Subspace

Let V be a vector space over a field F and let U be a subset of V. We say that U is a *subspace* of V if U is also a vector space over F under the operations of V.

EXAMPLE 6 The set $\{a_2x^2 + a_1x + a_0 \mid a_0, a_1, a_2 \in \mathbf{R}\}$ is a subspace of the vector space of all polynomials with real coefficients over $\mathbf{R}$.

EXAMPLE 7 Let V be a vector space over F and let $v_1, v_2, \ldots, v_n$ be (not necessarily distinct) elements of V. Then the subset

$$\langle v_1, v_2, \ldots, v_n \rangle = \{a_1v_1 + a_2v_2 + \cdots + a_nv_n \mid a_1, a_2, \ldots, a_n \in F\}$$

is called the *subspace of V spanned by* $v_1, v_2, \ldots, v_n$. Any sum of the form $a_1v_1 + a_2v_2 + \cdots + a_nv_n$ is called a *linear combination of* $v_1, v_2, \ldots, v_n$. If $\langle v_1, v_2, \ldots, v_n \rangle = V$, we say that $\{v_1, v_2, \ldots, v_n\}$ *spans* V. ∎

Linear Independence

The next definition is the heart of the theory.

Definition Linearly Dependent, Linearly Independent
A set S of vectors is said to be *linearly dependent* over the field F if there are vectors $v_1, v_2, \ldots, v_n$ from S and elements $a_1, a_2, \ldots, a_n$ from F, not all zero, such that $a_1v_1 + a_2v_2 + \cdots + a_nv_n = 0$. A set of vectors that is not linearly dependent over F is called *linearly independent* over F.

In other words, a set of vectors is linearly dependent over F if there is a nontrivial linear combination of them over F equal to 0.

EXAMPLE 8 In $\mathbf{R}^3$ the vectors (1, 0, 0), (1, 0, 1), and (1, 1, 1) are linearly independent over $\mathbf{R}$. To verify this, assume that there are real numbers a, b, and c such that $a(1, 0, 0) + b(1, 0, 1) + c(1, 1, 1) = (0, 0, 0)$. Then $(a + b + c, c, b + c) = (0, 0, 0)$. From this we see that $a = b = c = 0$. ∎

Certain kinds of linearly independent sets play a crucial role in the theory of vector spaces.

Definition Basis
Let V be a vector space over F. A subset B of V is called a *basis* for V if B is linearly independent over F and every element of V is a linear combination of elements of B.

The motivation for this definition is twofold. First, if B is a basis for a vector space V, then every member of V is a unique linear combination of the elements of B (see Exercise 19). Second, with every vector space spanned by finitely many vectors, we can use the notion of basis to associate a unique integer that tells us much about the vector space. (In fact, this integer and the field completely determine the vector space up to isomorphism—see Exercise 31.)

EXAMPLE 9 The set $V = \left\{ \begin{bmatrix} a & a+b \\ a+b & b \end{bmatrix} \middle| a, b \in \mathbf{R} \right\}$ is a vector space over **R** (see Exercise 17). We claim that the set $B = \left\{ \begin{bmatrix} 1 & 1 \\ 1 & 0 \end{bmatrix}, \begin{bmatrix} 0 & 1 \\ 1 & 1 \end{bmatrix} \right\}$ is a basis for V over **R**. To prove that the set B is linearly independent, suppose that there are real numbers a and b such that

$$a\begin{bmatrix} 1 & 1 \\ 1 & 0 \end{bmatrix} + b\begin{bmatrix} 0 & 1 \\ 1 & 1 \end{bmatrix} = \begin{bmatrix} 0 & 0 \\ 0 & 0 \end{bmatrix}.$$

This gives $\begin{bmatrix} a & a+b \\ a+b & b \end{bmatrix} = \begin{bmatrix} 0 & 0 \\ 0 & 0 \end{bmatrix}$, so that $a = b = 0$. On the other hand, since every member of V has the form

$$\begin{bmatrix} a & a+b \\ a+b & b \end{bmatrix} = a\begin{bmatrix} 1 & 1 \\ 1 & 0 \end{bmatrix} + b\begin{bmatrix} 0 & 1 \\ 1 & 1 \end{bmatrix},$$

we see that B spans V. ∎

We now come to the main result of this chapter.

Theorem 19.1 Invariance of Basis Size

If $\{u_1, u_2, \ldots, u_m\}$ and $\{w_1, w_2, \ldots, w_n\}$ are both bases of a vector space V over a field F, then $m = n$.

PROOF Suppose that $m \neq n$. To be specific, let us say that $m < n$. Consider the set $\{w_1, u_1, u_2, \ldots, u_m\}$. Since the u's span V, we know that w_1 is a linear combination of the u's, say, $w_1 = a_1u_1 + a_2u_2 + \cdots + a_mu_m$, where the a's belong to F. Clearly, not all the a's are 0. For convenience, say $a_1 \neq 0$. Then $\{w_1, u_2, \ldots, u_m\}$ spans V (see Exercise 21). Next, consider the set $\{w_1, w_2, u_2, \ldots, u_m\}$. This time, w_2 is a linear combination of $w_1, u_2, \ldots, u_m$, say, $w_2 = b_1w_1 + b_2u_2 + \cdots + b_mu_m$, where the b's belong to F. Then at least one of $b_2, \ldots, b_m$ is nonzero, for otherwise the w's are not linearly independent. Let us say $b_2 \neq 0$. Then $w_1, w_2, u_3, \ldots, u_m$ span V. Continuing in this fashion, we see that $\{w_1, w_2, \ldots, w_m\}$ spans V. But then w_{m+1} is a linear combination of $w_1, w_2, \ldots, w_m$ and, therefore, the set $\{w_1, \ldots, w_n\}$ is not linearly independent. This contradiction finishes the proof. ∎

Theorem 19.1 shows that any two finite bases for a vector space have the same size. Of course, not all vector spaces have finite bases. However, there is no vector space that has a finite basis and an infinite basis (see Exercise 25).

Definition Dimension

A vector space that has a basis consisting of *n* elements is said to have *dimension n*. For completeness, the trivial vector space {0} is said to be spanned by the empty set and to have dimension 0.

Although it requires a bit of set theory that is beyond the scope of this text, it can be shown that every vector space has a basis. A vector space that has a finite basis is called *finite dimensional*; otherwise, it is called *infinite dimensional.*

Exercises

Somebody who thinks logically is a nice contrast to the real world.

THE LAW OF THUMB

1. Verify that each of the sets in Examples 1–4 satisfies the axioms for a vector space. Find a basis for each of the vector spaces in Examples 1–4.
2. (Subspace Test) Prove that a nonempty subset U of a vector space V over a field F is a subspace of V if, for every u and u' in U and every a in F, $u + u' \in U$ and $au \in U$. (In words, a nonempty set U is a subspace of V if it is closed under the two operations of V.)
3. Verify that the set in Example 6 is a subspace. Find a basis for this subspace. Is $\{x^2 + x + 1, x + 5, 3\}$ a basis?
4. Verify that the set $\langle v_1, v_2, \ldots, v_n\rangle$ defined in Example 7 is a subspace.
5. Determine whether or not the set $\{(2, -1, 0), (1, 2, 5), (7, -1, 5)\}$ is linearly independent over $\mathbf{R}$.
6. Determine whether or not the set
$$\left\{\begin{bmatrix} 2 & 1 \\ 1 & 0 \end{bmatrix}, \begin{bmatrix} 0 & 1 \\ 1 & 2 \end{bmatrix}, \begin{bmatrix} 1 & 1 \\ 1 & 1 \end{bmatrix}\right\}$$
is linearly independent over Z_5.
7. If $\{u, v, w\}$ is a linearly independent subset of a vector space, show that $\{u, u + v, u + v + w\}$ is also linearly independent.
8. If $\{v_1, v_2, \ldots, v_n\}$ is a linearly dependent set of vectors, prove that one of these vectors is a linear combination of the other.
9. (Every spanning collection contains a basis.) If $\{v_1, v_2, \ldots, v_n\}$ spans a vector space V, prove that some subset of the v's is a basis for V.
10. (Every independent set is contained in a basis.) Let V be a finite-dimensional vector space and let $\{v_1, v_2, \ldots, v_n\}$ be a linearly

independent subset of V. Show that there are vectors $w_1, w_2, \ldots, w_m$ such that $\{v_1, v_2, \ldots, v_n, w_1, \ldots, w_m\}$ is a basis for V.

11. If V is a vector space over F of dimension 5 and U and W are subspaces of V of dimension 3, prove that $U \cap W \neq \{0\}$. Generalize.
12. Show that the solution set to a system of equations of the form

$$\begin{array}{c} a_{11}x_1 + \cdots + a_{1n}x_n = 0 \\ a_{21}x_1 + \cdots + a_{2n}x_n = 0 \\ \vdots \\ a_{m1}x_1 + \cdots + a_{mn}x_n = 0, \end{array}$$

where the a's are real, is a subspace of $\mathbf{R}^n$.

13. Let V be the set of all polynomials over Q of degree 2 together with the zero polynomial. Is V a vector space over Q?
14. Let $V = \mathbf{R}^3$ and $W = \{(a, b, c) \in V \mid a^2 + b^2 = c^2\}$. Is W a subspace of V? If so, what is its dimension?
15. Let $V = \mathbf{R}^3$ and $W = \{(a, b, c) \in V \mid a + b = c\}$. Is W a subspace of V? If so, what is its dimension?
16. Let $V = \left\{ \begin{bmatrix} a & b \\ b & c \end{bmatrix} \middle| \, a, b, c \in Q \right\}$. Prove that V is a vector space over Q, and find a basis for V over Q.
17. Verify that the set V in Example 9 is a vector space over $\mathbf{R}$.
18. Let $P = \{(a, b, c) \mid a, b, c \in \mathbf{R}, a = 2b + 3c\}$. Prove that P is a subspace of $\mathbf{R}^3$. Find a basis for P. Give a geometric description of P.
19. Let B be a subset of a vector space V. Show that B is a basis for V if and only if every member of V is a unique linear combination of the elements of B. (This exercise is referred to in this chapter and in Chapter 20.)
20. If U is a proper subspace of a finite-dimensional vector space V, show that the dimension of U is less than the dimension of V.
21. Referring to the proof of Theorem 19.1, prove that $\{w_1, u_2, \ldots, u_m\}$ spans V.
22. If V is a vector space of dimension n over the field Z_p, how many elements are in V?
23. Let $S = \{(a, b, c, d) \mid a, b, c, d \in \mathbf{R}, a = c, d = a + b\}$. Find a basis for S.
24. Let U and W be subspaces of a vector space V. Show that $U \cap W$ is a subspace of V and that $U + W = \{u + w \mid u \in U, w \in W\}$ is a subspace of V.

25. If a vector space has one basis that contains infinitely many elements, prove that every basis contains infinitely many elements. (This exercise is referred to in this chapter.)

26. Let $u = (2, 3, 1)$, $v = (1, 3, 0)$, and $w = (2, -3, 3)$. Since $(1/2)u - (2/3)v - (1/6)w = (0, 0, 0)$, can we conclude that the set $\{u, v, w\}$ is linearly dependent over Z_7?

27. Define the vector space analog of group homomorphism and ring homomorphism. Such a mapping is called a *linear transformation*. Define the vector space analog of group isomorphism and ring isomorphism.

28. Let T be a linear transformation from V to W. Prove that the image of V under T is a subspace of W.

29. Let T be a linear transformation of a vector space V. Prove that $\{v \in V \mid T(v) = 0\}$, the *kernel* of T, is a subspace of V.

30. Let T be a linear transformation of V onto W. If $\{v_1, v_2, \ldots, v_n\}$ spans V, show that $\{T(v_1), T(v_2), \ldots, T(v_n)\}$ spans W.

31. If V is a vector space over F of dimension n, prove that V is isomorphic as a vector space to $F^n = \{(a_1, a_2, \ldots, a_n) \mid a_i \in F\}$. (This exercise is referred to in this chapter.)

32. Let V be a vector space over an infinite field. Prove that V is not the union of finitely many proper subspaces of V.

Emil Artin

For Artin, to be a mathematician meant to participate in a great common effort, to continue work begun thousands of years ago, to shed new light on old discoveries, to seek new ways to prepare the developments of the future. Whatever standards we use, he was a great mathematician.

RICHARD BRAUER,
Bulletin of the American Mathematical Society

EMIL ARTIN was one of the leading mathematicians of the 20th century and a major contributor to linear algebra and abstract algebra. Artin was born on March 3, 1898, in Vienna, Austria, and grew up in what was recently known as Czechoslovakia. He received a Ph.D. in 1921 from the University of Leipzig. From 1923 until he emigrated to America in 1937, he was a professor at the University of Hamburg. After one year at Notre Dame, Artin went to Indiana University. In 1946 he moved to Princeton, where he stayed until 1958. The last four years of his career were spent where it began, at Hamburg.

Artin's mathematics is both deep and broad. He made contributions to number theory, group theory, ring theory, field theory, Galois theory, geometric algebra, algebraic topology, and the theory of braids—a field he invented. Artin received the American Mathematical Society's Cole Prize in number theory, and he solved one of the 23 famous problems posed by the eminent mathematician David Hilbert in 1900.

Artin was an outstanding teacher of mathematics at all levels, from freshman calculus to seminars for colleagues. Many of his Ph.D. students as well as his son Michael have become leading mathematicians. Through his research, teaching, and books, Artin exerted great influence among his contemporaries. He died of a heart attack, at the age of 64, in 1962.

For more information about Artin, visit:

http://www-groups.dcs.st-and.ac.uk/~history/

Olga Taussky-Todd

MaryAnn McLoughlin

"Olga Taussky-Todd was a distinguished and prolific mathematician who wrote about 300 papers."

EDITH LUCHINS AND MARY ANN McLOUGHLIN, *Notices of the American Mathematical Society, 1996*

OLGA TAUSSKY-TODD was born on August 30, 1906, in Olmütz in the Austro-Hungarian Empire. Taussky-Todd received her doctoral degree in 1930 from the University of Vienna. In the early 1930s she was hired as an assistant at the University of Göttingen to edit books on the work of David Hilbert. She also edited lecture notes of Emil Artin and assisted Richard Courant. She spent 1934 and 1935 at Bryn Mawr and the next two years at Girton College in Cambridge, England. In 1937, she taught at the University of London. In 1947, she moved to the United States and took a job at the National Bureau of Standards' National Applied Mathematics Laboratory. In 1957, she became the first woman to teach at the California Institute of Technology as well as the first woman to receive tenure and a full professorship in mathematics, physics, or astronomy there. Thirteen Caltech Ph.D. students wrote their Ph.D. theses under her direction.

In addition to her influential contributions to linear algebra, Taussky-Todd did important work in number theory.

Taussky-Todd received many honors and awards. She was elected a Fellow of the American Association for the Advancement of Science and vice president of the American Mathematical Society. In 1990, Caltech established an instructorship named in her honor. Taussky-Todd died on October 7, 1995, at the age of 89.

For more information about Taussky-Todd, visit:

http://www-groups.dcs.st-and.ac.uk/~history

http://www.agnesscott.edu/lriddle/women/women.htm

20 Extension Fields

In many respects this [Kronecker's Theorem] is the fundamental theorem of algebra.

RICHARD A. DEAN, *Elements of Abstract Algebra*

The Fundamental Theorem of Field Theory

In our work on rings, we came across a number of fields, both finite and infinite. Indeed, we saw that $Z_3[x]/\langle x^2 + 1\rangle$ is a field of order 9, whereas $\mathbf{R}[x]/\langle x^2 + 1\rangle$ is a field isomorphic to the complex numbers. In the next three chapters, we take up, in a systematic way, the subject of fields.

Definition Extension Field
A field E is an *extension field* of a field F if $F \subseteq E$ and the operations of F are those of E restricted to F.

Cauchy's observation in 1847 that $\mathbf{R}[x]/\langle x^2 + 1\rangle$ is a field that contains a zero of $x^2 + 1$ prepared the way for the following sweeping generalization of that fact.

Theorem 20.1 Fundamental Theorem of Field Theory (Kronecker's Theorem, 1887)

Let F be a field and let $f(x)$ be a nonconstant polynomial in $F[x]$. Then there is an extension field E of F in which $f(x)$ has a zero.

PROOF Since $F[x]$ is a unique factorization domain, $f(x)$ has an irreducible factor, say, $p(x)$. Clearly, it suffices to construct an extension field E of F in which $p(x)$ has a zero. Our candidate for E is $F[x]/\langle p(x)\rangle$. We already know that this is a field from Corollary 1 of Theorem 17.5. Also, since the mapping of ϕ: $F \to E$ given by $\phi(a) = a + \langle p(x)\rangle$ is one-to-one and preserves both operations, E has a subfield isomorphic to F. We may think of E as containing F if we simply identify the coset $a + \langle p(x)\rangle$ with its unique coset representative a that belongs to F [that is, think of $a + \langle p(x)\rangle$ as just a and vice versa; see Exercise 33 in Chapter 17].

Finally, to show that $p(x)$ has a zero in E, write

$$p(x) = a_n x^n + a_{n-1}x^{n-1} + \cdots + a_0.$$

Then, in E, $x + \langle p(x)\rangle$ is a zero of $p(x)$, because

$$\begin{aligned} p(x + \langle p(x)\rangle) &= a_n(x + \langle p(x)\rangle)^n + a_{n-1}(x + \langle p(x)\rangle)^{n-1} + \cdots + a_0 \\ &= a_n(x^n + \langle p(x)\rangle) + a_{n-1}(x^{n-1} + \langle p(x)\rangle) + \cdots + a_0 \\ &= a_n x^n + a_{n-1}x^{n-1} + \cdots + a_0 + \langle p(x)\rangle \\ &= p(x) + \langle p(x)\rangle = 0 + \langle p(x)\rangle. \end{aligned}$$

■

■ **EXAMPLE 1** Let $f(x) = x^2 + 1 \in Q[x]$. Then, viewing $f(x)$ as an element of $E[x] = (Q[x]/\langle x^2 + 1\rangle)[x]$, we have

$$\begin{aligned} f(x + \langle x^2 + 1\rangle) &= (x + \langle x^2 + 1\rangle)^2 + 1 \\ &= x^2 + \langle x^2 + 1\rangle + 1 \\ &= x^2 + 1 + \langle x^2 + 1\rangle \\ &= 0 + \langle x^2 + 1\rangle. \end{aligned}$$

Of course, the polynomial $x^2 + 1$ has the complex number $\sqrt{-1}$ as a zero, but the point we wish to emphasize here is that we have constructed a field that contains the rational numbers and a zero for the polynomial $x^2 + 1$ by using only the rational numbers. No knowledge of complex numbers is necessary. Our method utilizes only the field we are given. ■

■ **EXAMPLE 2** Let $f(x) = x^5 + 2x^2 + 2x + 2 \in Z_3[x]$. Then, the irreducible factorization of $f(x)$ over Z_3 is $(x^2 + 1)(x^3 + 2x + 2)$. So, to find an extension E of Z_3 in which $f(x)$ has a zero, we may take $E = Z_3[x]/\langle x^2 + 1\rangle$, a field with nine elements, or $E = Z_3[x]/\langle x^3 + 2x + 2\rangle$, a field with 27 elements. ■

Since every integral domain is contained in its field of quotients (Theorem 15.6), we see that every nonconstant polynomial with coefficients from an integral domain always has a zero in some field containing the ring of coefficients. The next example shows that this is not true for commutative rings in general.

■ **EXAMPLE 3** Let $f(x) = 2x + 1 \in Z_4[x]$. Then $f(x)$ has no zero in any ring containing Z_4 as a subring, because if β were a zero in such a ring, then $0 = 2\beta + 1$, and therefore $0 = 2(2\beta + 1) = 2(2\beta) + 2 = (2 \cdot 2)\beta + 2 = 0 \cdot \beta + 2 = 2$. But $0 \neq 2$ in Z_4. ∎

Splitting Fields

To motivate the next definition and theorem, let's return to Example 1 for a moment. For notational convenience, in $Q[x]/\langle x^2 + 1\rangle$, let $\alpha = x + \langle x^2 + 1\rangle$. Then, since α and $-\alpha$ are both zeros of $x^2 + 1$ in $(Q[x]/\langle x^2 + 1\rangle)[x]$, it should be the case that $x^2 + 1 = (x - \alpha)(x + \alpha)$. Let's check this out. First note that

$$(x - \alpha)(x + \alpha) = x^2 - \alpha^2 = x^2 - (x^2 + \langle x^2 + 1\rangle).$$

At the same time,

$$x^2 + \langle x^2 + 1\rangle = -1 + \langle x^2 + 1\rangle$$

and we have agreed to identify -1 and $-1 + \langle x^2 + 1\rangle$, so

$$(x - \alpha)(x + \alpha) = x^2 - (-1) = x^2 + 1.$$

This shows that $x^2 + 1$ can be written as a product of linear factors in some extension of Q. That was easy and you might argue coincidental. The polynomial given in Example 2 presents a greater challenge. Is there an extension of Z_3 in which that polynomial factors as a product of linear factors? Yes, there is. But first some notation and a definition.

Let F be a field and let $a_1, a_2, \ldots, a_n$ be elements of some extension E of F. We use $F(a_1,a_2, \ldots, a_n)$ to denote the smallest subfield of E that contains F and the set $\{a_1, a_2, \ldots, a_n\}$. We leave it as an exercise (Exercise 35) to show that $F(a_1, a_2, \ldots, a_n)$ is the intersection of all subfields of E that contain F and the set $\{a_1, a_2, \ldots, a_n\}$.

Definition Splitting Field

Let E be an extension field of F and let $f(x) \in F[x]$ with degree at least 1. We say that $f(x)$ *splits* in E if there are elements $a \in F$ and $a_1, a_2, \ldots, a_n \in E$ such that

$$f(x) = a(x - a_1)(x - a_2) \cdots (x - a_n).$$

We call E a *splitting field for $f(x)$ over F* if

$$E = F(a_1, a_2, \cdots, a_n).$$

Note that a splitting field of a polynomial over a field depends not only on the polynomial but on the field as well. Indeed, a splitting field of $f(x)$ over F is just a smallest extension field of F in which $f(x)$ splits.

The next example illustrates how a splitting field of a polynomial $f(x)$ over field F depends on F.

EXAMPLE 4 Consider the polynomial $f(x) = x^2 + 1 \in Q[x]$. Since $x^2 + 1 = (x + \sqrt{-1})(x - \sqrt{-1})$, we see that $f(x)$ splits in **C**, but a splitting field over Q is $Q(i) = \{r + si \mid r, s \in Q\}$. A splitting field for $x^2 + 1$ over **R** is **C**. Likewise, $x^2 - 2 \in Q[x]$ splits in **R**, but a splitting field over Q is $Q(\sqrt{2}) = \{r + s\sqrt{2} \mid r, s \in Q\}$. ∎

There is a useful analogy between the definition of a splitting field and the definition of an irreducible polynomial. Just as it makes no sense to say "$f(x)$ is irreducible," it makes no sense to say "E is a splitting field for $f(x)$." In each case, the underlying field must be specified; that is, one must say "$f(x)$ is irreducible over F" and "E is a splitting field for $f(x)$ over F."

Our notation in Example 4 appears to be inconsistent with the notation that we used in earlier chapters. For example, we denoted the set $\{a + b\sqrt{2} \mid a, b \in Z\}$ by $Z[\sqrt{2}]$ and the set $\{a + b\sqrt{2} \mid a, b \in Q\}$ by $Q(\sqrt{2})$. The difference is that $Z[\sqrt{2}]$ is merely a ring, whereas $Q(\sqrt{2})$ is a field. In general, parentheses are used when one wishes to indicate that the set is a field, although no harm would be done by using, say, $Q[\sqrt{2}]$ to denote $\{a + b\sqrt{2} \mid a, b \in Q\}$ if we were concerned with its ring properties only. Using parentheses rather than brackets simply conveys a bit more information about the set.

Theorem 20.2 Existence of Splitting Fields

Let F be a field and let f(x) be a nonconstant element of F[x]. Then there exists a splitting field E for f(x) over F.

PROOF We proceed by induction on $\deg f(x)$. If $\deg f(x) = 1$, then $f(x)$ is linear. Now suppose that the statement is true for all fields and all polynomials of degree less than that of $f(x)$. By Theorem 20.1, there is an extension E of F in which $f(x)$ has a zero, say, a_1. Then we may write $f(x) = (x - a_1)g(x)$, where $g(x) \in E[x]$. Since $\deg g(x) < \deg f(x)$, by induction, there is a field K that contains E and all the zeros of $g(x)$, say, $a_2, \ldots, a_n$. Clearly, then, a splitting field for $f(x)$ over F is $F(a_1, a_2, \ldots, a_n)$. ∎

EXAMPLE 5 Consider

$$f(x) = x^4 - x^2 - 2 = (x^2 - 2)(x^2 + 1)$$

over Q. Obviously, the zeros of $f(x)$ in $\mathbf{C}$ are $\pm\sqrt{2}$ and $\pm i$. So a splitting field for $f(x)$ over Q is

$$Q(\sqrt{2}, i) = Q(\sqrt{2})(i) = \{\alpha + \beta i \mid \alpha, \beta \in Q(\sqrt{2})\}$$
$$= \{(a + b\sqrt{2}) + (c + d\sqrt{2})i \mid a, b, c, d \in Q\}. \quad ▮$$

▮ EXAMPLE 6 Consider $f(x) = x^2 + x + 2$ over Z_3. Then $Z_3(i) = \{a + bi \mid a, b \in Z_3\}$ (see Example 9 in Chapter 13) is a splitting field for $f(x)$ over Z_3 because

$$f(x) = [x - (1 + i)][x - (1 - i)].$$

At the same time, we know by the proof of Kronecker's Theorem that the element $x + \langle x^2 + x + 2\rangle$ of

$$F = Z_3[x]/\langle x^2 + x + 2\rangle$$

is a zero of $f(x)$. Since $f(x)$ has degree 2, it follows from the Factor Theorem (Corollary 2 of Theorem 16.2) that the other zero of $f(x)$ must also be in F. Thus, $f(x)$ splits in F, and because F is a two-dimensional vector space over Z_3, we know that F is also a splitting field of $f(x)$ over Z_3. But how do we factor $f(x)$ in F? Factoring $f(x)$ in F is confusing because we are using the symbol x in two distinct ways: It is used as a placeholder to write the polynomial $f(x)$, and it is used to create the coset representatives of the elements of F. This confusion can be avoided by simply identifying the coset $1 + \langle x^2 + x + 2\rangle$ with the element 1 in Z_3 and denoting the coset $x + \langle x^2 + x + 2\rangle$ by β. With this identification, the field $Z_3[x]/\langle x^2 + x + 2\rangle$ can be represented as $\{0, 1, 2, \beta, 2\beta, \beta + 1, 2\beta + 1, \beta + 2, 2\beta + 2\}$. These elements are added and multiplied just as polynomials are, except that we use the observation that $x^2 + x + 2 + \langle x^2 + x + 2\rangle = 0$ implies that $\beta^2 + \beta + 2 = 0$, so that $\beta^2 = -\beta - 2 = 2\beta + 1$. For example, $(2\beta + 1)(\beta + 2) = 2\beta^2 + 5\beta + 2 = 2(2\beta + 1) + 5\beta + 2 = 9\beta + 4 = 1$. To obtain the factorization of $f(x)$ in F, we simply long divide, as follows:

$$\begin{array}{r l}
 & x + (\beta + 1) \\ \cline{2-2}
x - \beta & \big) \, x^2 + x + 2 \\
 & x^2 - \beta x \\ \cline{2-2}
 & (\beta + 1)x + 2 \\
 & (\beta + 1)x - (\beta + 1)\beta \\ \cline{2-2}
 & (\beta + 1)\beta + 2 = \beta^2 + \beta + 2 = 0.
\end{array}$$

So, $x^2 + x + 2 = (x - \beta)(x + \beta + 1)$. Thus, we have found two splitting fields for $x^2 + x + 2$ over Z_3, one of the form $F(a)$ and one of the form $F[x]/\langle p(x)\rangle$ [where $F = Z_3$ and $p(x) = x^2 + x + 2$]. ▮

The next theorem shows how the fields $F(a)$ and $F[x]/\langle p(x)\rangle$ are related in the case where $p(x)$ is irreducible over F and a is a zero of $p(x)$ in some extension of F.

Theorem 20.3 $F(a) \approx F[x]/\langle p(x)\rangle$

Let F be a field and let $p(x) \in F[x]$ be irreducible over F. If a is a zero of $p(x)$ in some extension E of F, then $F(a)$ is isomorphic to $F[x]/\langle p(x)\rangle$. Furthermore, if $\deg p(x) = n$, then every member of $F(a)$ can be uniquely expressed in the form

$$c_{n-1}a^{n-1} + c_{n-2}a^{n-2} + \cdots + c_1 a + c_0,$$

where $c_0, c_1, \ldots, c_{n-1} \in F$.

PROOF Consider the function ϕ from $F[x]$ to $F(a)$ given by $\phi(f(x)) = f(a)$. Clearly, ϕ is a ring homomorphism. We claim that $\operatorname{Ker}\phi = \langle p(x)\rangle$. (This is Exercise 31 in Chapter 17.) Since $p(a) = 0$, we have $\langle p(x)\rangle \subseteq \operatorname{Ker}\phi$. On the other hand, we know by Theorem 17.5 that $\langle p(x)\rangle$ is a maximal ideal in $F[x]$. So, because $\operatorname{Ker}\phi \neq F[x]$ [it does not contain the constant polynomial $f(x) = 1$], we have $\operatorname{Ker}\phi = \langle p(x)\rangle$. At this point it follows from the First Isomorphism Theorem for Rings and Corollary 1 of Theorem 17.5 that $\phi(F[x])$ is a subfield of $F(a)$. Noting that $\phi(F[x])$ contains both F and a and recalling that $F(a)$ is the smallest such field, we have $F[x]/\langle p(x)\rangle \approx \phi(F[x]) = F(a)$.

The final assertion of the theorem follows from the fact that every element of $F[x]/\langle p(x)\rangle$ can be expressed uniquely in the form

$$c_{n-1}x^{n-1} + \cdots + c_0 + \langle p(x)\rangle,$$

where $c_0, \ldots, c_{n-1} \in F$ (see Exercise 25 in Chapter 16), and the natural isomorphism from $F[x]/\langle p(x)\rangle$ to $F(a)$ carries $c_k x^k + \langle p(x)\rangle$ to $c_k a^k$. ∎

As an immediate corollary of Theorem 20.3, we have the following attractive result.

Corollary $F(a) \approx F(b)$

Let F be a field and let $p(x) \in F[x]$ be irreducible over F. If a is a zero of $p(x)$ in some extension E of F and b is a zero of $p(x)$ in some extension E′ of F, then the fields $F(a)$ and $F(b)$ are isomorphic.

PROOF From Theorem 20.3, we have

$$F(a) \approx F[x]/\langle p(x)\rangle \approx F(b).$$ ∎

Recall that a basis for an n-dimensional vector space over a field F is a set of n vectors $v_1, v_2, \ldots, v_n$ with the property that every member of the vector space can be expressed uniquely in the form $a_1v_1 + a_2v_2 + \cdots + a_nv_n$, where the a's belong to F (Exercise 19 in Chapter 19). So, in the language of vector spaces, the latter portion of Theorem 20.3 says that if a is a zero of an irreducible polynomial over F of degree n, then the set $\{1, a, \ldots, a^{n-1}\}$ is a basis for $F(a)$ over F.

Theorem 20.3 often provides a convenient way of describing the elements of a field.

EXAMPLE 7 Consider the irreducible polynomial $f(x) = x^6 - 2$ over Q. Since $\sqrt[6]{2}$ is a zero of $f(x)$, we know from Theorem 20.3 that the set $\{1, 2^{1/6}, 2^{2/6}, 2^{3/6}, 2^{4/6}, 2^{5/6}\}$ is a basis for $Q(\sqrt[6]{2})$ over Q. Thus,

$$Q(\sqrt[6]{2}) = \{a_0 + a_12^{1/6} + a_22^{2/6} + a_32^{3/6} + a_42^{4/6} + a_52^{5/6} \mid a_i \in Q\}.$$

This field is isomorphic to $Q[x]/\langle x^6 - 2\rangle$. ■

In 1882, Ferdinand von Lindemann (1852–1939) proved that π is not the zero of any polynomial in $Q[x]$. Because of this important result, Theorem 20.3 does not apply to $Q(\pi)$ (see Exercise 11). Fields of the form $F(a)$ where a is in some extension field of F but not the zero of an element of $F(x)$ are discussed in the next chapter.

In Example 6, we produced two splitting fields for the polynomial $x^2 + x + 2$ over Z_3. Likewise, it is an easy exercise to show that both $Q[x]/\langle x^2 + 1\rangle$ and $Q(i) = \{r + si \mid r, s \in Q\}$ are splitting fields of the polynomial $x^2 + 1$ over Q. But are these different-looking splitting fields algebraically different? Not really. We conclude our discussion of splitting fields by proving that splitting fields are unique up to isomorphism. To make it easier to apply induction, we will prove a more general result.

We begin by observing first that any ring isomorphism ϕ from F to F' has a natural extension from $F[x]$ to $F'[x]$ given by $c_nx^n + c_{n-1}x^{n-1} + \cdots + c_1x + c_0 \to \phi(c_n)x^n + \phi(c_{n-1})x^{n-1} + \cdots + \phi(c_1)x + \phi(c_0)$. Since this mapping agrees with ϕ on F, it is convenient and natural to use ϕ to denote this mapping as well.

Lemma

Let F be a field, let $p(x) \in F[x]$ be irreducible over F, and let a be a zero of $p(x)$ in some extension of F. If ϕ is a field isomorphism from F to F' and b is a zero of $\phi(p(x))$ in some extension of F', then there is an isomorphism from $F(a)$ to $F'(b)$ that agrees with ϕ on F and carries a to b.

PROOF First observe that since $p(x)$ is irreducible over F, $\phi(p(x))$ is irreducible over F'. It is straightforward to check that the mapping from $F[x]/\langle p(x)\rangle$ to $F'[x]/\langle \phi(p(x))\rangle$ given by

$$f(x) + \langle p(x)\rangle \to \phi(f(x)) + \langle \phi(p(x))\rangle$$

is a field isomorphism. By a slight abuse of notation, we denote this mapping by ϕ also. (If you object, put a bar over the ϕ.) From the proof of Theorem 20.3, we know that there is an isomorphism α from $F(a)$ to $F[x]/\langle p(x)\rangle$ that is the identity on F and carries a to $x + \langle p(x)\rangle$. Similarly, there is an isomorphism β from $F'[x]/\langle \phi(p(x))\rangle$ to $F'(b)$ that is the identity on F' and carries $x + \langle \phi(p(x))\rangle$ to b. Thus, $\beta\phi\alpha$ is the desired mapping from $F(a)$ to $F'(b)$. See Figure 20.1. ∎

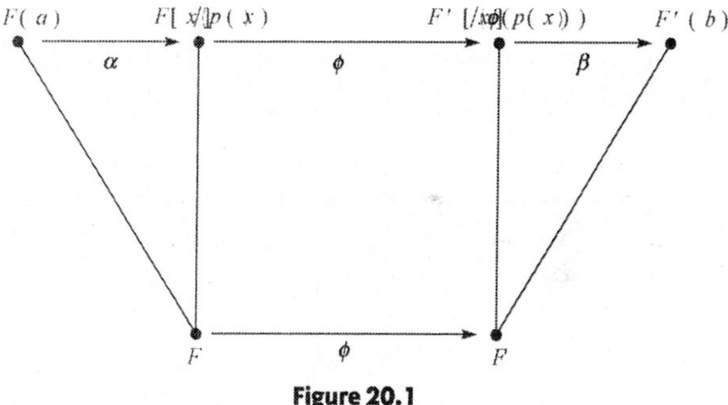

Figure 20.1

Theorem 20.4 Extending $\phi: F \to F'$

Let ϕ be an isomorphism from a field F to a field F' and let $f(x) \in F[x]$. If E is a splitting field for $f(x)$ over F and E' is a splitting field for $\phi(f(x))$ over F', then there is an isomorphism from E to E' that agrees with ϕ on F.

PROOF We use induction on $\deg f(x)$. If $\deg f(x) = 1$, then $E = F$ and $E' = F'$, so that ϕ itself is the desired mapping. If $\deg f(x) > 1$, let $p(x)$ be an irreducible factor of $f(x)$, let a be a zero of $p(x)$ in E, and let b be a zero of $\phi(p(x))$ in E'. By the preceding lemma, there is an isomorphism α from $F(a)$ to $F'(b)$ that agrees with ϕ on F and carries a to b. Now write $f(x) = (x - a)g(x)$, where $g(x) \in F(a)[x]$. Then E is a splitting field for $g(x)$ over $F(a)$ and E' is a splitting field for $\alpha(g(x))$ over $F'(b)$. Since $\deg g(x) < \deg f(x)$, there is an isomorphism from E to E' that agrees with α on $F(a)$ and therefore with ϕ on F. ∎

Corollary Splitting Fields Are Unique

Let F be a field and let $f(x) \in F[x]$. Then any two splitting fields of $f(x)$ over F are isomorphic.

PROOF Suppose that E and E' are splitting fields of $f(x)$ over F. The result follows immediately from Theorem 20.4 by letting ϕ be the identity from F to F. ∎

In light of the corollary above, we may refer to "the" splitting field of a polynomial over F without ambiguity.

Even though $x^6 - 2$ has a zero in $Q(\sqrt[6]{2})$, it does not split in $Q(\sqrt[6]{2})$. The splitting field is easy to obtain, however.

EXAMPLE 8 The Splitting Field of $x^n - a$ over Q

Let a be a positive rational number and let ω be a primitive nth root of unity (see Example 2 in Chapter 16). Then each of

$$a^{1/n}, \omega a^{1/n}, \omega^2 a^{1/n}, \ldots, \omega^{n-1} a^{1/n}$$

is a zero of $x^n - a$ in $Q(\sqrt[n]{a}, \omega)$. ∎

Zeros of an Irreducible Polynomial

Now that we know that every nonconstant polynomial over a field splits in some extension, we ask whether irreducible polynomials must split in some special way. Yes, they do. To discover how, we borrow something whose origins are in calculus.

Definition Derivative

Let $f(x) = a_n x^n + a_{n-1} x^{n-1} + \cdots + a_1 x + a_0$ belong to $F[x]$. The *derivative* of $f(x)$, denoted by $f'(x)$, is the polynomial $n a_n x^{n-1} + (n-1) a_{n-1} x^{n-2} + \cdots + a_1$ in $F[x]$.

Notice that our definition does not involve the notion of a limit. The standard rules for handling sums and products of functions in calculus carry over to arbitrary fields as well.

Lemma Properties of the Derivative

Let $f(x)$ and $g(x) \in F[x]$ and let $a \in F$. Then

1. $(f(x) + g(x))' = f'(x) + g'(x)$.
2. $(af(x))' = af'(x)$.
3. $(f(x)g(x))' = f(x)g'(x) + g(x)f'(x)$.

PROOF Properties 1 and 2 follow from straightforward applications of the definition. Using property 1 and induction on $\deg f(x)$, property 3 reduces to the special case in which $f(x) = a_n x^n$. This also follows directly from the definition. ▪

Before addressing the question of the nature of the zeros of an irreducible polynomial, we establish a general result concerning zeros of multiplicity greater than 1. Such zeros are called *multiple* zeros.

Theorem 20.5 Criterion for Multiple Zeros

A polynomial $f(x)$ over a field F has a multiple zero in some extension E if and only if $f(x)$ and $f'(x)$ have a common factor of positive degree in $F[x]$.

PROOF If a is a multiple zero of $f(x)$ in some extension E, then there is a $g(x)$ in $E[x]$ such that $f(x) = (x - a)^2 g(x)$. Since $f'(x) = (x - a)^2 g'(x) + 2(x - a)g(x)$, we see that $f'(a) = 0$. Thus, $x - a$ is a factor of both $f(x)$ and $f'(x)$ in the extension E of F. Now if $f(x)$ and $f'(x)$ have no common divisor of positive degree in $F[x]$, there are polynomials $h(x)$ and $k(x)$ in $F[x]$ such that $f(x)h(x) + f'(x)k(x) = 1$ (see Exercise 45 in Chapter 16). Viewing $f(x)h(x) + f'(x)k(x)$ as an element of $E[x]$, we see also that $x - a$ is a factor of 1. Since this is nonsense, $f(x)$ and $f'(x)$ must have a common divisor of positive degree in $F[x]$.

Conversely, suppose that $f(x)$ and $f'(x)$ have a common factor of positive degree. Let a be a zero of the common factor. Then a is a zero of $f(x)$ and $f'(x)$. Since a is a zero of $f(x)$, there is a polynomial $q(x)$ such that $f(x) = (x - a)q(x)$. Then $f'(x) = (x - a)q'(x) + q(x)$ and $0 = f'(a) = q(a)$. Thus, $x - a$ is a factor of $q(x)$ and a is a multiple zero of $f(x)$. ▪

Theorem 20.6 Zeros of an Irreducible

Let $f(x)$ be an irreducible polynomial over a field F. If F has characteristic 0, then $f(x)$ has no multiple zeros. If F has characteristic $p \neq 0$, then $f(x)$ has a multiple zero only if it is of the form $f(x) = g(x^p)$ for some $g(x)$ in $F[x]$.

PROOF If $f(x)$ has a multiple zero, then, by Theorem 20.5, $f(x)$ and $f'(x)$ have a common divisor of positive degree in $F[x]$. Since the only divisor of positive degree of $f(x)$ in $F[x]$ is $f(x)$ itself (up to associates), we see that $f(x)$ divides $f'(x)$. Because a polynomial over a field cannot divide a polynomial of smaller degree, we must have $f'(x) = 0$.

Now what does it mean to say that $f'(x) = 0$? If we write $f(x) = a_nx^n + a_{n-1}x^{n-1} + \cdots + a_1x + a_0$, then $f'(x) = na_nx^{n-1} + (n-1)a_{n-1}x^{n-2} + \cdots + a_1$. Thus, $f'(x) = 0$ only when $ka_k = 0$ for $k = 1, \ldots, n$.

So, when char $F = 0$, we have $f(x) = a_0$, which is not an irreducible polynomial. This contradicts the hypothesis that $f(x)$ is irreducible over F. Thus, $f(x)$ has no multiple zeros.

When char $F = p \neq 0$, we have $a_k = 0$ when p does not divide k. Thus, the only powers of x that appear in the sum $a_nx^n + \cdots + a_1x + a_0$ are those of the form $x^{pj} = (x^p)^j$. It follows that $f(x) = g(x^p)$ for some $g(x) \in F[x]$. [For example, if $f(x) = x^{4p} + 3x^{2p} + x^p + 1$, then $g(x) = x^4 + 3x^2 + x + 1$.] ∎

Theorem 20.6 shows that an irreducible polynomial over a field of characteristic 0 cannot have multiple zeros. The desire to extend this result to a larger class of fields motivates the following definition.

Definition Perfect Field

A field F is called *perfect* if F has characteristic 0 or if F has characteristic p and $F^p = \{a^p \mid a \in F\} = F$.

The most important family of perfect fields of characteristic p is the finite fields.

Theorem 20.7 Finite Fields Are Perfect

Every finite field is perfect.

PROOF Let F be a finite field of characteristic p. Consider the mapping ϕ from F to F defined by $\phi(x) = x^p$ for all $x \in F$. We claim that ϕ is a field automorphism. Obviously, $\phi(ab) = (ab)^p = a^pb^p = \phi(a)\phi(b)$. Moreover, $\phi(a + b) = (a + b)^p = a^p + \binom{p}{1} a^{p-1}b + \binom{p}{2} a^{p-2}b^2 + \cdots + \binom{p}{p-1} ab^{p-1} + b^p = a^p + b^p$, since each $\binom{p}{i}$ is divisible by p. Finally, since $x^p \neq 0$ when $x \neq 0$, Ker $\phi = \{0\}$. Thus, ϕ is one-to-one and, since F is finite, ϕ is onto. This proves that $F^p = F$. ∎

Theorem 20.8 Criterion for No Multiple Zeros

If $f(x)$ is an irreducible polynomial over a perfect field F, then $f(x)$ has no multiple zeros.

PROOF The case where F has characteristic 0 has been done. So let us assume that $f(x) \in F[x]$ is irreducible over a perfect field F of characteristic p and that $f(x)$ has multiple zeros. From Theorem 20.6 we know that $f(x) = g(x^p)$ for some $g(x) \in F[x]$, say, $g(x) = a_n x^n + a_{n-1}x^{n-1} + \cdots + a_1 x + a_0$. Since $F^p = F$, each a_i in F can be written in the form $b_i^{\,p}$ for some b_i in F. So, using Exercise 49a in Chapter 13, we have

$$\begin{aligned} f(x) = g(x^p) &= b_n^{\,p}x^{pn} + b_{n-1}^{\,p}x^{p(n-1)} + \cdots + b_1^{\,p}x^p + b_0^{\,p} \\ &= (b_n x^n + b_{n-1}x^{n-1} + \cdots + b_1 x + b_0)^p = (h(x))^p, \end{aligned}$$

where $h(x) \in F[x]$. But then $f(x)$ is not irreducible. ∎

The next theorem shows that when an irreducible polynomial does have multiple zeros, there is something striking about the multiplicities.

Theorem 20.9 Zeros of an Irreducible over a Splitting Field

Let $f(x)$ be an irreducible polynomial over a field F and let E be a splitting field of $f(x)$ over F. Then all the zeros of $f(x)$ in E have the same multiplicity.

PROOF Let a and b be distinct zeros of $f(x)$ in E. If a has multiplicity m, then in $E[x]$ we may write $f(x) = (x - a)^m g(x)$. It follows from the lemma preceding Theorem 20.4 and from Theorem 20.4 that there is a field isomorphism ϕ from E to itself that carries a to b and acts as the identity on F. Thus,

$$f(x) = \phi(f(x)) = (x - b)^m \phi(g(x)),$$

and we see that the multiplicity of b is greater than or equal to the multiplicity of a. By interchanging the roles of a and b, we observe that the multiplicity of a is greater than or equal to the multiplicity of b. So, we have proved that a and b have the same multiplicity. ∎

As an immediate corollary of Theorem 20.9 we have the following appealing result.

Corollary Factorization of an Irreducible over a Splitting Field

Let $f(x)$ be an irreducible polynomial over a field F and let E be a splitting field of $f(x)$. Then $f(x)$ has the form

$$a(x - a_1)^n(x - a_2)^n \cdots (x - a_t)^n,$$

where $a_1, a_2, \ldots, a_t$ are distinct elements of E and $a \in F$.

We conclude this chapter by giving an example of an irreducible polynomial over a field that does have a multiple zero. In particular, notice that the field we use is not perfect.

EXAMPLE 9 Let $F = Z_2(t)$ be the field of quotients of the ring $Z_2[t]$ of polynomials in the indeterminate t with coefficients from Z_2. (We must introduce a letter other than x, since the members of F are going to be our coefficients for the elements in $F[x]$.) Consider $f(x) = x^2 - t \in F[x]$. To see that $f(x)$ is irreducible over F, it suffices to show that it has no zeros in F. Well, suppose that $h(t)/k(t)$ is a zero of $f(x)$. Then $(h(t)/k(t))^2 = t$, and therefore $(h(t))^2 = t(k(t))^2$. Since $h(t), k(t) \in Z_2[t]$, we then have $h(t^2) = tk(t^2)$ (see Exercise 49 in Chapter 13). But deg $h(t^2)$ is even, whereas deg $tk(t^2)$ is odd. So, $f(x)$ is irreducible over F.

Finally, since t is a constant in $F[x]$ and the characteristic of F is 2, we have $f'(x) = 0$, so that $f'(x)$ and $f(x)$ have $f(x)$ as a common factor. So, by Theorem 20.5, $f(x)$ has a multiple zero in some extension of F. (Indeed, it has a single zero of multiplicity 2 in $K = F[x]/\langle x^2 - t\rangle$.) ■

Exercises

I have yet to see any problem, however complicated, which, when you looked at it in the right way, did not become still more complicated.

PAUL ANDERSON, *New Scientist*

1. Describe the elements of $Q(\sqrt[3]{5})$.
2. Show that $Q(\sqrt{2}, \sqrt{3}) = Q(\sqrt{2} + \sqrt{3})$.
3. Find the splitting field of $x^3 - 1$ over Q. Express your answer in the form $Q(a)$.
4. Find the splitting field of $x^4 + 1$ over Q.
5. Find the splitting field of

$$x^4 + x^2 + 1 = (x^2 + x + 1)(x^2 - x + 1)$$

over Q.

6. Let $a, b \in \mathbf{R}$ with $b \neq 0$. Show that $\mathbf{R}(a + bi) = \mathbf{C}$.
7. Find a polynomial $p(x)$ in $Q[x]$ such that $Q(\sqrt{1 + \sqrt{5}})$ is ring-isomorphic to $Q[x]/\langle p(x)\rangle$.
8. Let $F = Z_2$ and let $f(x) = x^3 + x + 1 \in F[x]$. Suppose that a is a zero of $f(x)$ in some extension of F. How many elements does $F(a)$ have? Express each member of $F(a)$ in terms of a. Write out a complete multiplication table for $F(a)$.
9. Let $F(a)$ be the field described in Exercise 8. Express each of a^5, a^{-2}, and a^{100} in the form $c_2a^2 + c_1a + c_0$.
10. Let $F(a)$ be the field described in Exercise 8. Show that a^2 and $a^2 + a$ are zeros of $x^3 + x + 1$.
11. Describe the elements in $Q(\pi)$.
12. Let $F = Q(\pi^3)$. Find a basis for $F(\pi)$ over F.
13. Write $x^7 - x$ as a product of linear factors over Z_3. Do the same for $x^{10} - x$.
14. Find all ring automorphisms of $Q(\sqrt[3]{5})$.
15. Let F be a field of characteristic p and let $f(x) = x^p - a \in F[x]$. Show that $f(x)$ is irreducible over F or $f(x)$ splits in F.
16. Suppose that β is a zero of $f(x) = x^4 + x + 1$ in some extension field E of Z_2. Write $f(x)$ as a product of linear factors in $E[x]$.
17. Find a, b, c in Q such that

$$(1 + \sqrt[3]{4})/(2 - \sqrt[3]{2}) = a + b\sqrt[3]{2} + c\sqrt[3]{4}.$$

Note that such a, b, c exist, since

$$(1 + \sqrt[3]{4})/(2 - \sqrt[3]{2}) \in Q(\sqrt[3]{2}) = \{a + b\sqrt[3]{2} + c\sqrt[3]{4} \mid a, b, c \in Q\}.$$

18. Express $(3 + 4\sqrt{2})^{-1}$ in the form $a + b\sqrt{2}$, where $a, b \in Q$.
19. Show that $Q(4 - i) = Q(1 + i)$, where $i = \sqrt{-1}$.
20. Let F be a field, and let a and b belong to F with $a \neq 0$. If c belongs to some extension of F, prove that $F(c) = F(ac + b)$. (F "absorbs" its own elements.)
21. Let $f(x) \in F[x]$ and let $a \in F$. Show that $f(x)$ and $f(x + a)$ have the same splitting field over F.
22. Recall that two polynomials $f(x)$ and $g(x)$ from $F[x]$ are said to be relatively prime if there is no polynomial of positive degree in $F[x]$ that divides both $f(x)$ and $g(x)$. Show that if $f(x)$ and $g(x)$ are relatively prime in $F[x]$, they are relatively prime in $K[x]$, where K is any extension of F.
23. Determine all of the subfields of $Q(\sqrt{2})$.

24. Let E be an extension of F and let a and b belong to E. Prove that $F(a, b) = F(a)(b) = F(b)(a)$.
25. Write $x^3 + 2x + 1$ as a product of linear polynomials over some extension field of Z_3.
26. Express $x^8 - x$ as a product of irreducibles over Z_2.
27. Prove or disprove that $Q(\sqrt{3})$ and $Q(\sqrt{-3})$ are ring-isomorphic.
28. For any prime p, find a field of characteristic p that is not perfect.
29. If β is a zero of $x^2 + x + 2$ over Z_5, find the other zero.
30. Show that $x^4 + x + 1$ over Z_2 does not have any multiple zeros in any extension field of Z_2.
31. Show that $x^{21} + 2x^8 + 1$ does not have multiple zeros in any extension of Z_3.
32. Show that $x^{21} + 2x^9 + 1$ has multiple zeros in some extension of Z_3.
33. Let F be a field of characteristic $p \neq 0$. Show that the polynomial $f(x) = x^{p^n} - x$ over F has distinct zeros.
34. Find the splitting field for $f(x) = (x^2 + x + 2)(x^2 + 2x + 2)$ over $Z_3[x]$. Write $f(x)$ as a product of linear factors.
35. Let F be a field and E an extension field of F that contains $a_1, a_2, \ldots, a_n$. Prove that $F(a_1, a_2, \ldots, a_n)$ is the intersection of all subfields of E that contain F and the set $\{a_1, a_2, \ldots, a_n\}$. (This exercise is referred to in this chapter.)
36. Suppose that a is algebraic over a field F. Show that a and $1 + a^{-1}$ have the same degree over F.
37. Suppose that $f(x)$ is a fifth-degree polynomial that is irreducible over Z_2. Prove that every nonidentity element is a generator of the cyclic group $(Z_2[x]/\langle f(x)\rangle)^*$.
38. Show that $Q(\sqrt{7}, i)$ is the splitting field for $x^4 - 6x^2 - 7$.
39. Let p be a prime, $F = Z_p(t)$ (the field of quotients of the ring $Z_p[x]$), and $f(x) = x^p - t$. Prove that $f(x)$ is irreducible over F and has a multiple zero in $K = F[x]/\langle x^p - t\rangle$.
40. Let $f(x)$ be an irreducible polynomial over a field F. Prove that the number of distinct zeros of $f(x)$ in a splitting field divides $\deg f(x)$.

Leopold Kronecker

He [Kronecker] wove together the three strands of his greatest interests—the theory of numbers, the theory of equations and elliptic functions—into one beautiful pattern.

E. T. BELL

LEOPOLD KRONECKER was born on December 7, 1823, in Liegnitz, Prussia. As a schoolboy, he received special instruction from the great algebraist Kummer. Kronecker entered the University of Berlin in 1841 and completed his Ph.D. dissertation in 1845 on the units in a certain ring.

Kronecker devoted the years 1845–1853 to business affairs, relegating mathematics to a hobby. Thereafter, being well-off financially, he spent most of his time doing research in algebra and number theory. Kronecker was one of the early advocates of the abstract approach to algebra. He innovatively applied rings and fields in his investigations of algebraic numbers, established the Fundamental Theorem of Finite Abelian Groups, and was the first mathematician to master Galois's theory of fields.

Kronecker advocated constructive methods for all proofs and definitions. He believed that all mathematics should be based on relationships among integers. He went so far as to say to Lindemann, who proved that π is transcendental, that irrational numbers do not exist. His most famous remark on the matter was "God made the integers, all the rest is the work of man." Henri Poincaré once remarked that Kronecker was able to produce fine work in number theory and algebra only by temporarily forgetting his own philosophy.

Kronecker died on December 29, 1891, at the age of 68.

For more information about Kronecker, visit:

http://www-groups.dcs.st-and.ac.uk/~history/

21 Algebraic Extensions

Banach once told me, "Good mathematicians see analogies between theorems or theories, the very best ones see analogies between analogies."

S. M. ULAM, *Adventures of a Mathematician*

Characterization of Extensions

In Chapter 20, we saw that every element in the field $Q(\sqrt{2})$ has the particularly simple form $a + b\sqrt{2}$, where a and b are rational. On the other hand, the elements of $Q(\pi)$ have the more complicated form

$$(a_n\pi^n + a_{n-1}\pi^{n-1} + \cdots + a_0)/(b_m\pi^m + b_{m-1}\pi^{m-1} + \cdots + b_0),$$

where the a's and b's are rational. The fields of the first type have a great deal of algebraic structure. This structure is the subject of this chapter.

Definition **Types of Extensions**

Let E be an extension field of a field F and let $a \in E$. We call a *algebraic over F* if a is the zero of some nonzero polynomial in $F[x]$. If a is not algebraic over F, it is called *transcendental over F*. An extension E of F is called an *algebraic* extension of F if every element of E is algebraic over F. If E is not an algebraic extension of F, it is called a *transcendental* extension of F. An extension of F of the form $F(a)$ is called a *simple* extension of F.

Leonhard Euler used the term *transcendental* for numbers that are not algebraic because "they transcended the power of algebraic methods." Although Euler made this distinction in 1744, it wasn't until 1844 that the existence of transcendental numbers over Q was proved by Joseph Liouville. Charles Hermite proved that e is transcendental over Q in 1873, and Lindemann showed that π is transcendental over Q in 1882. To this day, it is not known whether $\pi + e$ is transcendental over Q. With a precise definition of "almost all," it can be shown that almost all real numbers are transcendental over Q.

Theorem 21.1 shows why we make the distinction between elements that are algebraic over a field and elements that are transcendental over a field. Recall that $F(x)$ is the field of quotients of $F[x]$; that is,

$$F(x) = \{f(x)/g(x) \mid f(x), g(x) \in F[x], g(x) \neq 0\}.$$

Theorem 21.1 Characterization of Extensions

Let E be an extension field of the field F and let $a \in E$. If a is transcendental over F, then $F(a) \approx F(x)$. If a is algebraic over F, then $F(a) \approx F[x]/\langle p(x)\rangle$, where $p(x)$ is a polynomial in $F[x]$ of minimum degree such that $p(a) = 0$. Moreover, $p(x)$ is irreducible over F.

PROOF Consider the homomorphism ϕ: $F[x] \to F(a)$ given by $f(x) \to f(a)$. If a is transcendental over F, then Ker $\phi = \{0\}$, and so we may extend ϕ to an isomorphism $\overline{\phi}$: $F(x) \to F(a)$ by defining $\overline{\phi}(f(x)/g(x)) = f(a)/g(a)$.

If a is algebraic over F, then Ker $\phi \neq \{0\}$; and, by Theorem 16.4, there is a polynomial $p(x)$ in $F[x]$ such that Ker $\phi = \langle p(x)\rangle$ and $p(x)$ has minimum degree among all nonzero elements of Ker ϕ. Thus, $p(a) = 0$ and, since $p(x)$ is a polynomial of minimum degree with this property, it is irreducible over F. ∎

The proof of Theorem 21.1 can readily be adapted to yield the next two results also. The details are left to the reader (see Exercise 1).

Theorem 21.2 Uniqueness Property

If a is algebraic over a field F, then there is a unique monic irreducible polynomial $p(x)$ in $F[x]$ such that $p(a) = 0$.

The polynomial with the property specified in Theorem 21.2 is called the *minimal polynomial for a over F.*

Theorem 21.3 Divisibility Property

Let a be algebraic over F, and let $p(x)$ be the minimal polynomial for a over F. If $f(x) \in F[x]$ and $f(a) = 0$, then $p(x)$ divides $f(x)$ in $F[x]$.

If E is an extension field of F, we may view E as a vector space over F (that is, the elements of E are the vectors and the elements of F are the scalars). We are then able to use such notions as dimension and basis in our discussion.

Finite Extensions

Definition Degree of an Extension
Let E be an extension field of a field F. We say that E *has degree n over F* and write $[E:F] = n$ if E has dimension n as a vector space over F. If $[E:F]$ is finite, E is called a *finite extension* of F; otherwise, we say that E is an *infinite extension* of F.

Figure 21.1 illustrates a convenient method of depicting the degree of a field extension over a field.

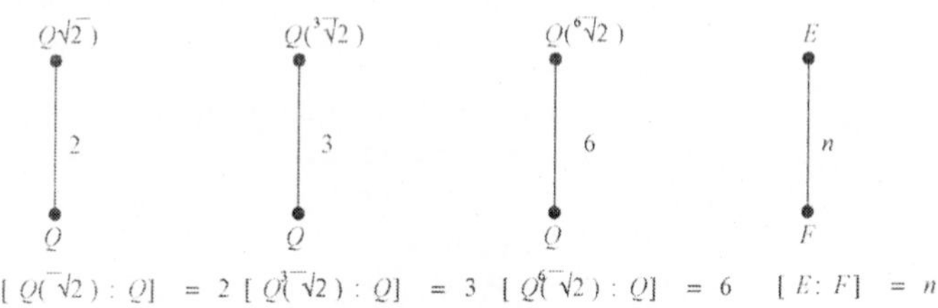

Figure 21.1

■ **EXAMPLE 1** The field of complex numbers has degree 2 over the reals, since $\{1, i\}$ is a basis. The field of complex numbers is an infinite extension of the rationals. ■

■ **EXAMPLE 2** If a is algebraic over F and its minimal polynomial over F has degree n, then, by Theorem 20.3, we know that $\{1, a, \ldots, a^{n-1}\}$ is a basis for $F(a)$ over F; and, therefore, $[F(a):F] = n$. In this case, we say that a has *degree n over F.* ■

■ Theorem 21.4 Finite Implies Algebraic

If E is a finite extension of F, then E is an algebraic extension of F.

PROOF Suppose that $[E:F] = n$ and $a \in E$. Then the set $\{1, a, \ldots, a^n\}$ is linearly dependent over F; that is, there are elements $c_0, c_1, \ldots, c_n$ in F, not all zero, such that

$$c_na^n + c_{n-1}a^{n-1} + \cdots + c_1a + c_0 = 0.$$

Clearly, then, a is a zero of the nonzero polynomial

$$f(x) = c_nx^n + c_{n-1}x^{n-1} + \cdots + c_1x + c_0.$$ ■

The converse of Theorem 21.4 is not true, for otherwise, the degrees of the elements of every algebraic extension of E over F would be bounded. But $Q(\sqrt{2}, \sqrt[3]{2}, \sqrt[4]{2}, \ldots)$ is an algebraic extension of Q that contains elements of every degree over Q (see Exercise 3).

The next theorem is the field theory counterpart of Lagrange's Theorem for finite groups. Like all counting theorems, it has far-reaching consequences.

Theorem 21.5 $[K:F] = [K:E][E:F]$

Let K be a finite extension field of the field E and let E be a finite extension field of the field F. Then K is a finite extension field of F and $[K:F] = [K:E][E:F]$.

PROOF Let $X = \{x_1, x_2, \ldots, x_n\}$ be a basis for K over E, and let $Y = \{y_1, y_2, \ldots, y_m\}$ be a basis for E over F. It suffices to prove that

$$YX = \{y_j x_i \mid 1 \leq j \leq m, 1 \leq i \leq n\}$$

is a basis for K over F. To do this, let $a \in K$. Then there are elements $b_1, b_2, \ldots, b_n \in E$ such that

$$a = b_1x_1 + b_2x_2 + \cdots + b_nx_n$$

and, for each $i = 1, \ldots, n$, there are elements $c_{i1}, c_{i2}, \ldots, c_{im} \in F$ such that

$$b_i = c_{i1}y_1 + c_{i2}y_2 + \cdots + c_{im}y_m.$$

Thus,

$$a = \sum_{i=1}^{n} b_i x_i = \sum_{i=1}^{n}\left(\sum_{j=1}^{m} c_{ij}y_j\right)x_i = \sum_{i,j} c_{ij}(y_jx_i).$$

This proves that YX spans K over F.

Now suppose there are elements c_{ij} in F such that

$$0 = \sum_{i,j} c_{ij}(y_jx_i) = \sum_i\left(\sum_j(c_{ij}y_j)\right)x_i.$$

Then, since each $\sum_j c_{ij}y_j \in E$ and X is a basis for K over E, we have

$$\sum_j c_{ij}y_j = 0$$

for each i. But each $c_{ij} \in F$ and Y is a basis for E over F, so each $c_{ij} = 0$. This proves that the set YX is linearly independent over F. ∎

Using the fact that for any field extension L of a field J, $[L:J] = n$ if and only if L is isomorphic to J^n as vector spaces (see Exercise 29), we

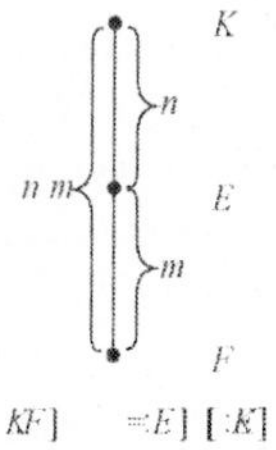

Figure 21.2

may give a concise conceptual proof of Theorem 21.5, as follows. Let $[K:E] = n$ and $[E:F] = m$. Then $K \approx E^n$ and $E \approx F^m$, so that $K \approx E^n \approx (F^m)^n \approx F^{mn}$. Thus, $[K:F] = mn$.

The content of Theorem 21.5 can be pictured as in Figure 21.2. Examples 3, 4, and 5 show how Theorem 21.5 is often utilized.

EXAMPLE 3 Since $\{1, \sqrt{3}\}$ is a basis for $Q(\sqrt{3}, \sqrt{5})$ over $Q(\sqrt{5})$ (see Exercise 7) and $\{1, \sqrt{5}\}$ is a basis for $Q(\sqrt{5})$ over Q, the proof of Theorem 21.5 shows that $\{1, \sqrt{3}, \sqrt{5}, \sqrt{15}\}$ is a basis for $Q(\sqrt{3}, \sqrt{5})$ over Q. (See Figure 21.3.) ∎

EXAMPLE 4 Consider $Q(\sqrt[3]{2}, \sqrt[4]{3})$. Then $[Q(\sqrt[3]{2}, \sqrt[4]{3}):Q] = 12$. For, clearly, $[Q(\sqrt[3]{2}, \sqrt[4]{3}):Q] = [Q(\sqrt[3]{2}, \sqrt[4]{3}):Q(\sqrt[3]{2})][Q(\sqrt[3]{2}):Q]$ and $[Q(\sqrt[3]{2}, \sqrt[4]{3}):Q] = [Q(\sqrt[3]{2}, \sqrt[4]{3}):Q(\sqrt[4]{3})][Q(\sqrt[4]{3}):Q]$ show that both $3 = [Q(\sqrt[3]{2}):Q]$ and $4 = [Q(\sqrt[4]{3}):Q]$ divide $[Q(\sqrt[3]{2}, \sqrt[4]{3}):Q]$. Thus, $[Q(\sqrt[3]{2}, \sqrt[4]{3}):Q] \geq 12$. On the other hand, $[Q(\sqrt[3]{2}, \sqrt[4]{3}):Q(\sqrt[3]{2})]$ is at most 4, since $\sqrt[4]{3}$ is a zero of $x^4 - 3 \in Q(\sqrt[3]{2})[x]$. Therefore, $[Q(\sqrt[3]{2}, \sqrt[4]{3}):Q] = [Q(\sqrt[3]{2}, \sqrt[4]{3}):Q(\sqrt[3]{2})][Q(\sqrt[3]{2}):Q] \leq 4 \cdot 3 = 12$. (See Figure 21.4.) ∎

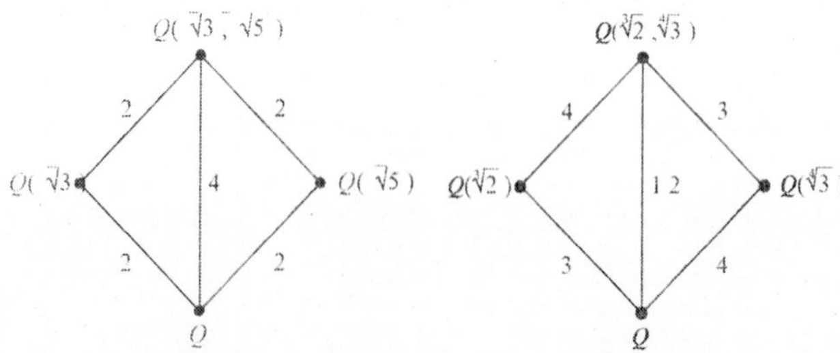

Figure 21.3

Figure 21.4

Theorem 21.5 can sometimes be used to show that a field does not contain a particular element.

EXAMPLE 5 Recall from Example 7 in Chapter 17 that $h(x) = 15x^4 - 10x^2 + 9x + 21$ is irreducible over Q. Let β be a zero of $h(x)$ in some extension of Q. Then, even though we don't know what β is, we can still prove that $\sqrt[3]{2}$ is not an element of $Q(\beta)$. For, if so, then $Q \subset Q(\sqrt[3]{2}) \subseteq Q(\beta)$ and $4 = [Q(\beta):Q] = [Q(\beta):Q(\sqrt[3]{2})][Q(\sqrt[3]{2}):Q]$ implies that 3 divides 4. Notice that this argument cannot be used to show that $\sqrt{2}$ is not contained in $Q(\beta)$. ∎

EXAMPLE 6 Consider $Q(\sqrt{3}, \sqrt{5})$. We claim that $Q(\sqrt{3}, \sqrt{5}) = Q(\sqrt{3} + \sqrt{5})$. The inclusion $Q(\sqrt{3} + \sqrt{5}) \subseteq Q(\sqrt{3}, \sqrt{5})$ is clear. Now note that since

$$(\sqrt{3} + \sqrt{5})^{-1} = \frac{1}{\sqrt{3} + \sqrt{5}} \cdot \frac{\sqrt{3} - \sqrt{5}}{\sqrt{3} - \sqrt{5}} = -\frac{1}{2}(\sqrt{3} - \sqrt{5}),$$

we know that $\sqrt{3} - \sqrt{5}$ belongs to $Q(\sqrt{3} + \sqrt{5})$. It follows that $[(\sqrt{3} + \sqrt{5}) + (\sqrt{3} - \sqrt{5})]/2 = \sqrt{3}$ and $[(\sqrt{3} + \sqrt{5}) - (\sqrt{3} - \sqrt{5})]/2 = \sqrt{5}$ both belong to $Q(\sqrt{3} + \sqrt{5})$, and therefore $Q(\sqrt{3}, \sqrt{5}) \subseteq Q(\sqrt{3} + \sqrt{5})$. ∎

EXAMPLE 7 It follows from Example 6 and Theorem 20.3 that the minimal polynomial for $\sqrt{3} + \sqrt{5}$ over Q has degree 4. How can we find this polynomial? We begin with $x = \sqrt{3} + \sqrt{5}$. Then $x^2 = 3 + 2\sqrt{15} + 5$. From this we obtain $x^2 - 8 = 2\sqrt{15}$ and, by squaring both sides, $x^4 - 16x + 64 = 60$. Thus, $\sqrt{3} + \sqrt{5}$ is a zero of $x^4 - 16x + 4$. We know that this is the minimal polynomial of $\sqrt{3} + \sqrt{5}$ over Q since it is monic and has degree 4. ∎

Example 6 shows that an extension obtained by adjoining two elements to a field can sometimes be obtained by adjoining a single element to the field. Our next theorem shows that, under certain conditions, this can always be done.

Theorem 21.6 Primitive Element Theorem (Steinitz, 1910)

If F is a field of characteristic 0, and a and b are algebraic over F, then there is an element c in F(a, b) such that F(a, b) = F(c).

PROOF Let $p(x)$ and $q(x)$ be the minimal polynomials over F for a and b, respectively. In some extension K of F, let $a_1, a_2, \ldots, a_m$ and $b_1, b_2, \ldots, b_n$ be the distinct zeros of $p(x)$ and $q(x)$, respectively, where $a = a_1$ and $b = b_1$. Among the infinitely many elements of F, choose an element d not equal to $(a_i - a)/(b - b_j)$ for all $i \geq 1$ and all $j > 1$. In particular, $a_i \neq a + d(b - b_j)$ for $j > 1$.

We shall show that $c = a + db$ has the property that $F(a, b) = F(c)$. Certainly, $F(c) \subseteq F(a, b)$. To verify that $F(a, b) \subseteq F(c)$, it suffices to prove that $b \in F(c)$, for then b, c, and d belong to $F(c)$ and $a = c - bd$. Consider the polynomials $q(x)$ and $r(x) = p(c - dx)$ [that is, $r(x)$ is obtained by substituting $c - dx$ for x in $p(x)$] over $F(c)$. Since both $q(b) = 0$ and $r(b) = p(c - db) = p(a) = 0$, both $q(x)$ and $r(x)$ are divisible by the minimal polynomial $s(x)$ for b over $F(c)$ (see Theorem 21.3). Because $s(x) \in F(c)[x]$, we may complete the proof by proving that $s(x) = x - b$. Since $s(x)$ is a common divisor of $q(x)$ and $r(x)$, the only possible zeros of $s(x)$ in K are the zeros of $q(x)$ that are also zeros of $r(x)$. But $r(b_j) = p(c - db_j) = p(a + db - db_j) = p(a + d(b - b_j))$ and d was chosen such that $a + d(b - b_j) \neq a_i$ for $j > 1$. It follows that b is the only zero of $s(x)$ in $K[x]$ and, therefore, $s(x) = (x - b)^u$. Since $s(x)$ is irreducible and F has characteristic 0, Theorem 20.6 guarantees that $u = 1$. ∎

In the terminology introduced earlier, it follows from Theorem 21.6 and induction that any finite extension of a field of characteristic 0 is a simple extension. An element a with the property that $E = F(a)$ is called a *primitive element* of E.

Properties of Algebraic Extensions

Theorem 21.7 Algebraic over Algebraic Is Algebraic

If K is an algebraic extension of E and E is an algebraic extension of F, then K is an algebraic extension of F.

PROOF Let $a \in K$. It suffices to show that a belongs to some finite extension of F. Since a is algebraic over E, we know that a is the zero of some irreducible polynomial in $E[x]$, say, $p(x) = b_nx^n + \cdots + b_0$. Now we construct a tower of extension fields of F, as follows:

$$F_0 = F(b_0),$$
$$F_1 = F_0(b_1), \ldots, F_n = F_{n-1}(b_n).$$

In particular,

$$F_n = F(b_0, b_1, \ldots, b_n),$$

so that $p(x) \in F_n[x]$. Thus, $[F_n(a):F_n] = n$; and, because each b_i is algebraic over F, we know that each $[F_{i+1}:F_i]$ is finite. So,

$$[F_n(a):F] = [F_n(a):F_n][F_n:F_{n-1}] \cdots [F_1:F_0][F_0:F]$$

is finite. (See Figure 21.5.) ∎

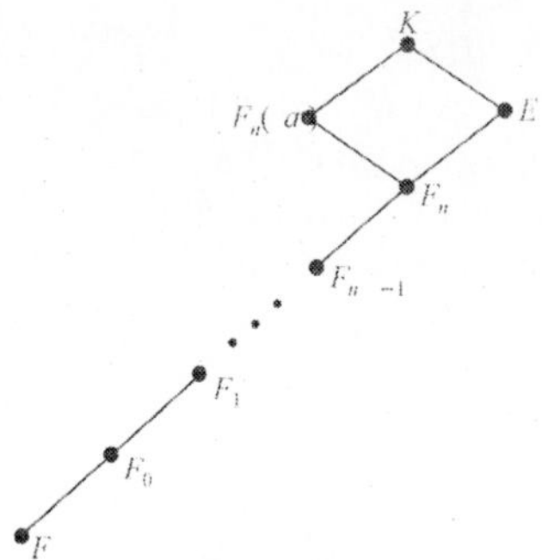

Figure 21.5

Corollary Subfield of Algebraic Elements

Let E be an extension field of the field F. Then the set of all elements of E that are algebraic over F is a subfield of E.

PROOF Suppose that a, $b \in E$ are algebraic over F and $b \neq 0$. To show that $a + b$, $a - b$, ab, and a/b are algebraic over F, it suffices to show that $[F(a, b):F]$ is finite, since each of these four elements belongs to $F(a, b)$. But note that

$$[F(a, b):F] = [F(a, b):F(b)][F(b):F].$$

Also, since a is algebraic over F, it is certainly algebraic over $F(b)$. Thus, both $[F(a, b):F(b)]$ and $[F(b):F]$ are finite. ∎

For any extension E of a field F, the subfield of E of the elements that are algebraic over F is called the *algebraic closure of F in E.*

One might wonder if there is such a thing as a maximal algebraic extension of a field F—that is, whether there is an algebraic extension E of F that has no proper algebraic extensions. For such an E to exist, it is necessary that every polynomial in $E[x]$ splits in E. Otherwise, it follows from Kronecker's Theorem that E would have a proper algebraic extension. This condition is also sufficient. If every member of $E[x]$ splits in E, and K is an algebraic extension of E, then every member of K is a zero of some element of $E[x]$. But the zeros of elements of $E[x]$ are in E. A field that has no proper algebraic extension is called *algebraically closed.* In 1910, Ernst Steinitz proved that every field F has a unique (up to isomorphism) algebraic extension that is algebraically closed. This field is called the *algebraic closure of F*. A proof of this result requires a sophisticated set theory background.

In 1799, Gauss, at the age of 22, proved that **C** is algebraically closed. This fact was considered so important at the time that it was called the Fundamental Theorem of Algebra. Over a 50-year period, Gauss found three additional proofs of the Fundamental Theorem. Today more than 100 proofs exist. In view of the ascendancy of abstract algebra in the 20th century, a more appropriate phrase for Gauss's result would be the Fundamental Theorem of Classical Algebra.

Exercises

It matters not what goal you seek
Its secret here reposes:
You've got to dig from week to week
To get Results or Roses.

EDGAR GUEST

1. Prove Theorem 21.2 and Theorem 21.3.
2. Let E be the algebraic closure of F. Show that every polynomial in $F[x]$ splits in E.
3. Prove that $Q(\sqrt{2}, \sqrt[3]{2}, \sqrt[4]{2}, \ldots)$ is an algebraic extension of Q but not a finite extension of Q. (This exercise is referred to in this chapter.)
4. Let E be an algebraic extension of F. If every polynomial in $F[x]$ splits in E, show that E is algebraically closed.
5. Suppose that F is a field and every irreducible polynomial in $F[x]$ is linear. Show that F is algebraically closed.
6. Suppose that $f(x)$ and $g(x)$ are irreducible over F and that $\deg f(x)$ and $\deg g(x)$ are relatively prime. If a is a zero of $f(x)$ in some extension of F, show that $g(x)$ is irreducible over $F(a)$.
7. Let a and b belong to Q with $b \neq 0$. Show that $Q(\sqrt{a}) = Q(\sqrt{b})$ if and only if there exists some $c \in Q$ such that $a = bc^2$.
8. Find the degree and a basis for $Q(\sqrt{3} + \sqrt{5})$ over $Q(\sqrt{15})$. Find the degree and a basis for $Q(\sqrt{2}, \sqrt[3]{2}, \sqrt[4]{2})$ over Q.
9. Suppose that E is an extension of F of prime degree. Show that, for every a in E, $F(a) = F$ or $F(a) = E$.
10. Let a be a complex number that is algebraic over Q. Show that $\sqrt{a}$ is algebraic over Q. Why does this prove that $\sqrt[2n]{a}$ is algebraic over Q?
11. Suppose that E is an extension of F and $a, b \in E$. If a is algebraic over F of degree m, and b is algebraic over F of degree n, where m and n are relatively prime, show that $[F(a, b):F] = mn$.

12. Find an example of a field F and elements a and b from some extension field such that $F(a, b) \neq F(a)$, $F(a, b) \neq F(b)$, and $[F(a, b):F] < [F(a):F][F(b):F]$.

13. Let K be a field extension of F and let $a \in K$. Show that $[F(a):F(a^3)] \leq 3$. Find examples to illustrate that $[F(a):F(a^3)]$ can be 1, 2, or 3.

14. Find the minimal polynomial for $\sqrt{-3} + \sqrt{2}$ over Q.

15. Let K be an extension of F. Suppose that E_1 and E_2 are contained in K and are extensions of F. If $[E_1:F]$ and $[E_2:F]$ are both prime, show that $E_1 = E_2$ or $E_1 \cap E_2 = F$.

16. Find the minimal polynomial for $\sqrt[3]{2} + \sqrt[3]{4}$ over Q.

17. Let E be a finite extension of $\mathbf{R}$. Use the fact that $\mathbf{C}$ is algebraically closed to prove that $E = \mathbf{C}$ or $E = \mathbf{R}$.

18. Suppose that $[E:Q] = 2$. Show that there is an integer d such that $E = Q(\sqrt{d})$ where d is not divisible by the square of any prime.

19. Suppose that $p(x) \in F[x]$ and E is a finite extension of F. If $p(x)$ is irreducible over F, and deg $p(x)$ and $[E:F]$ are relatively prime, show that $p(x)$ is irreducible over E.

20. Let E be an extension field of F. Show that $[E:F]$ is finite if and only if $E = F(a_1, a_2, \ldots, a_n)$, where $a_1, a_2, \ldots, a_n$ are algebraic over F.

21. If α and β are real numbers and α and β are transcendental over Q, show that either $\alpha\beta$ or $\alpha + \beta$ is also transcendental over Q.

22. Let $f(x)$ be a nonconstant element of $F[x]$. If a belongs to some extension of F and $f(a)$ is algebraic over F, prove that a is algebraic over F.

23. Let $f(x) = ax^2 + bx + c \in Q[x]$. Find a primitive element for the splitting field for $f(x)$ over Q.

24. Find the splitting field for $x^4 - x^2 - 2$ over Z_3.

25. Let $f(x) \in F[x]$. If deg $f(x) = 2$ and a is a zero of $f(x)$ in some extension of F, prove that $F(a)$ is the splitting field for $f(x)$ over F.

26. Let a be a complex zero of $x^2 + x + 1$ over Q. Prove that $Q(\sqrt{a}) = Q(a)$.

27. If F is a field and the multiplicative group of nonzero elements of F is cyclic, prove that F is finite.

28. Let a be a complex number that is algebraic over Q and let r be a rational number. Show that a^r is algebraic over Q.

29. Prove that, if K is an extension field of F, then $[K:F] = n$ if and only if K is isomorphic to F^n as vector spaces. (See Exercise 27 in Chapter 19 for the appropriate definition. This exercise is referred to in this chapter.)

30. Let a be a positive real number and let n be an integer greater than 1. Prove or disprove that $[Q(a^{1/n}):Q] = n$.

31. Let a and b belong to some extension field of F and let b be algebraic over F. Prove that $[F(a, b):F(a)] \leq [F(a, b):F]$.

32. Let $f(x)$ and $g(x)$ be irreducible polynomials over a field F and let a and b belong to some extension E of F. If a is a zero of $f(x)$ and b is a zero of $g(x)$, show that $f(x)$ is irreducible over $F(b)$ if and only if $g(x)$ is irreducible over $F(a)$.

33. Let β be a zero of $f(x) = x^5 + 2x + 4$ (see Example 8 in Chapter 17). Show that none of $\sqrt{2}, \sqrt[3]{2}, \sqrt[4]{2}$ belongs to $Q(\beta)$.

34. Prove that $Q(\sqrt{2}, \sqrt[3]{2}) = Q(\sqrt[6]{2})$.

35. Let a and b be rational numbers. Show that $Q(\sqrt{a}, \sqrt{b}) = Q(\sqrt{a} + \sqrt{b})$.

36. Let F, K, and L be fields with $F \subseteq K \subseteq L$. If L is a finite extension of F and $[L:F] = [L:K]$, prove that $F = K$.

37. Let F be a field and K a splitting field for some nonconstant polynomial over F. Show that K is a finite extension of F.

38. Prove that **C** is not the splitting field of any polynomial in $Q[x]$.

39. Prove that $\sqrt{2}$ is not an element of $Q(\pi)$.

40. Let $\alpha = \cos\frac{2\pi}{7} + i\sin\frac{2\pi}{7}$ and $\beta = \cos\frac{2\pi}{5} + i\sin\frac{2\pi}{5}$. Prove that β is not in $Q(\alpha)$.

41. Suppose that a is algebraic over a field F. Show that a and $1 + a^{-1}$ have the same degree over F.

42. Suppose K is an extension of F of degree n. Prove that K can be written in the form $F(x_1, x_2, \ldots, x_n)$ for some $x_1, x_2, \ldots, x_n$ in K.

Suggested Readings

R. L. Roth, "On Extensions of Q by Square Roots," *American Mathematical Monthly* 78 (1971): 392–393.

In this paper, it is proved that if $p_1, p_2, \ldots, p_n$ are distinct primes, then $[Q(\sqrt{p_1}, \sqrt{p_2}, \ldots, \sqrt{p_n}):Q] = 2^n$.

Paul B. Yale, "Automorphisms of the Complex Numbers," *Mathematics Magazine* 39 (1966): 135–141.

This award-winning expository paper is devoted to various results on automorphisms of the complex numbers.

Irving Kaplansky

American Mathematical Society

He got to the top of the heap by being a first-rate doer and expositor of algebra.

PAUL R. HALMOS, *I Have a Photographic Memory*

IRVING KAPLANSKY was born on March 22, 1917, in Toronto, Canada, a few years after his parents emigrated from Poland. Although his parents thought he would pursue a career in music, Kaplansky knew early on that mathematics was what he wanted to do. As an undergraduate at the University of Toronto, Kaplansky was a member of the winning team in the first William Lowell Putnam Competition, a mathematical contest for United States and Canadian college students. Kaplansky received a B.A. degree from Toronto in 1938 and an M.A. in 1939. In 1939, he entered Harvard University to earn his doctorate as the first recipient of a Putnam Fellowship. After receiving his Ph.D. from Harvard in 1941, Kaplansky stayed on as Benjamin Peirce Instructor until 1944. After one year at Columbia University, he went to the University of Chicago, where he remained until his retirement in 1984. He then became the director of the Mathematical Sciences Research Institute at the University of California, Berkeley.

Kaplansky's interests were broad, including areas such as ring theory, group theory, field theory, Galois theory, ergodic theory, algebras, metric spaces, number theory, statistics, and probability.

Among the many honors Kaplansky received are election to both the National Academy of Sciences and the American Academy of Arts and Sciences, election to the presidency of the American Mathematical Society, and the 1989 Steele Prize for cumulative influence from the American Mathematical Society. The Steele Prize citation says, in part, ". . . he has made striking changes in mathematics and has inspired generations of younger mathematicians." Kaplansky died on June 25, 2006, at the age of 89.

For more information about Kaplansky, visit:

http://www-groups.dcs.st-and.ac.uk/~history/

22 Finite Fields

This theory [of finite fields] is of considerable interest in its own right and it provides a particularly beautiful example of how the general theory of the preceding chapters fits together to provide a rather detailed description of all finite fields.

RICHARD A. DEAN, *Elements of Abstract Algebra*

Classification of Finite Fields

In this, our final chapter on field theory, we take up one of the most beautiful and important areas of abstract algebra—finite fields. Finite fields were first introduced by Galois in 1830 in his proof of the unsolvability of the general quintic equation. When Cayley invented matrices a few decades later, it was natural to investigate groups of matrices over finite fields. To this day, matrix groups over finite fields are among the most important classes of groups. In the past 50 years, there have been important applications of finite fields in computer science, coding theory, information theory, and cryptography. But, besides the many uses of finite fields in pure and applied mathematics, there is yet another good reason for studying them. They are just plain fun!

The most striking fact about finite fields is the restricted nature of their order and structure. We have already seen that every finite field has prime-power order (Exercise 51 in Chapter 13). A converse of sorts is also true.

Theorem 22.1 Classification of Finite Fields

For each prime p and each positive integer n, there is, up to isomorphism, a unique finite field of order p^n.

PROOF Consider the splitting field E of $f(x) = x^{p^n} - x$ over Z_p. We will show that $|E| = p^n$. Since $f(x)$ splits in E, we know that $f(x)$ has exactly p^n zeros in E, counting multiplicity. Moreover, by Theorem 20.5, every zero of $f(x)$ has multiplicity 1. Thus, $f(x)$ has p^n distinct zeros in E.

On the other hand, the set of zeros of $f(x)$ in E is closed under addition, subtraction, multiplication, and division by nonzero elements (see Exercise 37), so that the set of zeros of $f(x)$ is itself an extension field of Z_p in which $f(x)$ splits. Thus, the set of zeros of $f(x)$ is E and, therefore, $|E| = p^n$.

To show that there is a unique field for each prime-power, suppose that K is any field of order p^n. Then K has a subfield isomorphic to Z_p (generated by 1), and, because the nonzero elements of K form a multiplicative group of order $p^n - 1$, every element of K is a zero of $f(x) = x^{p^n} - x$ (see Exercise 27). So, K must be a splitting field for $f(x)$ over Z_p. By the corollary to Theorem 20.4, there is only one such field up to isomorphism. ∎

The existence portion of Theorem 22.1 appeared in the works of Galois and Gauss in the first third of the 19th century. Rigorous proofs were given by Dedekind in 1857 and by Jordan in 1870 in his classic book on group theory. The uniqueness portion of the theorem was proved by E. H. Moore in an 1893 paper concerning finite groups. The mathematics historian E. T. Bell once said that this paper by Moore marked the beginning of abstract algebra in America.

Because there is only one field for each prime-power p^n, we may unambiguously denote it by GF(p^n), in honor of Galois, and call it the *Galois field of order p^n*.

Structure of Finite Fields

The next theorem tells us the additive and multiplicative group structure of a field of order p^n.

Theorem 22.2 Structure of Finite Fields

As a group under addition, GF(p^n) ***is isomorphic to***

$$\underbrace{Z_p \oplus Z_p \oplus \cdots \oplus Z_p}_{n \text{ factors}}.$$

As a group under multiplication, the set of nonzero elements of GF(p^n) ***is isomorphic to*** Z_{p^n-1} ***(and is, therefore, cyclic).***

PROOF Since GF(p^n) has characteristic p (Theorem 13.3), every nonzero element of GF(p^n) has additive order p. Then by the Fundamental Theorem of Finite Abelian Groups, GF(p^n) under addition is isomorphic to a direct product of n copies of Z_p.

To see that the multiplicative group $GF(p^n)^*$ of nonzero elements of $GF(p^n)$ is cyclic, we first note that by the Fundamental Theorem of Finite Abelian Groups (Theorem 11.1), $GF(p^n)^*$ is isomorphic to a direct product of the form $Z_{n_1} \oplus Z_{n_2} \oplus \cdots \oplus Z_{n_m}$. If the orders of these components are pairwise relatively prime, then it follows from Corollary 1 of Theorem 8.2 that $GF(p^n)^*$ is cyclic. Hence we may assume that there is an integer $d > 1$ that divides the orders of two of the components. From the Fundamental Theorem of Cyclic Groups (Theorem 4.3) we know that each of these components has a subgroup of order d. This means that $GF(p^n)^*$ has two distinct subgroups of order d, call them H and K. But then every element of H and K is a zero of $x^d - 1$, which contradicts the fact that a polynomial of degree d over a field can have at most d zeros (Corollary 3 of Theorem 16.2). ∎

Some students misinterpret Theorem 22.2 to mean that $Z_p \oplus Z_p \oplus \cdots \oplus Z_p$ is a field of order p^n. Since any element of $Z_p \oplus Z_p \oplus \cdots \oplus Z_p$ that has at least one coordinate equal to 0 cannot have an inverse, it is not a field.

Since $Z_p \oplus Z_p \oplus \cdots \oplus Z_p$ is a vector space over Z_p with $\{(1, 0, \ldots, 0), (0, 1, 0, \ldots, 0), \ldots, (0, 0, \ldots, 1)\}$ as a basis, we have the following useful and aesthetically appealing formula.

Corollary 1

$$[GF(p^n):GF(p)] = n$$

Corollary 2 GF(p^n) Contains an Element of Degree n

Let a be a generator of the group of nonzero elements of* GF(p^n) *under multiplication. Then a is algebraic over* GF(p) *of degree n.

PROOF Observe that $[GF(p)(a):GF(p)] = [GF(p^n):GF(p)] = n$. ∎

EXAMPLE 1 Let's examine the field GF(16) in detail. Since $x^4 + x + 1$ is irreducible over Z_2, we know that

$$GF(16) \approx \{ax^3 + bx^2 + cx + d + \langle x^4 + x + 1\rangle \mid a, b, c, d \in Z_2\}.$$

Thus, we may think of GF(16) as the set

$$F = \{ax^3 + bx^2 + cx + d \mid a, b, c, d \in Z_2\},$$

where addition is done as in $Z_2[x]$, but multiplication is done modulo $x^4 + x + 1$. For example,

$$(x^3 + x^2 + x + 1)(x^3 + x) = x^3 + x^2,$$

since the remainder upon dividing

$$(x^3 + x^2 + x + 1)(x^3 + x) = x^6 + x^5 + x^2 + x$$

by $x^4 + x + 1$ in $Z_2[x]$ is $x^3 + x^2$. An easier way to perform the same calculation is to observe that in this context $x^4 + x + 1$ is 0, so

$$\begin{aligned} x^4 &= -x - 1 = x + 1, \\ x^5 &= x^2 + x, \\ x^6 &= x^3 + x^2. \end{aligned}$$

Thus,

$$x^6 + x^5 + x^2 + x = (x^3 + x^2) + (x^2 + x) + x^2 + x = x^3 + x^2.$$

Another way to simplify the multiplication process is to make use of the fact that the nonzero elements of GF(16) form a cyclic group of order 15. To take advantage of this, we must first find a generator of this group. Since any element F^* must have a multiplicative order that divides 15, all we need to do is find an element α in F^* such that $\alpha^3 \neq 1$ and $\alpha^5 \neq 1$. Obviously, x has these properties. So, we may think of GF(16) as the set $\{0, 1, x, x^2, \ldots, x^{14}\}$, where $x^{15} = 1$. This makes multiplication in F trivial, but, unfortunately, it makes addition more difficult. For example, $x^{10} \cdot x^7 = x^{17} = x^2$, but what is $x^{10} + x^7$? So, we face a dilemma. If we write the elements of F^* in the additive form $ax^3 + bx^2 + cx + d$, then addition is easy and multiplication is hard. On the other hand, if we write the elements of F^* in the multiplicative form x^i, then multiplication is easy and addition is hard. Can we have the best of both? Yes, we can. All we need to do is use the relation $x^4 = x + 1$ to make a two-way conversion table, as in Table 22.1.

So, we see from Table 22.1 that

$$\begin{aligned} x^{10} + x^7 &= (x^2 + x + 1) + (x^3 + x + 1) \\ &= x^3 + x^2 = x^6 \end{aligned}$$

and

$$\begin{aligned} (x^3 + x^2 + 1)(x^3 + x^2 + x + 1) &= x^{13} \cdot x^{12} \\ &= x^{25} = x^{10} = x^2 + x + 1. \quad ■ \end{aligned}$$

Don't be misled by the preceding example into believing that the element x is always a generator for the cyclic multiplicative group of nonzero elements. It is not. (See Exercise 19.) Although any two

Table 22.1 Conversion Table for Addition and Multiplication in GF(16)

Multiplicative Form to Additive Form		Additive Form to Multiplicative Form	
1	1	1	1
x	x	x	x
x^2	x^2	$x+1$	x^4
x^3	x^3	x^2	x^2
x^4	$x+1$	x^2+x	x^5
x^5	x^2+x	x^2+1	x^8
x^6	x^3+x^2	x^2+x+1	x^{10}
x^7	x^3+x+1	x^3	x^3
x^8	x^2+1	x^3+x^2	x^6
x^9	x^3+x	x^3+x	x^9
x^{10}	x^2+x+1	x^3+1	x^{14}
x^{11}	x^3+x^2+x	x^3+x^2+x	x^{11}
x^{12}	x^3+x^2+x+1	x^3+x^2+1	x^{13}
x^{13}	x^3+x^2+1	x^3+x+1	x^7
x^{14}	x^3+1	x^3+x^2+x+1	x^{12}

irreducible polynomials of the same degree over $Z_p[x]$ yield isomorphic fields, some are better than others for computational purposes.

EXAMPLE 2 Consider $f(x) = x^3 + x^2 + 1$ over Z_2. We will show how to write $f(x)$ as the product of linear factors. Let $F = Z_2[x]/\langle f(x)\rangle$ and let a be a zero of $f(x)$ in F. Then $|F| = 8$ and $|F^*| = 7$. So, by Corollary 2 to Theorem 7.1, we know that $|a| = 7$. Thus, by Theorem 20.3,

$$\begin{aligned} F &= \{0, 1, a, a^2, a^3, a^4, a^5, a^6\} \\ &= \{0, 1, a, a + 1, a^2, a^2 + a + 1, a^2 + 1, a^2 + a\}. \end{aligned}$$

We know that a is one zero of $f(x)$, and we can test the other elements of F to see if they are zeros. We can simplify the calculations by using the fact that $a^3 + a^2 + 1 = 0$ to make a conversion table for the two forms of writing the elements of F. Because char $F = 2$, we know that $a^3 = a^2 + 1$. Then,

$$\begin{aligned} a^4 &= a^3 + a = (a^2 + 1) + a = a^2 + a + 1, \\ a^5 &= a^3 + a^2 + a = (a^2 + 1) + a^2 + a = a + 1, \\ a^6 &= a^2 + a, \\ a^7 &= 1. \end{aligned}$$

Now let's see whether a^2 is a zero of $f(x)$.

$$\begin{aligned} f(a^2) &= (a^2)^3 + (a^2)^2 + 1 = a^6 + a^4 + 1 \\ &= (a^2 + a) + (a^2 + a + 1) + 1 = 0. \end{aligned}$$

So, yes, it is. Next we try a^3.

$$f(a^3) = (a^3)^3 + (a^3)^2 + 1 = a^9 + a^6 + 1$$
$$= a^2 + (a^2 + a) + 1 = a + 1 \neq 0.$$

Now a^4.

$$f(a^4) = (a^4)^3 + (a^4)^2 + 1 = a^{12} + a^8 + 1$$
$$= a^5 + a + 1 = (a + 1) + a + 1 = 0.$$

So, a^4 is our remaining zero. Thus, $f(x) = (x - a)(x - a^2)(x - a^4) = (x + a)(x + a^2)(x + a^4)$, since char $F = 2$.

We may check this factorization by expanding the product and using a conversion table to obtain $f(x) = x^3 + x^2 + 1$. ∎

Subfields of a Finite Field

Theorem 22.1 gives us a complete description of all finite fields. The following theorem gives us a complete description of all the subfields of a finite field. Notice the close analogy between this theorem and Theorem 4.3, which describes all the subgroups of a finite cyclic group.

Theorem 22.3 Subfields of a Finite Field

***For each divisor m of n,* GF(p^n) *has a unique subfield of order* p^m*. Moreover, these are the only subfields of* GF(p^n).**

PROOF To show the existence portion of the theorem, suppose that m divides n. Then, since

$$p^n - 1 = (p^m - 1)(p^{n-m} + p^{n-2m} + \cdots + p^m + 1),$$

we see that $p^m - 1$ divides $p^n - 1$. For simplicity, write $p^n - 1 = (p^m - 1)t$. Let $K = \{x \in \text{GF}(p^n) \mid x^{p^m} = x\}$. We leave it as an easy exercise for the reader to show that K is a subfield of GF(p^n) (Exercise 25). Since the polynomial $x^{p^m} - x$ has at most p^m zeros in GF(p^n), we have $|K| \leq p^m$. Let $\langle a \rangle = \text{GF}(p^n)^*$. Then $|a^t| = p^m - 1$, and since $(a^t)^{p^m-1} = 1$, it follows that $a^t \in K$. So, K is a subfield of GF(p^n) of order p^m.

The uniqueness portion of the theorem follows from the observation that if GF(p^n) had two distinct subfields of order p^m, then the polynomial $x^{p^m} - x$ would have more than p^m zeros in GF(p^n). This contradicts Corollary 3 of Theorem 16.2.

Finally, suppose that F is a subfield of GF(p^n). Then F is isomorphic to GF(p^m) for some m and, by Theorem 21.5,

$$\begin{aligned} n &= [\mathrm{GF}(p^n):\mathrm{GF}(p)] \\ &= [\mathrm{GF}(p^n):\mathrm{GF}(p^m)][\mathrm{GF}(p^m):\mathrm{GF}(p)] \\ &= [\mathrm{GF}(p^n):\mathrm{GF}(p^m)]m. \end{aligned}$$

Thus, m divides n. ∎

Theorems 22.2 and 22.3, together with Theorem 4.3, make the task of finding the subfields of a finite field a simple exercise in arithmetic.

EXAMPLE 3 Let F be the field of order 16 given in Example 1. Then there are exactly three subfields of F, and their orders are 2, 4, and 16. Obviously, the subfield of order 2 is $\{0, 1\}$ and the subfield of order 16 is F itself. To find the subfield of order 4, we merely observe that the three nonzero elements of this subfield must be the cyclic subgroup of $F^* = \langle x \rangle$ of order 3. So the subfield of order 4 is

$$\{0, 1, x^5, x^{10}\} = \{0, 1, x^2 + x, x^2 + x + 1\}.$$ ∎

EXAMPLE 4 If F is a field of order $3^6 = 729$ and α is a generator of F^*, then the subfields of F are

1. $\mathrm{GF}(3) = \{0\} \cup \langle \alpha^{364} \rangle = \{0, 1, 2\}$,
2. $\mathrm{GF}(9) = \{0\} \cup \langle \alpha^{91} \rangle$,
3. $\mathrm{GF}(27) = \{0\} \cup \langle \alpha^{28} \rangle$,
4. $\mathrm{GF}(729) = \{0\} \cup \langle \alpha \rangle$. ∎

EXAMPLE 5 The subfield lattice of GF(2^{24}) is the following.

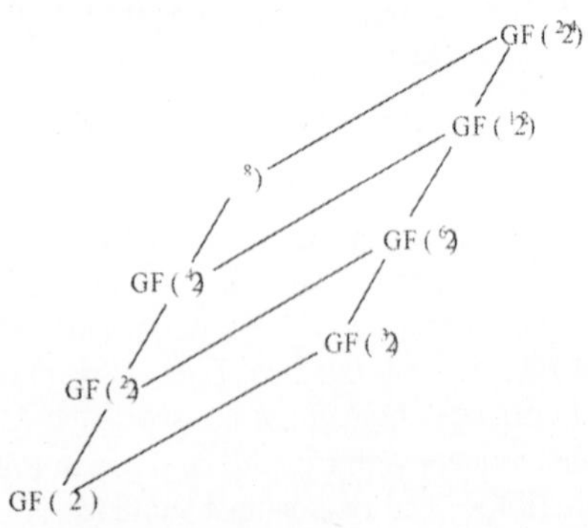

∎

Exercises

No pressure, no diamonds.

MARY CASE

1. Find [GF(729):GF(9)] and [GF(64):GF(8)].
2. If m divides n, show that $[GF(p^n):GF(p^m)] = n/m$.
3. Draw the lattice of subfields of GF(64).
4. Let α be a zero of $x^3 + x^2 + 1$ in some extension field of Z_2. Find the multiplicative inverse of $\alpha + 1$ in $Z_2[\alpha]$.
5. Let α be a zero of $f(x) = x^2 + 2x + 2$ in some extension field of Z_3. Find the other zero of $f(x)$ in $Z_3[\alpha]$.
6. Let α be a zero of $f(x) = x^3 + x + 1$ in some extension field of Z_2. Find the other zeros of $f(x)$ in $Z_2[\alpha]$.
7. Let K be a finite extension field of a finite field F. Show that there is an element a in K such that $K = F(a)$.
8. How many elements of the cyclic group GF(81)* are generators?
9. Let $f(x)$ be a cubic irreducible over Z_2. Prove that the splitting field of $f(x)$ over Z_2 has order 8.
10. Prove that the rings $Z_3[x]/\langle x^2 + x + 2\rangle$ and $Z_3[x]/\langle x^2 + 2x + 2\rangle$ are isomorphic.
11. Show that the *Frobenius mapping* ϕ: $GF(p^n) \to GF(p^n)$, given by $a \to a^p$, is a ring automorphism of order n (that is, ϕ^n is the identity mapping). (This exercise is referred to in Chapter 32.)
12. Determine the possible finite fields whose largest proper subfield is $GF(2^5)$.
13. Prove that the degree of any irreducible factor of $x^8 - x$ over Z_2 is 1 or 3.
14. Find the smallest field that has exactly 6 subfields.
15. Find the smallest field of characteristic 2 that contains an element whose multiplicative order is 5 and the smallest field of characteristic 3 that contains an element whose multiplicative order is 5.
16. Verify that the factorization for $f(x) = x^3 + x^2 + 1$ over Z_2 given in Example 2 is correct by expanding.
17. Show that x is a generator of the cyclic group $(Z_3[x]/\langle x^3 + 2x + 1\rangle)^*$.
18. Suppose that $f(x)$ is a fifth-degree polynomial that is irreducible over Z_2. Prove that x is a generator of the cyclic group $(Z_2[x]/\langle f(x)\rangle)^*$.

19. Show that x is not a generator of the cyclic group $(Z_3[x]/\langle x^3 + 2x + 2\rangle)^*$. Find one such generator.
20. If $f(x)$ is a cubic irreducible polynomial over Z_3, prove that either x or $2x$ is a generator for the cyclic group $(Z_3[x]/\langle f(x)\rangle)^*$.
21. Prove the uniqueness portion of Theorem 22.3 using a group theoretic argument.
22. Suppose that α and β belong to GF(81)*, with $|\alpha| = 5$ and $|\beta| = 16$. Show that $\alpha\beta$ is a generator of GF(81)*.
23. Construct a field of order 9 and carry out the analysis as in Example 1, including the conversion table.
24. Show that any finite subgroup of the multiplicative group of a field is cyclic.
25. Show that the set K in the proof of Theorem 22.3 is a subfield.
26. If $g(x)$ is irreducible over GF(p) and $g(x)$ divides $x^{p^n} - x$, prove that deg $g(x)$ divides n.
27. Use a purely group theoretic argument to show that if F is a field of order p^n, then every element of F^* is a zero of $x^{p^n} - x$. (This exercise is referred to in the proof of Theorem 22.1.)
28. Draw the subfield lattices of GF(3^{18}) and of GF(2^{30}).
29. How does the subfield lattice of GF(2^{30}) compare with the subfield lattice of GF(3^{30})?
30. If $p(x)$ is a polynomial in $Z_p[x]$ with no multiple zeros, show that $p(x)$ divides $x^{p^n} - x$ for some n.
31. Suppose that p is a prime and $p \neq 2$. Let a be a nonsquare in GF(p)—that is, a does not have the form b^2 for any b in GF(p). Show that a is a nonsquare in GF(p^n) if n is odd and that a is a square in GF(p^n) if n is even.
32. Let $f(x)$ be a cubic irreducible over Z_p, where p is a prime. Prove that the splitting field of $f(x)$ over Z_p has order p^3 or p^6.
33. Show that every element of GF(p^n) can be written in the form a^p for some unique a in GF(p^n).
34. Suppose that F is a field of order 1024 and $F^* = \langle\alpha\rangle$. List the elements of each subfield of F.
35. Suppose that F is a field of order 125 and $F^* = \langle\alpha\rangle$. Show that $\alpha^{62} = -1$.
36. Show that no finite field is algebraically closed.
37. Let E be the splitting field of $f(x) = x^{p^n} - x$ over Z_p. Show that the set of zeros of $f(x)$ in E is closed under addition, subtraction, multiplication, and division (by nonzero elements). (This exercise is referred to in the proof of Theorem 22.1.)

38. Suppose that L and K are subfields of GF(p^n). If L has p^s elements and K has p^t elements, how many elements does $L \cap K$ have?

39. Give an example to show that the mapping $a \to a^p$ need not be an automorphism for arbitrary fields of prime characteristic p.

40. In the field GF(p^n), show that for every positive divisor d of n, $x^{p^n} - x$ has an irreducible factor over GF(p) of degree d.

41. Let a be a primitive element for the field GF(p^n), where p is an odd prime and n is a positive integer. Find the smallest positive integer k such that $a^k = p - 1$.

42. Let a be a primitive element for the field GF(5^n), where n is a positive integer. Find the smallest positive integer k such that $a^k = 2$.

43. Let p be a prime such that $p \bmod 4 = 1$. How many elements of order 4 are in GF(p^n)*?

44. Let p be a prime such that $p \bmod 4 = 3$. How many elements of order 4 are in GF(p^n)*?

Computer Exercises

Software for the computer exercises in this chapter is available at the website:

http://www.d.umn.edu/~jgallian

Suggested Reading

Judy L. Smith and J. A. Gallian, "Factoring Finite Factor Rings," *Mathematics Magazine* 58 (1985): 93–95.

This paper gives an algorithm for finding the group of units of the ring $F[x]/\langle g(x)^m \rangle$.

L .E. Dickson

American Mathematical Society

One of the books [written by L. E. Dickson] is his major, three-volume *History of the Theory of Numbers* which would be a life's work by itself for a more ordinary man.

A. A. ALBERT,
Bulletin of the American Mathematical Society

LEONARD EUGENE DICKSON was born in Independence, Iowa, on January 22, 1874. In 1896, he received the first Ph.D. to be awarded in mathematics at the University of Chicago. After spending a few years at the University of California and the University of Texas, he was appointed to the faculty at Chicago and remained there until his retirement in 1939.

Dickson was one of the most prolific mathematicians of the 20th century, writing 267 research papers and 18 books. His principal interests were matrix groups, finite fields, algebra, and number theory.

Dickson had a disdainful attitude toward applicable mathematics; he would often say, "Thank God that number theory is unsullied by any applications." He also had a sense of humor. Dickson would often mention his honeymoon: "It was a great success," he said, "except that I only got two research papers written."

Dickson received many honors in his career. He was the first to be awarded the prize from the American Association for the Advancement of Science for the most notable contribution to the advancement of science, and the first to receive the Cole Prize in algebra from the American Mathematical Society. The University of Chicago has research instructorships named after him. Dickson died on January 17, 1954.

For more information about Dickson, visit:

http://www-groups.dcs.st-and.ac.uk/~history/

23 Geometric Constructions

At the age of eleven, I began Euclid. . . . This was one of the great events of my life, as dazzling as first love.

BERTRAND RUSSELL

Historical Discussion of Geometric Constructions

The ancient Greeks were fond of geometric constructions. They were especially interested in constructions that could be achieved using only a straightedge without markings and a compass. They knew, for example, that any angle can be bisected, and they knew how to construct an equilateral triangle, a square, a regular pentagon, and a regular hexagon. But they did not know how to trisect every angle or how to construct a regular seven-sided polygon (heptagon). Another problem that they attempted was the duplication of the cube—that is, given any cube, they tried to construct a new cube having twice the volume of the given one using only an unmarked straightedge and a compass. Legend has it that the ancient Athenians were told by the oracle at Delos that a plague would end if they constructed a new altar to Apollo in the shape of a cube with double the volume of the old altar, which was also a cube. Besides "doubling the cube," the Greeks also attempted to "square the circle"—to construct a square with area equal to that of a given circle. They knew how to solve all these problems using other means, such as a compass and a straightedge with two marks, or an unmarked straightedge and a spiral, but they could not achieve any of the constructions with a compass and an unmarked straightedge alone. These problems vexed mathematicians for over 2000 years. The resolution of these perplexities was made possible when they were transferred from questions of geometry to questions of algebra in the 19th century.

The first of the famous problems of antiquity to be solved was that of the construction of regular polygons. It had been known since Euclid that regular polygons with a number of sides of the form 2^k, $2^k \cdot 3$, $2^k \cdot 5$, and $2^k \cdot 3 \cdot 5$ could be constructed, and it was believed that no others were possible. In 1796, while still a teenager, Gauss proved that the 17-sided regular polygon is constructible. In 1801, Gauss asserted that a regular polygon of n sides is constructible if and only if n has the form $2^k p_1 p_2 \cdots p_i$, where the p's are distinct primes of the form $2^{2^s} + 1$. We provide a proof of this statement in Theorem 33.5.

Thus, regular polygons with 3, 4, 5, 6, 8, 10, 12, 15, 16, 17, and 20 sides are possible to construct, whereas those with 7, 9, 11, 13, 14, 18, and 19 sides are not. How these constructions can be effected is another matter. One person spent 10 years trying to determine a way to construct the 65,537-sided polygon.

Gauss's result on the constructibility of regular n-gons eliminated another of the famous unsolved problems, because the ability to trisect a 60° angle enables one to construct a regular 9-gon. Thus, there is no method for trisecting a 60° angle with an unmarked straightedge and a compass. In 1837, Wantzel proved that it was not possible to double the cube. The problem of the squaring of a circle resisted all attempts until 1882, when Ferdinand Lindemann proved that π is transcendental, since, as we will show, all constructible numbers are algebraic.

Constructible Numbers

With the field theory we now have, it is an easy matter to solve the following problem: Given an unmarked straightedge, a compass, and a unit length, what other lengths can be constructed? To begin, we call a real number α *constructible* if, by means of an unmarked straightedge, a compass, and a line segment of length 1, we can construct a line segment of length $|\alpha|$ in a finite number of steps. It follows from plane geometry that if α and β ($\beta \neq 0$) are constructible numbers, then so are $\alpha + \beta$, $\alpha - \beta$, $\alpha \cdot \beta$, and α/β. (See the exercises for hints.) Thus, the set of constructible numbers contains Q and is a subfield of the real numbers. What we desire is an algebraic characterization of this field. To derive such a characterization, let F be any subfield of the reals. Call the subset $\{(x, y) \in R^2 \mid x, y \in F\}$ of the real plane the *plane of* F, call any line joining two points in the plane of F a *line in* F, and call any circle whose center is in the plane of F and whose radius is in F a *circle in* F. Then a line in F has an equation of the form

$$ax + by + c = 0, \qquad \text{where } a, b, c \in F,$$

and a circle in F has an equation of the form

$$x^2 + y^2 + ax + by + c = 0, \qquad \text{where } a, b, c \in F.$$

In particular, note that to find the point of intersection of a pair of lines in F or the points of intersection of a line in F and a circle in F, one need only solve a linear or quadratic equation in F. We now come to the crucial question. Starting with points in the plane of some field F, which points in the real plane can be obtained with an unmarked straightedge and a compass? Well, there are only three ways to construct points, starting with points in the plane of F.

1. Intersect two lines in F.
2. Intersect a circle in F and a line in F.
3. Intersect two circles in F.

In case 1, we do not obtain any new points, because two lines in F intersect in a point in the plane of F. In case 2, the point of intersection is the solution to either a linear equation in F or a quadratic equation in F. So, the point lies in the plane of F or in the plane of $F(\sqrt{\alpha})$, where $\alpha \in F$ and α is positive. In case 3, no new points are obtained, because, if the two circles are given by $x^2 + y^2 + ax + by + c = 0$ and $x^2 + y^2 + a'x + b'y + c' = 0$, then we have $(a - a')x + (b - b')y + (c - c') = 0$, which is a line in F. So, the points of intersection are in F.

It follows, then, that the only points in the real plane that can be constructed from the plane of a field F are those whose coordinates lie in fields of the form $F(\sqrt{\alpha})$, where $\alpha \in F$ and α is positive. Of course, we can start over with $F_1 = F(\sqrt{\alpha})$ and construct points whose coordinates lie in fields of the form $F_2 = F_1(\sqrt{\beta})$, where $\beta \in F_1$ and β is positive. Continuing in this fashion, we see that a real number c is constructible if and only if there is a series of fields $Q = F_1 \subseteq F_2 \subseteq \cdots \subseteq F_n \subseteq \mathbf{R}$ such that $F_{i+1} = F_i(\sqrt{\alpha_i})$, where $\alpha_i \in F_i$ and $c \in F_n$. Since $[F_{i+1}:F_i] = 1$ or 2, we see by Theorem 21.5 that if c is constructible, then $[Q(c):Q] = 2^k$ for some nonnegative integer k.

We now dispatch the problems that plagued the Greeks. Consider doubling the cube of volume 1. The enlarged cube would have an edge of length $\sqrt[3]{2}$. But $[Q(\sqrt[3]{2}):Q] = 3$, so such a cube cannot be constructed.

Next consider the possibility of trisecting a 60° angle. If it were possible to trisect an angle of 60°, then cos 20° would be constructible. (See Figure 23.1.) In particular, $[Q(\cos 20°):Q] = 2^k$ for some k. Now, using the trigonometric identity $\cos 3\theta = 4\cos^3\theta - 3\cos\theta$, with $\theta = 20°$, we

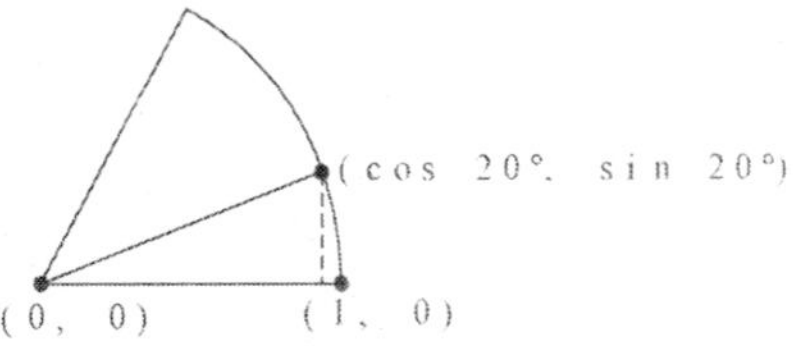

Figure 23.1

see that $1/2 = 4\cos^3 20^\circ - 3\cos 20^\circ$, so that $\cos 20^\circ$ is a zero of $8x^3 - 6x - 1$. But, since $8x^3 - 6x - 1$ is irreducible over Q (see Exercise 13), we must also have $[Q(\cos 20^\circ):Q] = 3$. This contradiction shows that trisecting a 60° angle is impossible.

The remaining problems are relegated to the reader as Exercises 14, 15, and 17.

Angle-Trisectors and Circle-Squarers

Down through the centuries, hundreds of people have claimed to have achieved one or more of the impossible constructions. In 1775, the Paris Academy, so overwhelmed with these claims, passed a resolution to no longer examine these claims or claims of machines purported to exhibit perpetual motion. Although it has been more than 100 years since the last of the constructions was shown to be impossible, there continues to be a steady parade of people who claim to have done one or more of them. Most of these people have heard that this is impossible but have refused to believe it. One person insisted that he could trisect any angle with a straightedge alone [2, p. 158]. Another found his trisection in 1973 after 12,000 hours of work [2, p. 80]. One got his from God [2, p. 73]. In 1971, a person with a Ph.D. in mathematics asserted that he had a valid trisection method [2, p. 127]. Many people have claimed the hat trick: trisecting the angle, doubling the cube, and squaring the circle. Two men who did this in 1961 succeeded in having their accomplishment noted in the *Congressional Record* [2, p. 110]. Occasionally, newspapers and magazines have run stories about "doing the impossible," often giving the impression that the construction may be valid. Many angle-trisectors and circle-squarers have had their work published at their own expense and distributed to colleges and universities. One had his printed in four languages! There are two delightful books written by mathematicians about their encounters with these people. The books are full of wit, charm, and humor ([1] and [2]).

Exercises

Only prove to me that it is impossible, and I will set about it this very evening.

Spoken by a member of the audience after De Morgan gave a lecture on the impossibility of squaring the circle.

1. If a and b are constructible numbers and $a \geq b > 0$, give a geometric proof that $a + b$ and $a - b$ are constructible.

2. If a and b are constructible, give a geometric proof that ab is constructible. (*Hint:* Consider the following figure. Notice that all segments in the figure can be made with an unmarked straightedge and a compass.)

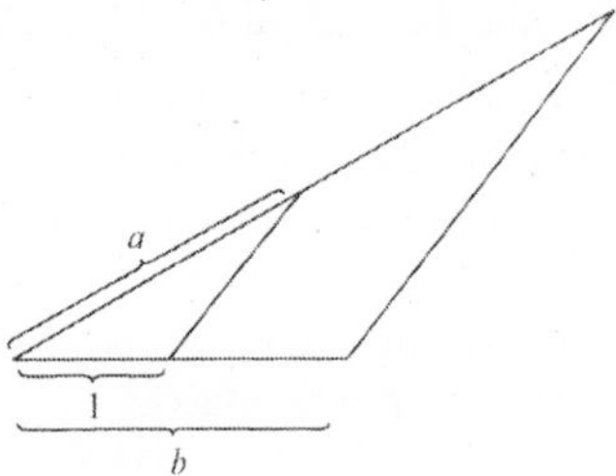

3. Prove that if c is a constructible number, then so is $\sqrt{|c|}$. (*Hint:* Consider the following semicircle with diameter $1 + |c|$.) (This exercise is referred to in Chapter 33.)

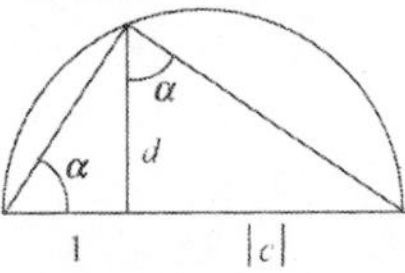

4. If a and b $(b \neq 0)$ are constructible numbers, give a geometric proof that a/b is constructible. (*Hint:* Consider the following figure.)

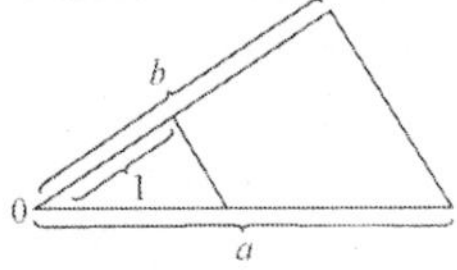

5. Prove that $\sin \theta$ is constructible if and only if $\cos \theta$ is constructible.

6. Prove that an angle θ is constructible if and only if $\sin \theta$ is constructible.

7. Prove that $\cos 2\theta$ is constructible if and only if $\cos \theta$ is constructible.

8. Prove that 30° is a constructible angle.

9. Prove that a 45° angle can be trisected with an unmarked straightedge and a compass.

10. Prove that a 40° angle is not constructible.

11. Show that the point of intersection of two lines in the plane of a field F lies in the plane of F.

12. Show that the points of intersection of a circle in the plane of a field F and a line in the plane of F are points in the plane of F or in the plane of $F(\sqrt{\alpha})$, where $\alpha \in F$ and α is positive. Give an example of a circle and a line in the plane of Q whose points of intersection are not in the plane of Q.
13. Prove that $8x^3 - 6x - 1$ is irreducible over Q.
14. Use the fact that $8\cos^3(2\pi/7) + 4\cos^2(2\pi/7) - 4\cos(2\pi/7) - 1 = 0$ to prove that a regular seven-sided polygon is not constructible with an unmarked straightedge and a compass.
15. Show that a regular 9-gon cannot be constructed with an unmarked straightedge and a compass.
16. Show that if a regular n-gon is constructible, then so is a regular $2n$-gon.
17. (Squaring the Circle) Show that it is impossible to construct, with an unmarked straightedge and a compass, a square whose area equals that of a circle of radius 1. You may use the fact that π is transcendental over Q.
18. Use the fact that $4\cos^2(2\pi/5) + 2\cos(2\pi/5) - 1 = 0$ to prove that a regular pentagon is constructible.
19. Can the cube be "tripled"?
20. Can the cube be "quadrupled"?
21. Can the circle be "cubed"?
22. If a, b, and c are constructible, show that the real roots of $ax^2 + bx + c$ are constructible.

References

1. Augustus De Morgan, *A Budget of Paradoxes*, Dover Publications, 1954, books.google.com.
2. Underwood Dudley, *A Budget of Trisections*, New York: Springer-Verlag, 1987.

Suggested Website

http://en.wikipedia.org/wiki/Squaring_the_circle

This website provides an excellent account of efforts to square the circle, and links for articles about trisecting the angle and doubling the cube.

Supplementary Exercises for Chapters 19–23

Difficulties strengthen the mind, as labor does the body.

SENECA

True/false questions for Chapters 19–23 are available on the Web at:

http://www.d.umn.edu/~jgallian/TF

1. Show that $x^{50} - 1$ has no multiple zeros in any extension of Z_3.
2. Suppose that $p(x)$ is a quadratic polynomial with rational coefficients and is irreducible over Q. Show that $p(x)$ has two zeros in $Q[x]/\langle p(x)\rangle$.
3. Let F be a finite field of order q and let a be a nonzero element in F. If n divides $q - 1$, prove that the equation $x^n = a$ has either no solutions in F or n distinct solutions in F.
4. Without using the Primitive Element Theorem, prove that if $[K:F]$ is prime, then K has a primitive element.
5. Let a be a zero of $x^2 + x + 1$. Express $(5a^2 + 2)/a$ in the form $c + ba$, where c and b are rational.
6. Describe the elements of the extension $Q(\sqrt[4]{2})$ over the field $Q(\sqrt{2})$.
7. If $[F(a):F] = 5$, find $[F(a^3):F]$. Does your argument apply equally well if a^3 is replaced with a^2 or a^4?
8. If $p(x) \in F[x]$ and $\deg p(x) = n$, show that the splitting field for $p(x)$ over F has degree at most $n!$.
9. Let a be a nonzero algebraic element over F of degree n. Show that a^{-1} is also algebraic over F of degree n.
10. Prove that $\pi^2 - 1$ is algebraic over $Q(\pi^3)$.
11. If ab is algebraic over F and $b \neq 0$, prove that a is algebraic over $F(b)$.
12. Let E be an algebraic extension of a field F. If R is a ring and $E \supseteq R \supseteq F$, show that R must be a field.
13. If a is transcendental over F, show that every element of $F(a)$ that is not in F is transcendental over F.
14. What is the order of the splitting field of $x^5 + x^4 + 1 = (x^2 + x + 1) \cdot (x^3 + x + 1)$ over Z_2?
15. Show that a finite extension of a finite field is a simple extension.
16. Let R be an integral domain that contains a field F as a subring. If R is finite dimensional when viewed as a vector space over F, prove that R is a field.

17. Show that it is impossible to find a basis for the vector space of $n \times n$ $(n > 1)$ matrices such that each pair of elements in the basis commutes under multiplication.
18. Let $P_n = \{a_n x^n + a_{n-1}x^{n-1} + \cdots + a_1 x + a_0 \mid \text{each } a_i \text{ is a real number}\}$. Is it possible to have a basis for P_n such that every element of the basis has x as a factor?
19. Find a basis for the vector space $\{f \in P_3 \mid f(0) = 0\}$. (See Exercise 18 for notation.)
20. Given that f is a polynomial of degree n in P_n, show that $\{f, f', f'', \ldots, f^{(n)}\}$ is a basis for P_n. ($f^{(k)}$ denotes the kth derivative of f.)
21. Suppose that K is an extension field of a field F of characteristic $p \neq 0$. Let $L = \{a \in K \mid a^{p^n} \in F \text{ for some nonnegative integer } n\}$. Prove that L is a subfield of K that contains F.
22. In which fields does $x^n - x$ have a multiple zero?

Index of Mathematicians

(Biographies appear on pages in boldface.)

Index of Terms

Cayley Tables

Cayley Table for the Alternating Group A_4 of Even Permutations of $\{1, 2, 3, 4\}$

(In this table, the permutations of A_4 are designated as $\alpha_1, \alpha_2, \ldots, \alpha_{12}$ and an entry k inside the table represents α_k. For example, $\alpha_3 \alpha_8 = \alpha_6$.)

	α_1	α_2	α_3	α_4	α_5	α_6	α_7	α_8	α_9	α_{10}	α_{11}	α_{12}
$(1) = \alpha_1$	1	2	3	4	5	6	7	8	9	10	11	12
$(12)(34) = \alpha_2$	2	1	4	3	6	5	8	7	10	9	12	11
$(13)(24) = \alpha_3$	3	4	1	2	7	8	5	6	11	12	9	10
$(14)(23) = \alpha_4$	4	3	2	1	8	7	6	5	12	11	10	9
$(123) = \alpha_5$	5	8	6	7	9	12	10	11	1	4	2	3
$(243) = \alpha_6$	6	7	5	8	10	11	9	12	2	3	1	4
$(142) = \alpha_7$	7	6	8	5	11	10	12	9	3	2	4	1
$(134) = \alpha_8$	8	5	7	6	12	9	11	10	4	1	3	2
$(132) = \alpha_9$	9	11	12	10	1	3	4	2	5	7	8	6
$(143) = \alpha_{10}$	10	12	11	9	2	4	3	1	6	8	7	5
$(234) = \alpha_{11}$	11	9	10	12	3	1	2	4	7	5	6	8
$(124) = \alpha_{12}$	12	10	9	11	4	2	1	3	8	6	5	7

Cayley Table for the Quaternion Group

	e	a	a^2	a^3	b	ba	ba^2	ba^3
e	e	a	a^2	a^3	b	ba	ba^2	ba^3
a	a	a^2	a^3	e	ba^3	b	ba	ba^2
a^2	a^2	a^3	e	a	ba^2	ba^3	b	ba
a^3	a^3	e	a	a^2	ba	ba^2	ba^3	b
b	b	ba	ba^2	ba^3	a^2	a^3	e	a
ba	ba	ba^2	ba^3	b	a	a^2	a^3	e
ba^2	ba^2	ba^3	b	ba	e	a	a^2	a^3
ba^3	ba^3	b	ba	ba^2	a^3	e	a	a^2

Cayley Tables

Cayley Table for the Dihedral Group of Order 6

	R_0	R_{120}	R_{240}	F	F'	F''
R_0	R_0	R_{120}	R_{240}	F	F'	F''
R_{120}	R_{120}	R_{240}	R_0	F'	F''	F
R_{240}	R_{240}	R_0	R_{120}	F''	F	F'
F	F	F''	F'	R_0	R_{240}	R_{120}
F'	F'	F	F''	R_{120}	R_0	R_{240}
F''	F''	F'	F	R_{240}	R_{120}	R_0

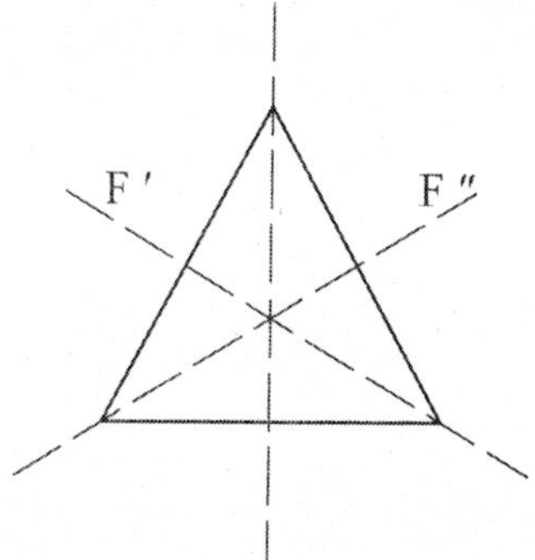

Cayley Table for the Dihedral Group of Order 8

	R_0	R_{90}	R_{180}	R_{270}	H	V	D	D'
R_0	R_0	R_{90}	R_{180}	R_{270}	H	V	D	D'
R_{90}	R_{90}	R_{180}	R_{270}	R_0	D'	D	H	V
R_{180}	R_{180}	R_{270}	R_0	R_{90}	V	H	D'	D
R_{270}	R_{270}	R_0	R_{90}	R_{180}	D	D'	V	H
H	H	D	V	D'	R_0	R_{180}	R_{90}	R_{270}
V	V	D'	H	D	R_{180}	R_0	R_{270}	R_{90}
D	D	V	D'	H	R_{270}	R_{90}	R_0	R_{180}
D'	D'	H	D	V	R_{90}	R_{270}	R_{180}	R_0

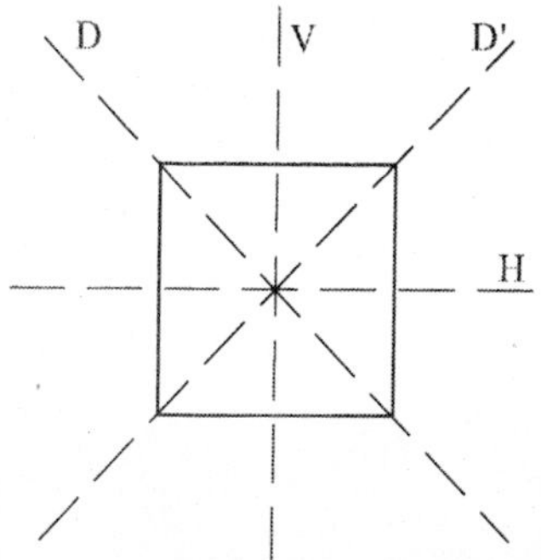